Programming Language Explorations

Programming Language Explorations

Ray Toal
Rachel Rivera
Alexander Schneider
Eileen Choe

CRC Press is an imprint of the
Taylor & Francis Group, an informa business
A CHAPMAN & HALL BOOK

CRC Press
Taylor & Francis Group
6000 Broken Sound Parkway NW, Suite 300
Boca Raton, FL 33487-2742

Printed on acid-free paper
Version Date: 20160822

International Standard Book Number-13: 978-1-4987-3846-0 (Paperback)

Visit the Taylor & Francis Web site at
http://www.taylorandfrancis.com

and the CRC Press Web site at
http://www.crcpress.com

To Noëlle and the girls
—RT

Contents

Preface

Much has happened since the question "Why would you want more than machine language?" was answered.

- 1950s–60s: Several computing professionals, most notably Grace Hopper, John McCarthy, and John Backus, create high-level, machine-independent programming languages. We soon have Fortran for scientific computing, COBOL for business, Algol for computing research, and Lisp for artificial intelligence. In the mid-1960s, PL/I and Algol 68 emerge for multi-purpose computing.
- 1960s–1970s: Industry learns that hard-to-read code prevents many projects from achieving success. The *structured programming* revolution begins. Languages providing *information hiding*, such as Modula, CLU, Simula, Mesa, and Euclid attempt to address the "software crisis." Structured and modular features are bolted on to earlier languages.
- 1980s: *Object-orientation* (OO) takes over the world. Though begun in the 1960s with Simula and refined in the 1970s at Xerox PARC with Smalltalk, OO—or approximations to it—explodes in the 1980s with C with Classes (since renamed C++), Objective-C, Eiffel, and Self. Earlier languages such as Lisp and Pascal gain OO features, becoming Common Lisp and Object Pascal, respectively.
- 1990s: The World Wide Web appears and Perl becomes popular. Java, JavaScript, and PHP are created with web applications in mind.
- 2000s: With machine speeds increasing, interest in established dynamic languages such as Ruby takes off. Scala shows that static languages can feel dynamic. Clojure arrives as a dynamic, modern Lisp, leveraging Java's virtual machine.
- 2010s: Old things become new again. Multicore processors and "Big Data" revive interest in functional programming. Older languages such as Python and R, and the newer Julia language, find use in data science. Net-centric computing and performance concerns make static typing popular again as Go, Rust, and Swift challenge C and C++ for native applications.

Each of the thirty-two languages we've mentioned, and the tens of thousands we did not, is created for some purpose. New creations may address a particular problem domain, improve upon old languages, or help us express solutions in a more efficient manner. Some introduce big ideas that give us new ways to *think about* programming. There's no single best language for all possible tasks, so we learn many of them.

This book aims to acquaint you with a number of programming languages in use today. For each, we'll provide a brief tour of the language's basic and advanced features augmented with dozens of runnable scripts. In doing so, we'll introduce many fundamental concepts transcending multiple languages, providing you a foundation for more effectively using your current favorite languages—and for learning new ones, too.

ORGANIZATION

In order to become proficient in multiple languages, you must learn fundamental language *concepts*. Conversely, mastering concepts requires that you learn multiple languages. While *both* directions of influence are important, we've chosen to proceed language-by-language, rather than concept-by-concept, so as to better convey each language's unique style and form. As we present each language, we'll introduce new concepts as they appear, and revisit familiar ones, comparing their implementation with those from languages we've seen in prior chapters. Our goal is to *present and explain common theoretical concepts of language design and usage*, illustrated in the context of practical language overviews.

After the obligatory introductory chapter outlining the book's goals and objectives, we present 12 chapters covering 12 languages. We introduce each with a common trio of example programs, take a brief tour of its basic elements, and touch on its type system, functional forms, scoping rules, concurrency patterns, and sometimes, metaprogramming facilities.

We've carefully chosen the order of languages to make a story out of the book. We start with JavaScript because it is extremely popular and full of modern and interesting features, including first-class functions, prototypes, and promises. It has influenced hundreds of successors, including our second language, CoffeeScript.

The next three languages, Lua, Python, and Ruby, are general-purpose scripting languages. They are followed by Julia, whose treatment of dynamic polymorphism via generic functions and multimethods contrasts nicely with Ruby's classes.

We then take a quick look at Java, a popular language in the enterprise computing space, and our first statically-typed language. Java was introduced along with the *Java platform*, which includes a virtual machine and a very extensive set of powerful libraries for just about every computing task imaginable. Hundreds of languages are targeted to this platform, including our next language, Clojure. Clojure will introduce us to a mostly-functional view of the world, with persistent data structures and software transactional memory. Next up is Elm, which, like Java, is statically typed, but with an incredibly powerful type inference mechanism. We continue with Erlang, a functional language like Clojure and Elm, specifically designed for building large-scale, highly reliable systems.

Our last two languages require a bit of memory management knowledge on the part of the programmer. First we visit Go, a language with explicit **pointers** and designed, like Erlang, with highly concurrent and distributed applications in mind. We close with Swift, which exemplifies many of the modern trends in programming language design—it is static, safe, and expressive—and introduces us to a different model of garbage collection (reference counting) than all of our previous languages.

Each language chapter ends with a summary, pointers to open source projects featuring the language, references to materials for further study, and a collection of exercises, designed as *further explorations*. We've added the exercises to motivate you to find the language's online playground or downloadable installation, and to try out various interesting snippets of code—and muse on what they say about the language. Occasionally we'll suggest that you write more substantial scripts, research various aspects of the language not covered in the chapter proper, or ask you to do some comparisons between languages.

Following our twelve featured language chapters, we provide a brief tour of a couple dozen or so additional languages, and a summary chapter bringing together many of the questions explored throughout the text.

A NOTE ON THE SELECTED LANGUAGES

No choice of languages can possibly satisfy everyone; you will undoubtably see your favorite languages missing, or wonder why the great influencers—Algol, Lisp, Smalltalk, Prolog, ML, etc.—do not appear. Our goal is to focus on modern languages in use today. After all, the influencers have modern descendants: Clojure is a modern Lisp; Ruby borrows much from Smalltalk; Elm is the latest of the venerable ML family, with a light Haskell influence. Python, Ruby, and Java, each over 20 years old, continue to evolve in significant ways. Recent languages, such as Julia, Go, and Swift, collect many of the best ideas of the last fifty years; we've selected them as a way to illustrate the breadth of the field.

Notice our emphasis on *ideas* rather than *features*. Alan Kay [71] has written:

> Programming languages can be categorized in a number of ways: imperative, applicative, logic-based, problem-oriented, etc. But they all seem to be either an "agglutination of features" or a "crystalization of style." COBOL, PL/I, Ada, etc., belong to the first kind; LISP, APL—and Smalltalk—are the second kind.

We haven't selected *only* languages of the second kind, but we will try to emphasize big ideas, insights, and styles in addition to features. And while we focus on today's languages, many of the great classics, along with a few other modern languages, and even a few esoterics, are discussed in a later chapter.

AUDIENCE

This book is targeted both to professionals and advanced college undergraduates looking to expand the range of languages and programming patterns they can apply in their work and studies. We've paid attention to modern programming practice, covered some cutting-edge languages and patterns, and provided many runnable examples, so readers will discover new skills to apply in their craft. In addition, we've slipped in a bit of academic terminology because foundational principles are required to grow from a hobbyist to a professional able to build large, scalable, efficient systems. However, the book remains unapologetically an *exploration* of high-level language practice, rather than a theoretical treatise based on formal syntax and semantics. The exploration style places this book between a tutorial and a reference: we make no attempt to teach any particular language from scratch, nor to provide complete descriptions of any language. Indeed, we've kept the focus *on the concepts and practices underlying programming language design and usage*, and you should find fairly complete coverage of these within the context of the different languages.

If you are an instructor looking for material to supplement a programming languages or software engineering course, you'll find our approach rather unconventional, but, we hope, a lot more fun. The traditional topics—binding, expressions, control flow, types, modularity, concurrency, and metaprogramming—all appear, but show up repeatedly within the context of whole-language overviews. Students thus have a chance to explore how the various features in a given language work together, rather than focusing too much on how a given feature is expressed in say, twelve different language fragments. You'll also find that we've chosen not to visit the same themes in every chapter. We instead highlight (say) scope *only* in those chapters with interesting scoping issues, and ditto for types, coroutines, metaprogramming, and so on. Don't go looking for any unifying themes across *all* languages; rather, look for themes appearing across *several* languages, and note which themes each language favors over others. In other words, we've targeted many, but not all, of the squares in a grid similar to the following:

	Binding/Scope	Funcs/Closures	Frames/ArgPassing	Prototypes/Classes	Extension	Dispatch	Operators	Control Flow	Matching	Type Systems	Type Inference	Immutability	Persistent Structs	Processes/Threads	Messaging	Synchronization	Pointers/Garbage	Metaprogramming
JavaScript																		
CoffeeScript																		
Lua																		
Python																		
Ruby																		
Julia																		
Java																		
Clojure																		
Elm																		
Erlang																		
Go																		
Swift																		

PREREQUISITIES

This book is not for beginning programmers. We're assuming you know at least two languages pretty well, and hopefully a couple more. You should know the basic concepts surrounding variables, expressions, operators, functions, and basic data structures, and have experience writing nontrivial applications—this book is a tour, not a tutorial. We will be covering many interesting (and often powerful) features, and purposely do so with code that is **often very dense, and sometimes cryptic**, even when presenting unfamiliar paradigms for the first time. Be on the lookout for new concepts that are introduced within the code examples themselves, rather than in the surrounding text.

You should also know your way around your file system and be skilled in using a command line interface to create and delete files, manage directories, and launch programs. You should also be pretty good with a text editor. Finally, you should be able to find and install software as needed to build and run the examples, and use a REPL or playground to practice.

SUPPLEMENTARY MATERIAL

All of the code from the book, together with test scripts that show how to run the examples, can be found in the online repository at `https://github.com/rtoal/ple`. We've also included notes in the repository for installing and working with the languages featured in the text, since this information changes too frequently to be mentioned in print. The repository also contains additional code of interest, and will grow and be updated as the programming languages evolve. Errata, additional exercises, and bonus material can be found at the companion website at `http://rtoal.github.io/ple/`.

ACKNOWLEDGEMENTS

We'd like to thank Loren Abrams, Facebook; David Pedowitz, Friendbuy; Andy Won, Amazon; Saturnino Garcia, University of San Diego; Caskey Dickson, Microsoft; B.J. Johnson, Claremont Graduate University; Matt Brown, UCLA; and Craig Reinhardt, California Lutheran University for their reviews of early drafts and many constructive comments. Zane Kansil, Jasmine Dahilig, Zoey Ho, Juan Carrillo, Stephen Smith, Andrew Akers, Trixie Roque, Matt Flickner, Andrés Buriticá, and Ed Bramanti reviewed portions of the text, were available for several discussions, and at times wrote some sample code. Shashi Kumar provided quite a bit of helpful LaTeX expertise.

We are grateful to the owners of the following trademarks and images for use in the production of this book:

- *The Unofficial JavaScript logo* by Chris Williams is used under the terms of the license at `https://github.com/voodootikigod/logo.js/blob/master/LICENSE`.
- *The CoffeeScript logo* is used by permission of Jeremy Ashkenas.
- *The Lua project logo*, © 1998, Lua.org, designed by Alexandre Nakonechnyj, is used according to the terms at `https://www.lua.org/images/`.
- *The Python logo*, a trademark of the Python Software Foundation, is used by permission of the Python Software Foundation Trademarks Committee.
- *The Ruby logo*, © 2006, Yukihiro Matsumoto, is used under the terms of the Creative Commons Attribution-ShareAlike 2.5 License (`https://www.ruby-lang.org/en/about/logo/`).
- *The logo for the Julia programming language*, by Stefan Karpinski, is used by permission of Jeff Bezanson.
- *The Java logo* is used by permission of Oracle Corporation. Oracle and Java are registered trademarks of Oracle and/or its affiliates.
- *Clojure* and *The Clojure logo* are copyright © 2007 Rich Hickey. The logo is used by permission of Mr. Hickey.
- *The Elm logo* is used by permission of Evan Czaplicki.
- *The Logo of the Programming Language Erlang* is used by permission of Ericsson.
- *The Go Gopher*, created by Renee French, is Creative Commons Attributions 3.0 licensed. (`https://blog.golang.org/gopher`)
- *Swift* and the *Swift Logo* are trademarks of Apple Inc., registered in the U.S. and other countries. The logo is used by permission of Apple Inc.

We are also grateful to the staff at Taylor & Francis, including Randi Cohen, Senior Acquisitions Editor and Marcus Fontaine, Project Coordinator, without whose hard work this book would not have been possible.

CHAPTER 0

Introduction

Hello! It appears that you might be interested in languages of the programming kind. Whether you are certain that you are, or are not so sure that you are, we'll try, in this chapter, to pique your interest and make the study of programming languages look not only fun, but worthwhile.

0.1 WHY STUDY PROGRAMMING LANGUAGES

Learning new programming languages will enable you to think about, and solve, problems in new and sometimes surprising ways.

How so? Try writing a function to sum the squares of the even numbers in an array. If all you know is C, you might think through this exercise as follows:

```
int sum_of_even_squares(int* a, unsigned int length) {
    int total = 0;
    for (unsigned int i = 0; i < length; i++) {
        if (a[i] % 2 == 0) {
            total += a[i] * a[i];
        }
    }
    return total;
}
```

But this is unsatisfying: *why is the variable* i *there?* Why do we care about the *indexes* of the array? In Swift, we can iterate over the array *elements* directly:

```
func sumOfEvenSquares(_ a: [Int]) -> Int {
    var sum = 0
    for x in a {
        if x % 2 == 0 {
            sum += x * x
        }
    }
    return sum
}
```

Slightly better, perhaps, but we can express a solution in a completely different way. Rather than the for-loop and its nested if-statement putting each individual element in charge of the overall computation, we can let the array do the work, by:

1. *Selecting*, or *filtering by*, its even elements,
2. *Mapping* the square operation over the selected elements, then finally
3. *Reducing* the squares to a single value by summing them together.

Many popular languages have these operations. Ruby does:

```
def sum_of_even_squares(a)
  a.select{|x| x % 2 == 0}.map{|x| x * x}.reduce(0, :+)
end
```

And so does Clojure:

```
(defn sum-of-even-squares [a]
  (->> a (filter even?) (map #(* % %)) (reduce +)))
```

We should be fair and show that Swift is just as capable:

```
func sumOfEvenSquares(_ a: [Int]) -> Int {
    return a.filter{$0 % 2 == 0}.map{$0 * $0}.reduce(0, combine: +)
}
```

Java, too, can filter, map, and reduce:

```
public static int sumOfEvenSquares(int[] a) {
    return IntStream.of(a).filter(x -> x%2==0).map(x -> x*x).sum();
}
```

Python expresses mapped and filtered sequences with a *generator expression*:

```
def sum_of_even_squares(a):
    return sum(x*x for x in a if x % 2 == 0)
```

K uses powerful array operators to yield a remarkably terse one-liner:

```
sumofevensquares: {+/x[&~x!2]^2}
```

Let's move from loops to assignment. What happens after executing `x = y + z`? The variable x gets updated on the spot and the old value is lost forever, right? Not so fast.... You may encounter languages in which the "value" of x is its entire *history* of values, or in which x automatically updates whenever y or z are subsequently changed. There are even languages that prohibit assignment altogether!

The fun doesn't stop there. In learning new languages, you'll find lists that aren't physically stored but produce their elements on demand. And models of the world where every piece of data, even small integers, can act as "little computers" that can send and receive messages. You may run into languages that can perform computation during type checking and languages that allow processes to restart after they die. You will even encounter languages in which you never code step-by-step algorithms; instead, you simply state the properties you expect of a solution, and let a built-in inference engine find the solution for you.

Exposure to new languages can help you write better code in the languages you *do* use every day, as you discover ways to simulate features that your language lacks. In addition, you'll acquire the foundation to learn new languages more easily, and even design your own language. Even if you don't create the next Python or C++, remember that language design happens at small scales too: you can create little languages to control robots or paintbrushes, invoke commands in a conversation-based game, describe formulas to input into a calculator or spreadsheet, or specify questions to ask of a search engine or database.

Steve McConnell articulates these benefits as well as anyone [82]:

> ...Mastering more than one language is often a watershed in the career of a professional programmer. Once a programmer realizes that programming principles transcend the syntax of any specific language, the doors swing open to knowledge that truly makes a difference in quality and productivity.

0.2 PROGRAMMING LANGUAGE ELEMENTS

What are some of the principles, or concepts, that transcend multiple languages? We'll answer this question throughout the book, but first, it helps to provide a foundation by introducing the basic linguistic elements that underlie these concepts. These elements will provide us with the working vocabulary we will need to describe programming principles and compare and contrast programming languages in a meaningful way.

A **value** is a unit of data. We have numeric values (e.g., *three*, π, *ninety-seven point eight*), values for truth and falsity, character and text values (strings), values containing other values (whose internal values may be named or numbered), and values that indicate missing or unknown information.

A **literal** is a representation of a value. A few examples follow:

- `95` is a literal for the value *ninety-five*. Alternate representations for this value include `0x5F` (`0x` is a common prefix for hexadecimal numerals), `0b1011111` (`b` means binary), and `0.0095E4` (`E` means "times-ten-to-the").
- `true` (sometimes `True` or `T`) is the literal representing truth; `false` (or `False` or `F`) is the literal representing falsity.
- The quoted sequence of characters `"Hello, how are you?"` is a **string** literal representing a common English greeting.
- `:alice` (rendered in some languages as just `alice`) is a kind of literal known as an **atom**, or **symbol**. Atoms stand only for themselves and nothing else. Unlike strings, they are not decomposed into character sequences.
- `[0, true, 98.6]` is literal for the sequence of three values: *zero*, *truth*, and *ninety-eight point six*.
- `{latitude: 29.9792, longitude: 31.1344}` is a literal representing the location of the Great Pyramid of Giza.
- `x => x / 2` is a literal representing a function that halves its argument. Alternate forms for this function include `{%0 / 2}`, `#(/ % 2)`, `lambda x: x / 2`, and `function (x) {return x / 2}`.
- `null` (sometimes `nil` or `None`) is a literal used to indicate the intentional absence of a value.
- `undefined` is a literal used to indicate some desired value is unknown or purposely not being divulged. Think of this as "I don't know," "I don't care," or "None of your business."

A **variable** is a name that refers to a value. Do not confuse variables and values: variables either *stand for* or *hold* values, they are not themselves values. In some languages, a variable is simply a name bound to a value; in others, a variable is a container into which different values can be placed at different times during program execution. Variables of the latter kind are more properly called **assignables**. [49] Regardless of the kind of variable, please note a *value never changes* (five is always five), but *which* value is bound to or held by a particular variable may vary.

An **expression** is a combination of literals, variables, and operators that is **evaluated** to produce a value. Some operators require all of their operands to be evaluated; some, like the popular `&&` ("and then") and `||` ("or else"), do not. An example of expression evaluation, assuming s holds the value `"car"`, y holds the value 100, and $found$ holds the value `true`, follows:

```
7 * s.indexOf('r') + Math.sqrt(y) / 2 <= 0 || !found
⇒ 7 * 2 + Math.sqrt(y) / 2 <= 0 || !found
⇒ 14 + Math.sqrt(y) / 2 <= 0 || !found
⇒ 14 + Math.sqrt(100) / 2 <= 0 || !found
⇒ 14 + 10 / 2 <= 0 || !found
⇒ 14 + 5 <= 0 || !found
⇒ 19 <= 0 || !found
⇒ false || !found
⇒ !found
⇒ !true
⇒ false
```

A **routine** is a (generally parameterized) unit of code. Routines that can only start from the beginning when called are known as **subroutines**; those able to resume from where they last yielded when called are known as **coroutines**. Some languages use the term **function** in place of subroutine; others use the term only for subroutines that return values, while using **procedure** for subroutines returning no values. Coroutines interleave their execution on a single sequence of instructions, explicitly yielding control to each other, while **threads** and **processes** run independently, preempted at times by the operating system when their number exceeds the number of available hardware execution units. Generally, we speak of threads as sharing memory, while processes maintain their own local memory and communicate exclusively via **messages**.

A **type**, roughly, determines a collection of values with some prescribed behavior. That a value v has type T means that only certain operations may be applied to v. The notion of "type" in programming language theory is extremely deep, and has filled entire books, for example [97]. Typing is central to the study of languages, as the point of a language's type system is *to provide (meaningful) constraints on what we can and cannot say.*

We have an intuitive sense that values belong to types—`true` to the boolean type, `0.3` to some kind of numeric type, and `"Hello"` to the string type—but exactly how these types are defined can vary significantly by language. Some languages use different types for integer and non-integer numeric values. Some arrange types into supertype-subtype relationships, where a value of a subtype can be used wherever a value of a supertype is expected. Some languages group all functions into a single type, while others type their functions based on the types of arguments or permitted return values:

Language	Literal	Type
Lua	`function (x) return x * x end`	`function`
JavaScript	`x => x * x`	`Object`
Elm	`\x -> x * x`	`number -> number`
Python	`lambda x: x * x`	`function`
Julia	`x -> x * x`	`Function`
Scala	`(x: Int) => x * x`	`Int => Int`
Go	`func (x int) int {return x*x}`	`func(int) int`

Closely related to the notion of type is that of **class**, which, roughly, is a kind of factory for instantiating objects that have a particular internal structure. Some languages have no classes, some conflate the notion of class and type, and some take great care to distinguish classes from types.

A **statement** is code that performs an action. Common types of statements include (1) *declaration statements* to bring variables into existence, (2) *expression statements* to evaluate expressions, (3) *assignment statements*, (4) *invocation statements* to run subroutines or start coroutines, (5) *conditional statements* (`if`, `unless`, `switch`, `match`) to execute code only under certain conditions, including the resumption of one of several coroutines based on the state of various guarded expressions, (6) *iteration statements* (`while`, `do while`, `for`, `repeat`) to execute code repeatedly, and (7) *disruptive statements* (`break`, `continue`, `retry`, `throw`, `return`, `yield`, `resume`) for changing the "normal" control flow.

Many languages provide a means for *throwing* or *raising* an **exception** when something has gone wrong, and a means to *catch*, or *handle*, thrown exceptions. Languages without an exception facility will often have a means to allow certain operations to return two values, one indicating success or failure, and the other for the expected value on success. You may also encounter **option values**, in which the value itself is tagged with an indication of it being a proper value or an error (or missing) value.

Modern languages support the notion of programming-in-the-large via **modules** or **packages**, which encapsulate strongly related entities of a library or application. The programmer may **export** some of the entities to the rest of the program, leaving the others private to the module. Modules may **import** (sometimes called load or require) entities from other modules. Often, but not always, modules correspond to files.

There are, of course, many more language elements that we need to learn. It is especially crucial to study concepts and language models which are not "mainline." In [123], Bret Victor shows how work in the 1960s and 1970s created programming models that manipulate data rather than code, are parallel rather than sequential, use goals instead of procedures, and are represented spatially rather than in text strings. It would be a tragedy, he says, if people were to miss these and other ideas, or worse, to simply master one way of programming and forget that they could ever have new ideas about programming models.

0.3 EVALUATING PROGRAMMING LANGUAGES

Why do people sometimes *argue* over their favorite languages, often claiming "their" language is better than "your" language? All languages have their flaws, which can be a source of great humor: Gary Bernhardt's *WAT* [12], Melissa Elliott's *PHP Manual Masterpieces* [29], and Eric Wastl's *PHP Sadness* [126]. But most languages do some things well and other things not so well. You might be able to rate a language's suitability for certain tasks, but an overall language rating is not worth looking for. To judge a language, you must know *why* it was designed. Let's take a look at a few of these reasons:

Fortran	Numeric computation
Lisp	Symbolic computation; Artificial intelligence
COBOL	Business computation
Algol 60	Algorithmic description
Algol 68	General-purpose computation
Pascal	Teaching structured programming
Modula	Modular programming
C	Systems programming
SQL	Database applications
Scheme	To be a simpler alternative to Lisp
Prolog	Expert systems; Natural language processing
Smalltalk	Personal computing
MATLAB	To be an alternative Fortran for numeric computation
Ada	Megaprogramming; Embedded systems
C++	Simulation
ML	Theorem proving
Perl	Scripts, with a focus on text processing
Erlang	Massively scalable, available, soft real-time systems
Python	Scripting, with an extensible language
Ruby	To be the language Matz[1] wanted, and to be better than Perl
Lua	General-purpose scripting
K	Ultra high-performance numerical (typically financial) analysis
R	Statistical and graphical computing
Java	Compact downloadable executable components
C#	Enterprise computing on the .NET platform
JavaScript	Client-side scripting for web applications
PHP	Server-side scripting for web applications
PostScript	Formatting of printed documents
Haskell	To bring together a number of existing functional languages
Io	Dynamic language experimentation
Scala	To support both functional and object oriented programming
F#	To solve complex problems in a simple way
Clojure	To be a modern Lisp dialect for the JVM
CoffeeScript	To be a cleaner JavaScript
Dart	Modern web applications
Go	Large distributed, interconnected systems
Parasail	Safe, secure, highly parallel applications
Rust	Large, safe, network clients and servers
TypeScript	To be a superset of JavaScript for large applications
Elm	Browser-based apps in a functional, reactive style
Elixir	To modernize and extend Erlang
Julia	High-performance scientific computing in a dynamic language
Hack	Web applications in a safe variant of PHP
Swift	iOS development with modern language features

[1]Yukihiro Matsumoto, the designer of Ruby

It's probably a good bet that a language designed expressly for the purpose of formatting printed documents might not be good for writing web servers. So what is the point of arguing whether PostScript is better than Go? The correct retort to "*A* is better than *B*" is "For what, exactly?"

In addition to understanding the kinds of problems a language was designed to solve, we need to understand *how* a language solves these problems. We can then classify languages in a way that enables us to compare and contrast. Unlike book outlines and company organization charts, we can't fit programming languages into a neat, single, hierarchical ontology. This isn't too surprising: in many knowledge disciplines, hierarchies simply don't exist. Clay Shirky makes this point beautifully in his essay *Ontology is Overrated* [109]. When there is no hierarchy, we employ *tags*. Here are a few tags we can use to label languages:

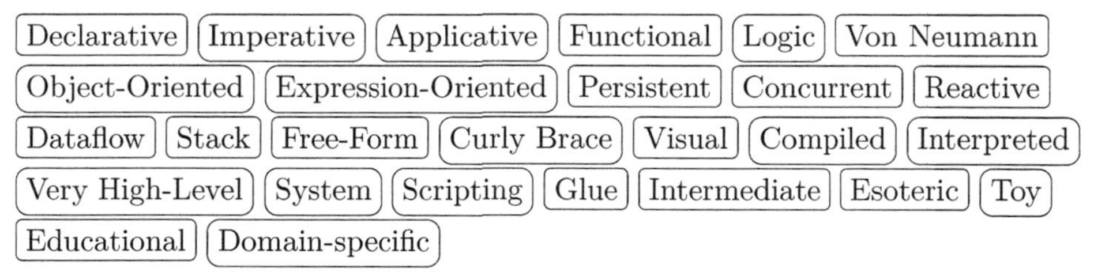

A given language will generally have quite a few tags. Java, for example, was introduced to the world as simple, object-oriented, familiar, robust, secure, architecture neutral, portable, high-performance, interpreted, threaded, and dynamic [42]. Scala, by design, is both functional and object-oriented. Understanding the various tags, or categories, provides important contextual information for language evaluation. But on exactly what criteria might we evaluate languages?

We may, for example, look at *technical* reasons:

- **Readability**. A significant portion of a programmer's time is spent trying to figure out existing code. Code must be understandable enough to be validated, fixed, or adapted to new requirements.

- **Writability**. Sometimes a language makes programming a chore, requiring one to fit a solution into the language's "opinionated" constraints. James Hague calls these *puzzle languages*, writing: "if it takes focused thought to phrase that solution into working code, you go down one path then back up, then give up, then try something completely different—then you're almost certainly using a puzzle language." [47]

- **Expressiveness**. Isn't it lovely when `A=B+C` just works—when A, B, and C are ... matrices? Or when we can avoid writing type expressions because the compiler infers the types for us? Or when we have high-level constructs to make concurrent programming safe without us having to remember to acquire and release locks? Or when we can simply express *what* we want and have the system figure out *how* to get it for us?

- **Guidance**. It can be very frustrating when a programmer uses a value of the wrong type, or forgets a case in a switch statement, or mis-indents, and the code happily executes—with surprising results. Some languages try to make common mistakes impossible to write.

- **Efficient compilation, efficient execution, or both**. Who wants to sit around waiting for code to compile,[2] or run?

[2]Obligatory xkcd link: `http://xkcd.com/303/`

But we may end up choosing a language for other, "softer," reasons:

- It might be the only language suitable for a specific problem.
- You just like it. We all have our own preferences. As they say, "there's no accounting for taste."
- You need to hire enough people that already know it. Face it, some languages are very popular; some are not.
- There is a wealth of development tools (IDEs, fast compilers), or resources (books, blog posts, activity on programming Q&A sites) for it.
- A big government, or a big company, created or backed it (IBM created PL/I, Sun—Java, Google—Go, Facebook—Hack, Mozilla—Rust, and Apple—Swift).
- Everyone else is using it. There's an old saying that goes: "No manager ever got fired for choosing Java."
- Your company already invested too much in it, and it's too expensive to change.
- You're sticking with it because you're too tired, or don't have enough time, to learn something new.

When you finally go out on a limb and proclaim your love of language A, or that your team should use language B for task C, you should be ready with an informed rationale, perhaps informed from the criteria above. If your decision is less than popular, you may be challenged with questions such as *why* your preferred language X does not have feature Y. The answer may be "because it has feature Z instead." For instance:

- The expressive power of dynamic typing, persistent data structures, functions as first-class values, higher-order functions, and closures (all topics we'll introduce in the book) saves development time and makes it easier to reason about program behavior, but may negatively impact runtime performance.
- Automatic garbage collection, where unreachable memory is automatically freed up without direct programmer intervention, may save thousands of hours of programmer time, but should not be used in embedded, life-critical, real-time systems. A collector might decide to free up space while your code is performing a time-sensitive complicated maneuver in a fighter plane or delivering radiation to a patient.

These tradeoffs, and others like them, tell us why there cannot be a best language for all situations. But this makes life better, because we get so many choices. The next twelve chapters will take you through twelve modern languages and the choices made by their designers. We'll see design tradeoffs time and time again. Have fun.

CHAPTER 1

JavaScript

We'll begin our tour with JavaScript, because it is, by *some* measures, the most popular programming language in the world.

First appeared 1995
Creator Brendan Eich
Notable versions ES3 (1999) • ES5 (2009) • ES2015 (2015)
Recognized for First-class functions, Weak typing, Prototypes
Notable uses Web application clients, Asynchronous servers
Tags Imperative, Functional, Dynamic, Prototypal
Six words or less "The assembly language of the web"

JavaScript was designed and implemented in ten days in 1995 by Brendan Eich, then at Netscape Communications Corporation, with the goal of creating an amateur-friendly scripting language embedded into a web browser. The syntax of the language was strongly influenced by C, with curly braces, assignment statements, and the ubiquitous `if`, `while`, and `for` statements. Semantically, however, JavaScript and C are worlds apart. JavaScript's influence here was the lesser-known language Scheme. Functions are **first-class values**: they can be assigned to variables, passed to functions, and returned from functions.

The goal of allowing novice programmers to write small scripts in web page markup led to some well-loved design choices, including array (e.g., `[10, 20, 30]`) and object (e.g., `{x:3, y:5}`) literals. Yet the attempt to keep the language simple led to several notorious features as well. **Weak typing**, where expressions of the wrong type are automatically coerced to "something that works," and **automatic semicolon insertion**, where the language will figure out where your statements begin and end when you are not explicit, save typing but sometimes produce utterly surprising behavior. Douglas Crockford [20] has catalogued these and a number of other "Bad Parts" and "Awful Parts," while at the same time praising JavaScript as "[having] some extraordinarily good parts. In JavaScript there is a beautiful, elegant, highly expressive language that is buried under a steaming pile of good intentions and blunders." [20, p. 2]

Though originally targeted to beginners, the language has grown a great deal over time and has been used in thousands of successful, sophisticated applications. Every major web browser comes with a JavaScript engine. Running applications written in languages other

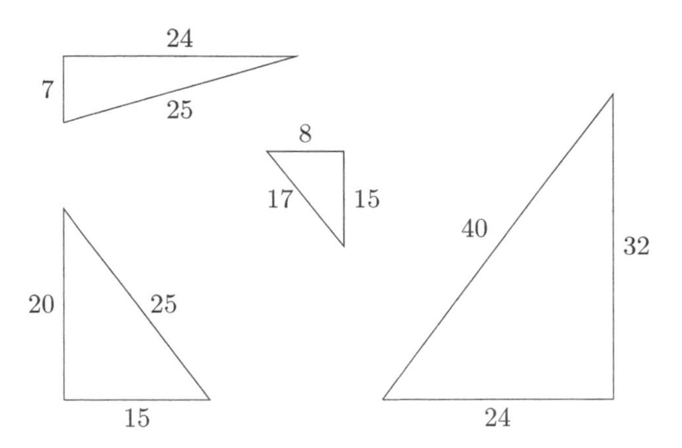

Figure 1.1 Integer right-triangle lengths

than JavaScript in a browser (including graphics-intensive applications written in C++ using the Unreal Engine [30]) is often done by first translating, or compiling, to JavaScript. But JavaScript's success is not limited to the browser; the language powers server-based applications supporting thousands of concurrent users.

As the first language on our tour, we'll use JavaScript to introduce several aspects of functions: argument passing, scope, the handling of free variables, closures, and anonymous and higher-order functions. We'll show how prototypes enable us to create sets of similar objects, and look at one of JavaScript's more interesting features: the all-purpose `this`-expression, which allows code to behave in certain ways based on context. We'll continue with a brief look at the concept of scope. We'll close the chapter with a short overview of asynchronous programming, one of many approaches to concurrent programming we'll encounter in this book.

1.1 HELLO JAVASCRIPT

Let's get a feel for JavaScript by generating some lucky numbers:

```
for (let c = 1; c <= 40; c++) {
  for (let b = 1; b < c; b++) {
    for (let a = 1; a < b; a++) {
      if (a * a + b * b === c * c) {
        console.log(`${a}, ${b}, ${c}`);
      }
    }
  }
}
```

We're terribly sorry for the flashback to high school math, but this code works pretty well as an opener, and is certainly more interesting than "Hello, world." It outputs all possible right-triangle measurements with integer values up to size 40 (see Figure 1.1 for examples) as follows:

```
3, 4, 5
6, 8, 10
5, 12, 13
9, 12, 15
8, 15, 17
12, 16, 20
15, 20, 25
7, 24, 25
10, 24, 26
20, 21, 29
18, 24, 30
16, 30, 34
21, 28, 35
12, 35, 37
15, 36, 39
24, 32, 40
```

As mentioned in the preface of this book, we're expecting you to have programming experience, so you can probably figure out the for-loops and the if-statement. Note the spelling of the equality operator: `x === y` is true iff x and y have the same value and the same type. We don't use JavaScript's `==` operator, as it only computes whether two objects are *similar* to, rather than equal to, each other, and sometimes behaves unexpectedly (e.g., `null == undefined` and `false == " \t "`). Within backquote-delimited strings, the construct `${`e`}` **interpolates** the value of expression e into the string.

Our second example writes the permutations of its **command line argument** to standard output. The language does not define how to access the command line, so we'll have to choose a specific implementation. We'll use the popular Node.js [70]:

```
function generatePermutations(a, n) {
  if (n === 0) {
    console.log(a.join(''));
  } else {
    for (let i = 0; i < n; i++) {
      generatePermutations(a, n - 1);
      const j = n % 2 === 0 ? 0 : i;
      [a[j], a[n]] = [a[n], a[j]];
    }
    generatePermutations(a, n - 1);
  }
}

if (process.argv.length !== 3) {
  console.error('Exactly one argument is required');
  process.exit(1);
}
const word = process.argv[2];
generatePermutations(word.split(''), word.length - 1);
```

Let's run this script:[1]

```
$ node --use-strict anagrams.js rat
rat
art
tra
rta
atr
tar
```

The script first checks that we have passed a single argument, and if not, terminates with an error message (written to standard error, *not* standard output) and non-zero exit code. Node's built-in variable `process.argv` contains the command line tokens invoking our script (excluding options), in our case:

```
process.argv === ['node', 'anagrams.js', 'rat']
```

A value of 3, therefore, indicates that one argument was passed to the script. Next, we call the recursive `generatePermutations` function of two arguments. We won't describe how the algorithm works; it's *Heap's algorithm*, which you can look up at Wikipedia. We will, however, note the script's use of `split` and `join`. JavaScript strings are **immutable**: you cannot make a string longer or shorter or change any of its constituent characters. Implementing Heap's algorithm requires us to split the string into an *array* of its characters, as arrays *are* mutable. We repeatedly rearrange the array in place through a series of swaps, each time applying the `join` operation to get a new string for output.

Let's move on to a more complex example. We'll read text from standard input and produce a report with the number of times each word appears. For now, words will consist only of the Latin letters A-Z and apostrophes; we'll remedy this deficiency later in the chapter. We'll again use Node.js, as we need a JavaScript runtime providing access to standard input:

```
const readline = require('readline');
const reader = readline.createInterface(process.stdin, null);
const counts = new Map();

reader.on('line', line => {
  for (let word of line.toLowerCase().match(/[a-z']+/g) || []) {
    counts.set(word, (counts.get(word) || 0) + 1);
  }
}).on('close', () => {
  for (let word of Array.from(counts.keys()).sort()) {
    console.log(`${word} ${counts.get(word)}`);
  }
});
```

To see the script in action, download, for fun, the plain text version of the complete works of Robert Burns from Project Gutenberg (`http://www.gutenberg.org/cache/epub/18500/pg18500.txt`) and call the file *burns.txt*. Store the script in *wordcount.js* and run:

[1] JavaScript scripts, and even certain portions of scripts, can be run in either strict mode or non-strict mode. The details of the two modes would fill many pages. For simplicity, all of the examples in this chapter assume strict mode, which can be triggered by invoking Node with the `-use-strict` option.

```
node --use-strict wordcount.js < burns.txt
```

This script uses Node's built-in `readline` module, and sets up a reader to read line-by-line from standard input. It then creates an empty map that we will fill with words and their counts. Next, the script ensures that whenever a line is ready (i.e., the system has fired a `line` event), it will be lowercased, broken up into words, and the counts for each word incremented. The script extracts from the lowercased line by matching against a **regular expression**[2] defining a word as a sequence of one or more letters in the range `a-z` and apostrophes. Finally, it arranges that when input has been fully read (via a `close` event), the script will write out each word, and its count, in sorted order.

1.2 THE BASICS

A JavaScript program is made up of scripts and modules, each containing a sequence of statements and function declarations. Statements include variable declarations, assignments, function calls, conditionals and loops, among others. Values have one of exactly 7 types:

- The type containing the sole value `undefined`.
- The type containing the sole value `null`.
- **Boolean**, containing the two values `true` and `false`.
- **Number**, the type of all numbers, including -98.88, 22.7×10^{100}, `Infinity`, `-Infinity`, and, strangely enough, `NaN`, the number meaning "not a number."
- **String**, roughly, the type of character sequences, but technically the type of sequences of UTF-16 code points.[3] String literals are delimited by either single quotes, double quotes, or backquotes, with the latter allowed to span lines and contain interpolated expressions.
- **Symbol**, the type of symbols (not covered in this chapter).
- **Object**, the type of all other values, including arrays and functions. Objects have named properties each holding a value, for example `{x: 3, y: 5}`. Properties of an array include `0`, `1`, `2`, and so on, and `length`.

The first six types are **primitive types**; Object is a **reference type**. The difference can be explained by picturing variables as boxes containing values. The *declarations* `let x = 1; let y = true; let z = 'so ' + y` create variables with initial values as follows:

x [1] y [`true`] z [`'so true'`]

After a variable is declared, an **assignment** puts a new value in its box; for example, `y = x` will *copy* the value currently in `x` into `y` (overwriting the contents of `y`):

x [1] y [1] z [`'so true'`]

Values of a primitive type are written directly inside the variable boxes, but object values are actually **references** to entities holding the object properties. This is best explained by analyzing the following script, and its visualization in Figure 1.2:

[2] Complete details regarding regular expressions are beyond the scope of this text.

[3] Details of code points can be found in Appendix B

```
const a = {x: 3, y: 5};  // creates an object
const b = a.y;           // simply puts 5 into b
const c = null;          // simply puts null into c
const d = {x: 3, y: 5};  // creates an object
const e = d;             // does not create an object
```

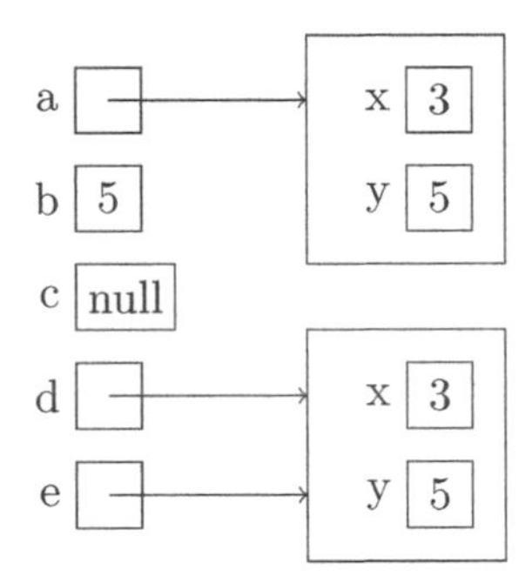

Figure 1.2 JavaScript Primitives and References

References allow multiple variables to refer to the same object. In our figure, `d.x` and `e.x` are **aliases** of each other—an assignment to one updates the other. References lead to multiple interpretations of what it means to copy: when nested objects exist, will copying only the references suffice (a **shallow copy**), or must *all* of the data be copied (a **deep copy**)? To perform a shallow copy, iterate through the properties of an object and assign their values to properties in the copy, or, for arrays, use JavaScript's `slice` method (see Figure 1.3):

```
const a = [{x:0, y:0}, {x:3, y:0}, {x: 3, y:4}];

const b = a;             // copies the reference, nothing more
const c = a.slice();     // makes a SHALLOW COPY of array elements
```

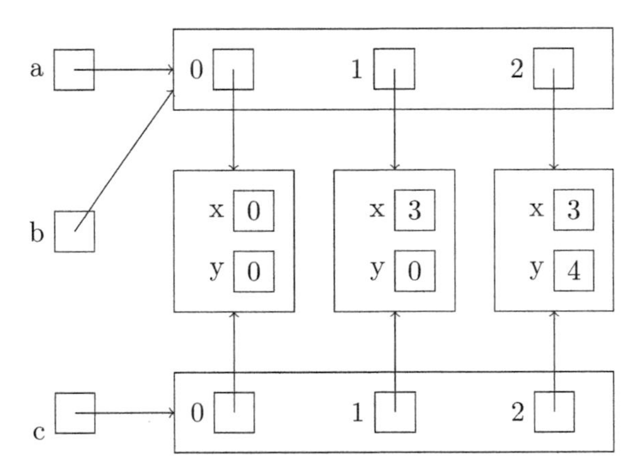

Figure 1.3 A Shallow Copy

To make a deep copy, iterate through the components of an object, copying primitives and recursively creating deep copies of objects.

JavaScript is a **weakly-typed language**, because more often than not, values of one type can appear where values of other types are expected.[4] For example:

- In `if` and `while` statements expecting a boolean condition, any value can appear. `0`, `null`, `undefined`, `false`, `NaN`, and the empty string act as false and are called **falsy**; all other values act as true and are called **truthy**.
- When a string is expected, `undefined` acts as `"undefined"`, `null` acts as `"null"`, `false` acts as `"false"`, `3` acts as `"3"`, and so on. To use an object x in a string context, JavaScript evaluates x`.toString()`.
- When a number is expected, `undefined` acts as `NaN`, `null` as `0`, `false` as `0`, `true` as `1`, and strings act as the number they "look like" or `NaN`. To use an object x in a numeric context, JavaScript evaluates x`.valueOf()`.

Table 1.1 provides concrete examples.

Value	as Boolean	as String	as Number
`undefined`	`false`	`'undefined'`	`NaN`
`null`	`false`	`'null'`	`0`
`false`	`false`	`'false'`	`0`
`true`	`true`	`'true'`	`1`
`0`	`false`	`'0'`	`0`
`858`	`true`	`'858'`	`858`
`NaN`	`false`	`'NaN'`	`NaN`
`'0'`	`true`	`'0'`	`0`
`'858'`	`true`	`'858'`	`858`
`''`	`false`	`''`	`0`
`'dog'`	`true`	`'dog'`	`NaN`
`Symbol('dog')`	`true`	`'Symbol(dog)'`	*throws* `TypeError`
any object x	`true`	*result of* x`.toString()`	*result of* x`.valueOf()`

Table 1.1 JavaScript Automatic Type Conversions

Implicit type conversion, also known as **coercion** tends to be the rule in JavaScript. `TypeErrors` are very rare, thrown when using a symbol as a number, `null` or `undefined` as an object, or trying to call a non-function.

JavaScript gives us a few options for specifying function values, including the arrow (`=>`), the construct `function` (*params*) `{` *body* `}`, and the function declaration. Functions can be called via the name bound to the function in its declaration, the variable to which the function value was assigned, or even **anonymously**. Functions that accept functions as parameters or return functions are called **higher-order functions**. Higher-order functions facilitate a style of coding—called **functional programming**—in which function composition replaces assignment statements and explicit loops. Examples follow:[5]

[4]Contrast this with a strongly-typed language, in which using values of the wrong type more often than not generates an error.

[5]In this book, we'll be illustrating many behaviors via runnable scripts featuring assertions. Assertions either succeed quietly or fail by throwing an exception. Node.js provides a number of assertions in its `assert` module. In particular: `assert(`*e*`)` tests whether *e* is truthy, `assert.strictEqual(`*e1*`,`*e2*`)` tests whether *e1* `===` *e2*, `assert.deepStrictEqual(`*e1*`,`*e2*`)` tests whether objects *e1* and *e2* have the same structure with equal values throughout, and `assert.throws(`*f*`)` tests whether function *f* will throw an exception when called.

```
const assert = require('assert');

// Function values can use `=>` or `function`
const square = x => x * x;
const odd = x => Math.abs(x % 2) === 1;
const lessThanTen = function (x) {return x < 10};
const twice = (f, x) => f(f(x));

// An anonymous function call
assert((x => x + 5)(10) === 15);

// We can pass function values to other functions
assert(twice(square, -3) === 81);
assert(twice(x => x + 1, 5) === 7);

// We can create and return new functions on the fly
function compose(f, g) {
  return x => f(g(x));
}
const isOddWhenSquared = compose(odd, square);
assert(isOddWhenSquared(7));
assert(!isOddWhenSquared(0));

// Array functions often take the place of loops
const a = [9, 7, 4, -1, 8];
assert(!a.every(odd));
assert(a.some(odd));
assert(a.every(lessThanTen));
assert.deepStrictEqual(a.filter(odd), [9, 7, -1]);
assert.deepStrictEqual(a.map(square), [81, 49, 16, 1, 64]);
```

In JavaScript, **arguments** are fully evaluated before the call and their values are assigned (copied) to the **parameters** left-to-right. If you pass too many arguments, the extras are ignored; pass too few and the extra parameters begin as `undefined`. You can, however, mark your final parameter with `...` to pack extra arguments into an array. You may also use `...` on the argument side, to unpack an array to pass into multiple parameters.

```
const assert = require('assert');

function f(x, y) {return [x, y]}
function g(x, ...y) {return [x, y]}

assert.deepStrictEqual(f(1), [1, undefined]);       // too few args
assert.deepStrictEqual(f(1, 2, 3), [1, 2]);         // too many args
assert.deepStrictEqual(g(1, 2, 3), [1, [2, 3]]);    // args packed
const args = [1, 2]
assert.deepStrictEqual(f(...args), [1, 2]);         // args unpacked
```

Each function call allocates a **frame**, or **activation record**, to hold parameters and **local variables** for this call. Local variables are those **declared** with `var`, `let`, or `const` inside the function. Let's illustrate how function calls are implemented by revisiting the permutations script from Section 1.1. Consider running the script with argument `rat`. The script calls `generatePermutations(['r','a','t'], 2)`, generating a new frame with local variable `i`. This activation makes a call to `generatePermutations(a,1)`, which calls `generatePermutations(a,0)`. Figure 1.4 shows a snapshot of program execution at this point, at the beginning of the third activation. The figure shows three important aspects of JavaScript function implementation: (1) variables live inside frames, while objects live outside frames, (2) recursive solutions work because each activation gets its own copies of local variable and parameters, and (3) passing objects is "cheap," as only references, and not object internals, are copied.

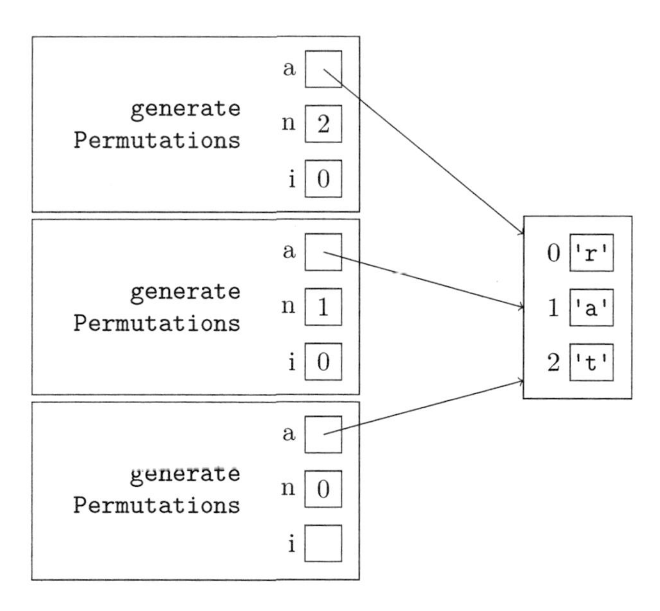

Figure 1.4 Function Call Execution Snapshot

Because parameters are always freshly allocated at call time, they are distinct variables from the arguments, so changing a parameter will not affect its corresponding argument. But keep in mind that since object values are just references, you can change the *properties* of a passed object through the parameter. Let's see how this works with a concrete example:

```
const assert = require('assert');

const x = [1,2,3];
const y = [4,5,6];

function f(a, b) {
  a = 300;                                    // Change *parameter*
  assert.deepStrictEqual(x, [1,2,3]);         // Argument still intact!
  b[1] = 400;                                 // Change *property*
  assert.deepStrictEqual(y, [4,400,6]); // See the change!
}

f(x, y);
```

Variables declared outside of any function or module are called **global variables** and are implemented in JavaScript as properties of the **global object**. Strictly speaking, properties are not variables at all: they live inside an object, not a frame! The global object contains dozens of properties including `isFinite`, `String`, `Object`, and `Date`, as well as a property that references the global object itself, called `global` or `window`, depending on the environment. And here's an interesting language design choice: referencing an undeclared variable throws a `ReferenceError`, while referencing a missing property produces `undefined`.

```
const assert = require('assert');

const p = {x: 3, y: 5}
assert(p.z === undefined);                              // There's no p.z
assert.throws(() => z, ReferenceError);                 // No variable z
```

1.3 CLOSURES

Variables used but not declared inside the function are called **free variables**. All of the functions we have seen so far operate only on parameters and local variables, so let's see how JavaScript handles free variables.

```
const x = 'OUTER';
function second() {console.log(x);}
function first() {const x = 'FIRST'; second();}
first();
```

The variable x within `second` is free. Does x take on the value from its caller, function `first`? Or from the outer x? Languages that use caller's values for free variables are **dynamically scoped**; those that look outward to textually enclosing regions are **statically scoped**. JavaScript is statically scoped.

Things become even more interesting when functions are nested inside other functions. Consider this script:

```
function second(f) {
  const name = 'new';
  f();
}

function first() {
  const name = 'old';
  const printName = () => console.log(name);
  second(printName);
}

first();
```

The function assigned to `printName` in `first` has a free variable; it is passed to `second` (as f) and then executed, logging the value of `name`. When the function was created, it sees `name` defined within `first`, but when called (as f), might it see the `name` variable in `second`? If a system binds the free variables of a passed function after passing, we speak of **shallow binding**; if bound where the function is defined, we have **deep binding**.

What does JavaScript do? When a nested function with free variables is sent outside its environment, either by being passed to or returned from another function, the function carries the bindings of those variables from the enclosing environment *of its definition* with it. These bindings "close over" the inner function, so the function, together with its bindings, is called a **lexical closure**, or **closure** for short.

Closures can be used to make **generators**.[6] A generator function produces a "next" value each time it is called. A generator for a sequence of squares would produce 0 on its first call, 1 on its second, then 4, 9, 16, and so on. A function that increments a global variable then returns its square would be **insecure** because other parts of the code could change the global variable, disrupting future calls to the generator! Fortunately, we can use the fact that a function's local variables are completely hidden from the outside:

```
const nextSquare = (() => {
  let previous = -1;
  return () => {
    previous++;
    return previous * previous;
  }
})();

const assert = require('assert');
assert(nextSquare() === 0);
assert(nextSquare() === 1);
assert(nextSquare() === 4);
```

The value assigned to `nextSquare` is the result of calling an anonymous function; we call the right-hand side of the assignment an **immediately invoked function expression**, or IIFE. The call returns a closure. The variable that holds the number to be squared is local to the enclosing function. The generator (`nextSquare`) can see this value, but no other parts of the code can. The generator is secure.

1.4 METHODS

An object can have properties whose values are functions:

```
const circle = {
  radius: 10,
  area: function () {return Math.PI * this.radius * this.radius},
  circumference: function () {return 2 * Math.PI * this.radius},
  expand: function (scale) {this.radius *= scale}
};
```

When we call a function via property access notation (e.g., `circle.area()`), we say the function is a **method** and the object is the **receiver**. Note that we've used the long syn-

[6] Generators are more commonly created with the `yield` statement, but we're using a low-level approach here simply to illustrate closures in action.

tax for function values: if we define the method value with the `function ( params ) { body }` syntax, the special expression `this` refers to the receiver. Functions defined with `( params ) => { body }` do not get a special `this`. You may wish to adopt the convention of using the `function` syntax for methods and the arrow notation in all other cases. If you prefer, there's a shorthand notation for the `function` syntax inside of object literals:

```
const circle = {
  radius: 10,
  area() {return Math.PI * this.radius * this.radius},
  circumference() {return 2 * Math.PI * this.radius},
  expand(scale) {this.radius *= scale}
};
```

The purpose of the special `this` expression is to allow *context-dependent code*. In the case of methods, `this` takes on the value of the method's receiver as determined at runtime. For example, if we copy a method defined in object A to object B and call the method through B, `this` will be B, not A. This **late binding** can be quite flexible, as we'll see in the next section.

JavaScript employs `this` in situations other than method calls. We can, for instance, force the value of `this` to take on the value of our choosing via `call`, `apply`, and `bind`:

```
function talkTo(message, suffix) {
  return message + ', ' + this.name + suffix;
}

const alice = {name: 'Alice', address: talkTo};
const bob = {name: 'Bob'};

const assert = require('assert');
assert(alice.address('Hello', '.') === 'Hello, Alice.');
assert(alice.address.call(bob, 'Yo', '!') === 'Yo, Bob!');
assert(alice.address.apply(bob, ['Bye', '...']) === 'Bye, Bob...');
assert(alice.address.bind(bob)('Right', '?') === 'Right, Bob?');
```

1.5 PROTOTYPES

Let's turn now from functions to objects. How do you efficiently create a number of similar objects? How do you define dozens, thousands, or millions of points, or of circles, or people, or votes, or airports, or web page index entries?

In JavaScript, we start with an initial (prototypical) object, then *derive* additional objects from it. These new objects have the original object as their **prototype**. What is a prototype? When we encounter the expression $q.x$, we look for an x property in q. If found, we produce the corresponding value; if not, we'll look in q's prototype (if it has one), and if necessary, in the prototype's prototype, and so on, until we find the property or reach the end of the "prototype chain." If no object on the chain has the property, the lookup produces `undefined`.

```
const unitCircle = {
  x: 0,
  y: 0,
  radius: 1,
  color: 'black',
  area() {return Math.PI * this.radius * this.radius},
  circumference() {return 2 * Math.PI * this.radius}
};

const c1 = Object.create(unitCircle);
c1.x = 3;
c1.color = 'green';

const c2 = Object.create(unitCircle);
c2.radius = 5;

const c3 = Object.create(unitCircle);

const assert = require('assert')
assert(c2.color === 'black' && c2.area() === 25 * Math.PI);
assert(c3.y === 0 && c3.area() === Math.PI);
```

The expression `Object.create(`*p*`)` creates a new object whose prototype is *p*. Our script creates a black circle of radius 1, centered at the origin, as the prototype of three other circles. Because of the way JavaScript **delegates** property lookup, we have `c1.x === 3` and `c1.color === 'green'` (obviously), as well as `c1.y === 0` and `c1.radius === 1`. In the object referenced by `c1`, `x` and `color` are called **own properties**, while `y` and `radius` are called **inherited properties**.

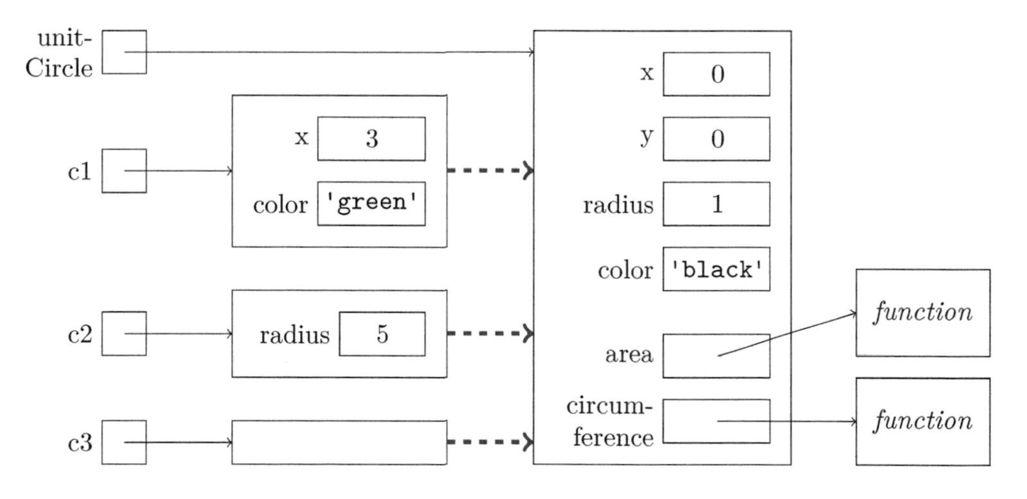

Figure 1.5 Objects sharing a prototype

Note that in each of the newly created objects, we store only those properties whose values *differ* from those in the prototypal circle. In particular, each circle inherits all of the methods from the prototype. Every circle computes its area and circumference the same way, so it would be wasteful to store copies of these functions in each circle.

Figure 1.5 shows the four circles and the prototype links to `unitCircle`. For simplicity, our diagram omits the prototype links from `unitCircle` to its prototype, and from the two functions to their prototype.[7]

In practice, programmers will want a function to construct each instance of a family of objects that each share a prototype. Interestingly, JavaScript provides a mechanism to do just that. We'll illustrate the technique with an example, and then discuss:

```
function Circle(centerX=0, centerY=0, radius=1, color='black') {
  this.x = centerX;
  this.y = centerY;
  this.radius = radius;
  this.color = color;
}

Circle.prototype.area = function () {
  return Math.PI * this.radius * this.radius;
};

Circle.prototype.circumference = function () {
  return 2 * Math.PI * this.radius;
};

const assert = require('assert');
const c = new Circle(1, 5);
assert.deepEqual(c, {x:1, y:5, radius:1, color:'black'})
assert(c.area() === Math.PI);
assert(c.circumference() === 2 * Math.PI);
assert(Object.getPrototypeOf(c) === Circle.prototype);
assert(c.constructor) === Circle;
assert(typeof(c) === 'object');
```

This simple looking script illustrates a lot of JavaScript magic. Every JavaScript function has two properties, `length` (the number of parameters) and `prototype`, the object that will be assigned as the prototype of all objects created by calling the function with the operator `new`. It's primed for you with a `constructor` property, referencing the function. This allows you to determine the function that created an object as a kind of run time "type check," since, as the last line of our example shows, `Circle` isn't a JavaScript type. That's a lot to take in, but studying Figure 1.6 may help!

The declaration `let c = new Circle(1, 5);` creates a new object with prototype `Circle.prototype`, passes it to the function `Circle` as `this`, and binds the now-initialized object to the variable *c*. The area and circumference methods are stored as properties in the prototype, just as in our earlier example. As an aside, we've taken the opportunity here to introduce **default parameter values**—parameters initialized to specified values, rather than `undefined`, when no argument is supplied.

Creating a constructor function to build **instances** of a user-defined type, and loading up the shared state and behavior into a common prototype occurs so often in JavaScript that there is a shorthand syntax for this pattern:

[7] One of the end-of-chapter exercises asks you to complete the figure.

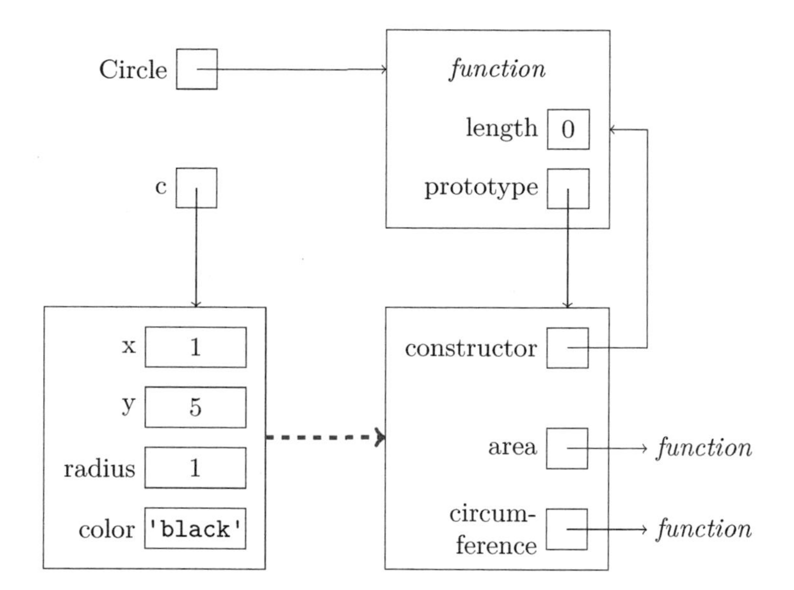

Figure 1.6 Construction of Objects using `new`

```
class Circle {
  constructor(centerX=0, centerY=0, radius=1, color='black') {
    this.x = centerX;
    this.y = centerY;
    this.radius = radius;
    this.color = color;
  }
  area() {return Math.PI * this.radius * this.radius;}
  circumference() {return 2 * Math.PI * this.radius;}
}

const assert = require('assert');
const c = new Circle(1, 5);
assert(c.circumference() === 2 * Math.PI);
assert(typeof Circle === 'function');
```

The last line of the script above shows us that the `class` keyword *does not* create a "class object": `Circle` is still a function! The class construct is **syntactic sugar**—a syntax that makes the standard form easier to read. It does no more and no less (in this case) than defining the function and assigning methods to the prototype.

1.6 SCOPE

A **binding** is an association of a name with an entity. The **scope of a binding** is the region of code where a particular binding is active. Let's take a look at two ways we can introduce bindings in JavaScript: `let` and `var` (`const` works here like `let`). Bindings introduced with `var` are scoped to the innermost function, and bindings introduced with `let` are scoped to the nearest block:

```
const assert = require('assert');

const a = 1, b = 2;

(function () {
  assert(a === undefined);      // the local `a` is in scope
  assert(b === 2);              // we see the outer `b`

  if (true) {
    var a = 100;                // scoped to whole function!
    let b = 200;                // scoped only inside this block
    const c = 300;                // scoped only inside this block
  }
  assert(a === 100);            // it's been initialized
  assert(b === 2);              // outer, because local used `let`

  assert.throws(() => c);       // there's no `c` out here at all
})()
```

Reading a `var`-declared variable in its scope but before its declaration produces `undefined`, as shown above. Reading a `let`-declared variable in its scope but before the `let` throws a `ReferenceError`.

Given that there are several other ways to create bindings (`const`, function declarations, class declarations, function parameters, etc.), the complete set of rules that determine scope would fill many pages. In general, we will not be detailing the complete scoping rules for any language in this book, though we will make time to show some of the more interesting design choices surrounding scope.

1.7 CALLBACKS AND PROMISES

Programmers often have to deal with uncertainties surrounding time. Users will, without any warning, click buttons, drag fingers across a surface, press and release keys, and move cursors in and out of regions on a display. Reading and writing files and databases, and exchanging information with programs running on different machines may take several seconds to a few minutes or more to complete. When an application stops and waits for external operations to complete, we call its behavior **synchronous**; if it can continue to do useful work until the long running operation finishes, we have **asynchronous** behavior.

An asynchronous architecture consists of code for *firing* and *responding to* **events**. Some events originate outside the program (button clicks, a cursor entering a canvas, data becoming ready from a file read) and some from **event emitters** that you write yourself. In either case, events are added to a task queue which the JavaScript engine repeatedly pulls from. An interactive application contains instructions saying "When event e is pulled from the queue, call function f with the data provided by e". Let's see how this looks in a browser. The following script creates a little canvas you can sketch in.[8]

[8]To run this script in a browser, save it in the file *sketch.js*, Create a new file, *sketch.html*, with the content `<script src="sketch.js"></script>`, and finally, open the HTML file in the browser.

```
window.addEventListener('load', e => {
  const canvas = document.createElement('canvas');
  const ctx = canvas.getContext('2d');
  let drawing = false;
  canvas.style.border = '2px solid purple';
  canvas.addEventListener('mousedown', e => {
    drawing = true;
    ctx.moveTo(e.clientX, e.clientY);
  });
  canvas.addEventListener('mousemove', e => {
    if (drawing) {
      ctx.lineTo(e.clientX, e.clientY);
      ctx.stroke();
    }
  });
  const stopDrawing = e => {drawing = false};
  canvas.addEventListener('mouseup', stopDrawing);
  canvas.addEventListener('mouseout', stopDrawing);
  document.body.appendChild(canvas);
});
```

The functions passed as the second argument of `addEventListener` are called **event handlers**, or **callbacks**, and the act of adding the listeners to the various objects known to the browser is called **registering** the callback. When an event is pulled from the queue, the browser passes an **event object**, containing data about the event, to your callback. Mouse events will contain the cursor position in the `clientX` and `clientY` properties; touch events for phones and tablets work in a similar fashion. Each of the callbacks is run to completion, one after the other.

To see how things work on the server side, let's extend our word count script from the beginning of the chapter. We'll allow words containing any letter from the Unicode character set, rather than limiting ourselves to the Basic Latin letters.[9]

```
const reader = require('readline').createInterface(process.stdin, null);
const XRegExp = require('xregexp').XRegExp;
const counts = new Map();

reader.on('line', line => {
  const wordPattern = XRegExp("[\\p{L}']+", 'g');
  for (let word of line.toLowerCase().match(wordPattern) || []) {
    counts.set(word, (counts.get(word) || 0) + 1);
  }
}).on('close', () => {
  for (let word of Array.from(counts.keys()).sort()) {
    console.log(`${word} ${counts.get(word)}`);
  }
});
```

[9]Full Unicode support for matching is not built-in to Node.js, you'll have to install an external module by invoking `npm install xregexp` on the command line.

Node's `process.stdin` is a file **stream** representing standard input. In Node, streams are event emitters; `stdin` will fire a `line` event when a line is ready to be read, and a `close` event when there is no more data to be read. We call the `on` method on the reader to register the callback to invoke when the event fires.

Node.js comes with a number of built-in modules that contain event emitters, such as `stream`, `fs` (for file system), `networking`, and `http`. Events are fired at all the expected times: when streams or files are opened or closed, when data from a file, stream, or socket becomes ready, or upon timeouts or errors.

As an alternative to writing asynchronous functions that take callbacks as parameters, you can write functions that return **promises** instead. A JavaScript promise is built from the `Promise` constructor of the standard library with a single argument, called the executor. The executor has two parameters, `resolve` and `reject`, running asynchronously, eventually calling `resolve` to indicate success or `reject` to indicate failure. You can build up a series of asynchronous calls by chaining promises with `then` (to capture successful resolutions) or `catch` (to capture rejections). The following example simulates a chain of three long running (three whole seconds each) asynchronous tasks. For simplicity, we illustrate potential "failure" only for the first task.

```
function initialize(configuration) {
  console.log('Initializing ' + configuration);
  return new Promise((resolve, reject) => {
    if (!configuration) {
      reject('Empty configuration');
    } else {
      setTimeout(() => resolve('the initialized data'), 3000);
    }
  });
}

function process(initialData) {
  console.log('Processing ' + initialData);
  return new Promise((resolve, reject) => {
    setTimeout(() => resolve('the processed data'), 3000);
  });
}

function report(output) {
  console.log('Reporting ' + output);
  return new Promise((resolve, reject) => {
    setTimeout(() => resolve('the reported data'), 3000);
  });
}

initialize('the configuration data')
  .then(text => process(text))
  .then(value => report(value))
  .then(value => console.log('Success: ' + value))
  .catch(reason => console.log('Error: ' + reason));
```

JavaScript's asynchronous callbacks and promises comprise only two ways to manage **concurrency**, the modeling and coordination of independent and distinct computing activities whose execution spans may overlap in time. We will see others throughout the book, including coroutines, threads, processes, actors, and functional reactive programming.

1.8 JAVASCRIPT WRAP UP

In this chapter we were introduced to JavaScript. We learned that:

- All modern web browsers run JavaScript; in addition, engines such as Node.js run JavaScript on the command line or server.
- JavaScript expressions manipulate values of exactly seven types: Undefined, Null, Boolean, Number, String, Symbol, and Object. The first six are primitive types; Object is a reference type. Arrays and functions are objects. Values of reference types are actually pointers, so assignment of these values creates aliasing, or sharing.
- JavaScript is weakly-typed, meaning that in most situations a value e of type t can be used in a context in which a value of type t' is expected. The runtime will find a coercion of e to a roughly-equivalent expression e' of type t'.
- Values that coerce to false are called *falsy*; all other values are called *truthy*. The only falsy values in JavaScript are `undefined`, `null`, `0`, `NaN`, the empty string, and the empty object.
- When calling functions, arguments are fully evaluated and then passed to parameters by copying values. Extra arguments are ignored; extra parameters are assigned `undefined`, or a default value if specified. Parameters and local variables live only during the function activation and are invisible to outer scopes.
- When functions with free variables are passed into, or copied into, different scopes, the free variables continue to refer to the variables in their originally enclosing scopes. Functions taking advantage of this ability are called closures.
- JavaScript is statically scoped and uses deep binding. It supports both function-scoped (`var`) and block-scoped (`let`, `const`) entities.
- Object properties are either *own properties* or *inherited properties*. Property lookup traverses the prototype chain, if required.
- The value of the expression `this` is context-dependent. It refers to (1) the global object when used in a global context, (2) the receiver of a method in a method call, provided the method is *not* defined with the fat arrow, (3) the newly created object when used with operator `new`, or (4) an object designated by the programmer in certain methods, such as `apply`, `call`, and `bind`.
- The keyword `class` provides sugar for defining a constructor function and methods that populate a prototype object.
- The design of JavaScript facilitates asynchronous, event-driven programming with callbacks. The standard library provides promises, which can be used instead of callbacks in many situations.

To continue your study of JavaScript beyond the introductory material of this chapter, you may wish to find and research the following:

- **Language features not covered in this chapter**. Symbols, regular expressions, property attributes, destructuring, generators, template strings, typed arrays, proxies, freezing, and the built-in objects. Some of these topics not covered here will appear later in the book in the context of other languages.
- **Open source projects using JavaScript**. Studying, and contributing to, open source projects is an excellent way to improve your proficiency in any language. Of the thousands of projects using JavaScript, you may enjoy jQuery (`https://github.com/jquery/jquery`), d3 (`https://github.com/mbostock/d3`), Backbone (`https://github.com/jashkenas/backbone`), and Underscore (`https://github.com/jashkenas/underscore`).
- **Reference manuals, tutorials, and books**. Strictly speaking, JavaScript is an implementation of the language ECMAScript, whose official specification is [28]. Douglas Crockford's *JavaScript: The Good Parts* [20] is a popular text covering many aspects of the language and its usage, while Axel Rauschmayer's *Speaking JavaScript* [102] and *Exploring ES6* [103] cover the language in depth. *NodeJS in Action* [15] will help you use the popular NodeJS and its ecosystem to build complex applications.

You may also be interested in several relatives of JavaScript, including the low-level *asm.js* [55], the web "compilation target format" WebAssembly [128], and the JavaScript superset language TypeScript [85].

EXERCISES

Now it's your turn. Continue exploring JavaScript with the activities and research questions below, and feel free to branch out on your own.

1.1 Find out how to execute JavaScript code in your web browser's "Developer Tools."

1.2 If you have not already done so, install Node.js on your machine. Try out the REPL (do a web search for "node repl" if you do not know what one is). Evaluate simple expressions such as `2+2` and `true || false`.

1.3 Evaluate the following JavaScript expressions in the Node REPL or browser console: (a) `"16" == 16`, (b) `16 == "0x10"`, and (c) `"0x10" == "16"`. Were the results surprising? Why or why not? If you said no, why might a reasonable person be surprised? (Hint: Consider transitivity.) What happens when you replace the `==` operator with the `===` operator?

1.4 Learn about standard output and standard error (Wikipedia has information at [139]). Why is it so important to write error messages to standard error rather than standard input?

1.5 Why is it so important that processes produce a return code of 0 on success, and non-zero on failure?

1.6 For the first three examples in this chapter, identify their literals, variables, operators, expressions and statements. (A precise understanding of this vocabulary is essential to being able to design, analyze, and implement languages.)

1.7 Research and explain the difference between `let` and `const`.

1.8 Write a function to produce a deep copy of an array of points, where each point is an object with three properties (x, y, and z) with numeric values.

1.9 Find out how to list all of the properties of the global object in your favorite JavaScript environment.

1.10 We saw, but did not explain, the array methods `every`, `some`, `filter`, and `map`. Each have an optional second argument. What is this second argument and why is it needed? Write a script that illustrates the need.

1.11 Try out this (presumably incorrect) script in a browser:

```
// Rookie mistake using var: alerts 10 for every button.
for (var i = 0; i < 10; i++) {
  const button = document.createElement("button");
  button.innerHTML = i;
  button.addEventListener('click', e => {alert(i);});
  document.body.appendChild(button);
}
```

What happens when each button is pressed? Why? Does changing `var` to `let` change the behavior? Why or why not?

1.12 If the previous example were rewritten to invoke the `forEach` method on the array `[0,1,2,3,4,5,6,7,8,9]`, would the surprising behavior occur?

1.13 Research the issue of static vs. dynamic scoping, and list the advantages and disadvantages of each. Do any modern languages use dynamic scoping? If so, which ones?

1.14 JavaScript uses static scoping and deep binding. Do you think it could have been designed to use shallow binding instead? What would be the difficulties in implementing shallowing binding, if any?

1.15 `Let`-declarations in `for` loops are special: there's a new binding for *each execution* of the loop. Given this JavaScript fact, try to determine the output of this script, before executing it:

```
let a = [], b = [];
for (var x of [1, 2, 3]) {
  a[x] = () => x;
}
for (let y of [1, 2, 3]) {
  b[y] = () => y;
}
console.log(a[1]());
console.log(b[1]());
```

Execute the script to see if your analysis was correct.

1.16 Rewrite the IIFE for `nextSquare` on page 19 so that the local variable `previous` is a parameter, and is initialized via the argument in the invocation. Why does this alternative implementation work? Do you find this version more or less readable than the original?

1.17 It turns out that JavaScript has a more direct way of implementing generators, using the keyword `function*` and the `yield` statement. Research the topic of generator functions and rewrite the `nextSquare` example to use this mechanism.

1.18 Research the details of the `call`, `apply`, and `bind` methods that we mentioned, but did not explain, in this chapter. Explain the behavior of each in your own words.

1.19 For a function f, object o, and expression e, show that the expressions `f.call(o,e)` and `f.bind(o)(e)` produce the same result. What, then, can `bind` do that `call` cannot? Hint: Look up the term "partial application."

1.20 What object is assigned as the prototype of all objects created with the object literal syntax (e.g., `{}`)? Which object is assigned as the prototype of all functions? What properties do these objects have? What are the prototypes of those prototypes?

1.21 Complete Figure 1.5 to show the prototypes of the unit circle object and its two methods. Show the prototypes of the prototypes where they exist.

1.22 In the example at the end of Section 1.5, we see a constructor function designed to be called with operator `new`. What happens if it is called without `new`?

1.23 Using the `lodash` module from npm, write a function that outputs a shuffled deck of (traditional) playing cards. Represent the deck as an array of 52 cards, each with one of thirteen ranks and one of four suits. (Hint: look for a method called `shuffle` in lodash.)

1.24 (Challenge) We have included the apostrophe in the set of "word characters" in our word count example from this chapter, allowing us to correctly pick up as words the following from the *War and Peace* text file from Project Gutenberg: `who'll`, `i've`, and `zdrzhinski's`. However, the following are picked up as "words" too: `'did`, `already'`, and the lone `'`, since the text uses apostrophes for quoted text within quoted text. Do some research on the topic of **regular expressions** and rewrite the script so that the pattern picks up apostrophes as word characters *only* if they are immediately preceded and followed by a letter.

1.25 Explore the introductory JavaScript course at Khan Academy. [72] Create and share an animation of your own design.

1.26 Rewrite the final example in the chapter to use callbacks instead of promises.

CHAPTER 2

CoffeeScript

CoffeeScript attempts to "expose the good parts of JavaScript in a simple way." [10]

First appeared 2009
Creator Jeremy Ashkenas
Notable versions 1.0 (2010) • 1.10 (2015)
Recognized for Transpiling to JavaScript, Expressiveness
Notable uses Web application clients
Tags Functional, Expression-Oriented, Dynamic
Six words or less "It's Just JavaScript"

CoffeeScript appeared in late 2009 with the goal of making web application clients easier to write. Because web browsers run JavaScript natively, CoffeeScript code is universally translated into JavaScript for execution. In fact the **golden rule of CoffeeScript** is "It's just JavaScript."

CoffeeScript exposes JavaScript's Good Parts only, and introduces a handful of features that were not present in JavaScript when CoffeeScript was first written, among them destructuring assignment, default parameter values, splats, and the `class` syntax for creating constructors and assigning prototype properties. It also features a few concepts not (yet!) in JavaScript, including the existential operators, the `do` construct, and comprehensions. It is an **expression-oriented language**: where JavaScript has `while`, `for`, and `if` *statements*, these constructs are *expressions* in CoffeeScript.

CoffeeScript, by design, removes a lot of the "noise" in JavaScript. It does away with curly braces for bracketing (preferring indentation), parentheses for arguments in function calls (though you can use them if you like), and allows object literals to forego braces and use indented lines for each property. Many of the messy parts of JavaScript have no translation at all. It wisely throws out JavaScript's `==` completely: writing `==`, or its synonym `is`, **transpiles** (translates at the source level) to JavaScript's `===`.

In this chapter, we will treat CoffeeScript as a language in its own right, but because of its golden rule, we'll make extensive comparisons with the language of the previous chapter.

2.1 HELLO COFFEESCRIPT

We begin with the same three introductory programs that began Chapter 1. First up is a script to generate integer right triangle measurements.

```
for c in [1..40]
  for b in [1...c]
    for a in [1...b]
      console.log "#{a}, #{b}, #{c}" if a * a + b * b is c * c
```

CoffeeScript expresses program structure by indentation rather than braces. Its `for-in`-loop iterates through the *values* of an array, and is equivalent to JavaScript's `for-of` construct.[1] Two dots give a range an inclusive upper bound; three dots generate an exclusive range. String interpolation uses `#{...}` and works only inside strings delimited with double quotes, not single quotes. The `if`-clause may *follow* an expression to be conditionally executed, useful when you wish to emphasize the action being taken rather than the condition under which it needs to be taken.

Let's translate our command line permutations example from last chapter's JavaScript version:[2]

```
generatePermutations = (a, n) ->
  if n is 0
    console.log a.join ''
  else
    for i in [0...n]
      generatePermutations a, n-1
      j = if n % 2 is 0 then 0 else i
      [a[j], a[n]] = [a[n], a[j]]
    generatePermutations a, n-1

if process.argv.length isnt 3
  console.error 'Exactly one argument is required'
  process.exit 1
word = process.argv[2]
generatePermutations word.split(''), word.length-1
```

Like JavaScript, CoffeeScript function expressions can be written in two ways. CoffeeScript's *thin arrow* syntax, (*params*) `->` *body*, is equivalent to JavaScript's `function (` *params* `) {` *body* `}`, while the *fat arrow* syntax works like JavaScript's fat arrow. Both languages have a destructuring assignment that simplifies swapping array elements.

In line with its philosophy of minimal syntactic noise, parentheses are optional in function calls, allowing you to write `f x, y` for `f(x, y)` and `f x y` for `f(x(y))`. Without parentheses, the parser will slurp up as many arguments as it can, meaning that `f x, g y, z` parses as `f(x, g(y, z))` rather than `f(x, g(y), z)`.

[1]In JavaScript `for-in` will iterate through the *indexes* of an array.

[2]This script is designed to be run with Node.js; running `npm install -g coffee-script` will get you a command line interpreter and program runner.

Now for our third introductory program:[3]

```
reader = require('readline').createInterface process.stdin, null
{XRegExp} = require 'xregexp'
counts = new Map()

reader.on 'line', (line) ->
  wordPattern = XRegExp("[\\p{L}']+", 'g')
  for word in (line.toLowerCase().match(wordPattern) or [])
    counts.set word, (counts.get(word) or 0) + 1

reader.on 'close', ->
  for word in Array.from(counts.keys()).sort()
    console.log "#{word} #{counts.get word}"
```

What's new from JavaScript? We see that functions with zero parameters can be written beginning with an arrow without the need for an explicit empty parameter list. The operator `or` can be used in place of `||` (and as you probably guessed, `and` for `&&` and `not` for `!`).

Did you notice one other difference between CoffeeScript and JavaScript that we had not mentioned: the lack of a `var` or `let` keyword? Keep this in mind, we'll have a *lot* to say about it in an upcoming section.

2.2 THE BASICS

Although CoffeeScript can be considered a dialect of JavaScript, the two languages "look" rather different. JavaScript is a *curly-brace language*, using `{` and `}` to express code structure, while CoffeeScript uses indentation. Indentation gives structure not only to compound statements, but also to objects:

```
circle =
  radius: 3
  center:
    x: 5
    y: 4
  color: 'green'
  fillColor: 'pink'
  thickness: 2
  lineStyle: 'dashed'
```

Since the line breaks and the number of spaces matter in determining the structure of a script, we say CoffeeScript has a **significant whitespace** syntax, as opposed to a **free-form** syntax. The moniker "free-form" refers to the way that multiple statements can appear on a line, or a single statement can be broken up across multiple lines, with very little, if any, constraints.

CoffeeScript employs whitespace-awareness to facilitate writing long strings, a task that many languages make surprisingly clumsy. Strings delimited by apostrophes or double

[3]This script makes use of an external module. Run `npm install xregexp` to make it available to your script.

quotes are allowed to span multiple lines; each line will be joined by a *single* space, even though you've kept your code pretty with proper left margins. For strings that *must* contain line breaks, use triple quotes (`'''` or `"""`); CoffeeScript will suppress the initial whitespace on each line to keep everything clean:

```
example =
  multi: "This is
    really a one line
    string"
  block: """
    One
    Two
      TwoPointFive
    Three
    """

console.log example.multi
console.log example.block
```

This script outputs:

```
This is really a one line string
One
Two
  TwoPointFive
Three
```

CoffeeScript expressions, types, and variables mimic their JavaScript counterparts. Functions do too: function values can be assigned to variables, passed as parameters to other functions, and even called directly without being named:

```
square = (x) -> x ** 2
squares = [1..5].map(square)  # pass function by name
result = ((x) -> x * 5) 16    # call anonymous function

assert = require 'assert'
assert.deepStrictEqual(squares, [1, 4, 9, 16, 25])
assert result is 80
```

Similarly, arguments are passed by value, extra arguments are ignored, and extra parameters start off `undefined`. It's also possible to specify a default value for a parameter when no corresponding argument is supplied:

```
f = (x, y=1, z=0) -> x * y + z

assert = require 'assert'
assert f(2) is 2             # 2*1+0 = 2
assert f(3, 5) is 15         # 3*5+0 = 15
assert f(2, 8, 1) is 17      # 2*8+1 = 17
```

You can even pass multiple arguments to a function, but have them all *packed* into a single parameter, using a **splat**:

```
average = (a...) ->
  (a.reduce ((x, y) -> x + y), 0) / a.length

assert = require 'assert'
result = average 7.5, -10, 50.5
assert result is 16
assert isNaN average()
```

Within the function body, the splat `a` is just an array. We can also use splats in calls to *unpack* an array before passing:

```
medianOfThree = (x, y, z) ->
  x + y + z - (Math.max x, y, z) - Math.min x, y, z

assert = require 'assert'
numbers = [80, 20, 55]
middle = medianOfThree numbers...
assert middle is 55
```

CoffeeScript has a `this` keyword, which operates exactly like JavaScript's `this`. You have two nice shorthands in CoffeeScript: `@` for `this`, and `@`*property* for `this.`*property*:

```
circle =
  radius: 10,
  area: -> Math.PI * @radius * @radius
  circumference: -> 2 * Math.PI * @radius

assert = require 'assert'
assert circle.area() is 100 * Math.PI
assert circle.circumference() is 20 * Math.PI
```

This script works because functions defined with the thin arrow (`->`) set the value of `this` to the receiver when used as methods. But CoffeeScript also has JavaScript's fat arrow (`=>`), which does *not* set the value of `this`. This is useful for those cases where we have functions nested inside of methods, and we do not want the value of `this` in the nested functions hijacked by the inner function.

The following example should help to illustrate the difference between the thin and fat arrows. A person wants to say hello after a one second delay. The global `setTimeout` function takes a function and a duration in milliseconds, and executes the function after the duration is passed. The code will only work with the fat arrow, since the thin arrow function hijacks the value of `this`:

```
person =
  name: 'Alice'
  tryToSayHelloButFail: (delay) ->
    setTimeout (() -> console.log "Hi from #{@name} :("), delay
  sayHello: (delay) ->
    setTimeout (() => console.log "Hi from #{@name} :)"), delay

person.tryToSayHelloButFail 1000          # Hi from undefined :(
person.sayHello 1000                      # Hi from Alice :)
```

CoffeeScript carries over many other features of JavaScript, including `Object.create` and the `new` operator. Prototypes and the expression `this` (and its alias `@`) work the same way. The `class` expression generates constructors and prototypes, just as in JavaScript. However, CoffeeScript comes with a slick way to handle property initialization:

```
class Circle
  constructor: (@x=0, @y=0, @radius=1, @color='black') ->
  area: -> Math.PI * @radius * @radius
  circumference: -> 2 * Math.PI * @radius
  expand: (scale) -> @radius *= scale

circles = [(new Circle 3, 5, 10, 'blue'), (new Circle)]

assert = require 'assert'
assert circles[0].color is 'blue'
assert circles[1].circumference() is 2 * Math.PI
```

As in JavaScript, the CoffeeScript `class` construct generates a function with the parameters of the constructor. The declarations inside the class—with the exception of the constructor—become properties of the associated prototype. There's some lovely syntactic sugar here: tagging a constructor parameter with an `@` allows you elide the oft-seen (boilerplate) property initialization within constructor bodies.

The class syntax makes it convenient to express **IS-A** relationships, which we can illustrate with some animals. All animals have a name, and when told to speak, they say their name and make a species-specific sound. Animals come in three kinds: cows, horses, and sheep. A language should make it convenient to capture what is common to all kinds of animals (e.g., having a name, speaking) in one place, and isolating the kind-specific behavior (the particular sounds made by each species).

Study the CoffeeScript solution:

```
class Animal
  constructor: (@name) ->
  speak: -> "#{@name} says #{@sound()}"

class Cow extends Animal
  sound: -> 'moooo'

class Horse extends Animal
  sound: -> 'neigh'

class Sheep extends Animal
  sound: -> 'baaaa'

assert = require 'assert'
h = new Horse 'CJ'
assert h.speak() is 'CJ says neigh'
c = new Cow 'Bessie'
assert c.speak() is 'Bessie says moooo'
assert new Sheep('Little Lamb').speak() is 'Little Lamb says baaaa'
```

Read the `extends` keyword as "is a"—that `Horse` extends `Animal` means a horse *is an* animal. Since an animal can speak, so can a horse. This works because the class extension mechanism makes `Animal.prototype` the prototype of `Horse.prototype`. The implementation is shown in Figure 2.1.

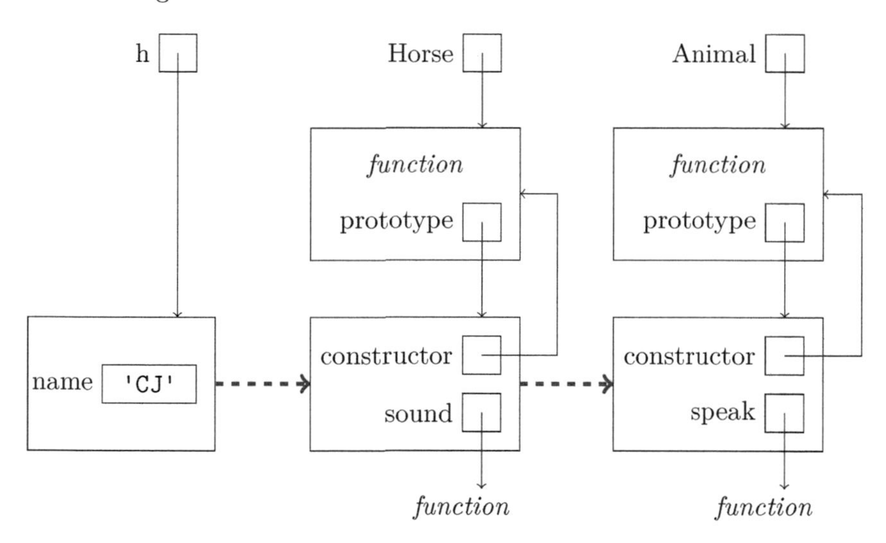

Figure 2.1 Implementation of IS-A in CoffeeScript

2.3 NO SHADOWING?!

In programming language theory, the **scope of a binding** is the region of the code where the binding (of a name to an entity) is in force. JavaScript and CoffeeScript both use lexical scoping: scopes are determined by looking only at the source code and not relying on any runtime behavior. JavaScript, as we saw in the last chapter, uses the `let`, `var`, or `const` keyword to explicitly create a new variable in an inner scope:

```
// JavaScript illustration of local variables
var a = 0, b = 1, c = 2;
(() => {
  var a = 100;     // Local, shadows the outer a
  b = 200;         // Forgot var, overwrites outer b!
  var d = 300;     // Local, will not exist after return
})();

const assert = require('assert')
assert.deepStrictEqual([a,b,c], [0, 200, 2])
assert.throws(() => d, ReferenceError)
```

Here the local variable *a* **shadows** the outer variable *a*. The two variables are distinct. As long as you remember to explicitly declare the inner variable with `var` or `let`, the outer variable is safe. If you forget the declaring keyword, you'll clobber an outer variable of the same name if one is present, or throw a `ReferenceError` if one is not present.[4] CoffeeScript, though, *never* shadows: writing `a = 100` inside a function will simply create a local variable *a* unless there is an outer *a*, in which case it will assign to the outer variable!

```
[a, b, c] = [0, 1, 2]
(() ->
  a = 100        # Overwrites outer: there's an a out there
  b = 200        # Overwrites outer: there's a b out there
  d = 300)()     # Local, because there's no d out there

assert = require 'assert'
assert.deepStrictEqual([a,b,c], [100,200,2])
assert.throws (-> d), ReferenceError
```

This design decision is controversial: a CoffeeScript function cannot, without getting fancy, declare a guaranteed local variable. Adding a new top-level variable to a file can turn a local variable non-local! While this is unlikely to happen when good coding standards are in place, (e.g., small modules, few globals, good naming conventions), the design has generated much discussion in the CoffeeScript community.[5]

Interestingly, parameters, unlike "local" variables, *do* shadow, so you can force the equivalent of a JavaScript-style local variable with an IIFE. Another "trick" is to use the `do` keyword, which sugars an IIFE. Let's see both of these in action:

```
[a, b, c] = [1, 2, 3]

bad = ->
  a = 10                      # clobbers outer a

okay = ->
  ((b) -> b = 20)()           # writes to local b

better = ->
  do (c=30) ->                # writes to local c

bad(); okay(); better()
assert = require 'assert'
assert.deepStrictEqual([a,b,c], [10,2,3])
```

2.4 COMPREHENSIONS

Programming languages vary in their ability to express values. Consider the common array. How can we initialize an array whose elements are generated via a known rule? We can always create an empty, mutable array, then fill it using a `for`-loop to generate candidate elements, and an `if`-statement to add only the desired elements. In CoffeeScript, we express such an array directly, using an array **comprehension**. Here we both get the names of

[4]For backward-compatibility reasons, JavaScript lets you run scripts in a "non-strict mode" in which the attempt to write to a non-declared variable creates a new global variable. This behavior has been identified as a cause of errors in real-world systems [53].

[5]`https://github.com/jashkenas/coffeescript/issues/712`

all employees earning a salary above some threshold, and get the employees with short names:

```
employees = [
  {name: 'alice', salary: 85000}
  {name: 'bob', salary: 77500}
  {name: 'chi', salary: 58200}
  {name: 'dinh', salary: 99259}
  {name: 'ekaterina', salary: 105882}
  {name: 'fahima', salary: 79999}]

assert = require 'assert'
highEarners = (e.name for e in employees when e.salary > 80000)
shortNames = (e.name for e in employees when e.name.length < 4)
assert.deepStrictEqual highEarners, ['alice', 'dinh', 'ekaterina']
assert.deepStrictEqual shortNames, ['bob', 'chi']
```

A comprehension is an expression that produces an array. No statements are required.

2.5 DESTRUCTURING

How do names take on values in a program? For many novice programmers, the answer "Assignment statements!" is the first thing to pop into their heads. With a little thought, passing arguments to parameters in function calls may come up, too. But assignment and parameter passing are simply specific instances of the generalized concept of **binding**. Once you begin to think in generalized terms, you may start asking questions such as "Why does assignment go right to left (`x=3`) instead of left-to-right (`3=x`)?" Or "Why can't I assign multiple values at once?" You might even combine those two questions and realize that binding might just be carried out via a generalized pattern match. Consider the following expression, from a hypothetical programming language:

```
{x:1, y: [z, "hello"]} <=> {y: [2, q], x: p}
```

The idea is to match up a variable (on either side) with the corresponding value on the other side. Our example would bind 1 to p (look at the values of the x keys on both sides), 2 to z (first element of the y-arrays), and `"hello"` to q. CoffeeScript doesn't go quite so far as binding variables on both sides, but it does allow expressive *patterns* on the left. The entire right-hand side is evaluated first (as is the case in classic assignment), then the variables on the left become bound. Some examples follow:

```
[x,y] = [10,20]                 # x gets 10 and y gets 20
[x,y] = [y,x]                   # yes, a swap!
{a, b} = {a: 5, b: 3}           # a nice idiom

{place: {name: mountain, loc: [lat, lon]}} = {
    place: {name: 'Everest', loc: [27.9881,86.9253]}}

assert = require 'assert'
assert.deepStrictEqual [x, y, a, b, mountain, lat, lon],
  [20, 10, 5, 3, 'Everest', 27.9881, 86.9253]
```

This mechanism is called **destructuring assignment** because the language is breaking up the complex objects for you. Contrast this with the more verbose *explicit* destructuring:

```
# NOTE: This is NOT idiomatic CoffeeScript!
# Use destructing assignment instead!

destination = {place: {name: 'Everest', loc: [27.9881,86.9253]}}
mountain = destination.place.name
lat = destination.place.loc[0]
lon = destination.place.loc[1]

assert = require 'assert'
assert.deepStrictEqual [lat,lon,mountain], [27.9881,86.9253,'Everest']
```

2.6 EXISTENTIAL OPERATORS

CoffeeScript, again like JavaScript, uses `null` for the intentional absence of information, and `undefined` to indicate no information is known. All other values are *known values*, and can act like objects—in fact, trying to access properties of `null` and `undefined` are among the few places a `TypeError` can ever be thrown:

```
assert = require 'assert'
assert 4.toFixed(2) is '4.00'
assert true.toString() is 'true'
assert 'abcde'.length is 5
assert [5,3,9,4,6].indexOf(3) is 1
assert.throws((-> null.toString()), TypeError)
assert.throws((-> undefined.toString()), TypeError)
```

The postfix operator `?` reports whether an expression is known, that is, not `null` and not `undefined`:

```
assert = require 'assert'

assert 78.8? is true
assert false? is true
assert []? is true
assert undefined? is false
assert null? is false
x = 9;
assert x? is true
```

The question mark combines with several operators to do some slick things. The most common use is

```
c = employee?.supervisor?.city?.name
```

which is a much nicer way to write:

```
c =
  if employee?
    _supervisor = employee.supervisor
    if _supervisor?
      _city = _supervisor.city
      if _city?
        _city.name
      else
        undefined
    else
      undefined
  else
    undefined
```

We also have the forms `a?[]` and `a?()`. And there's a `?=` which is a bit different than `or=`, and sometimes more useful. The expression `x ?= 1` assigns 1 to x only if x is `undefined` or `null`, whereas `x or= 1` will assign 1 if x is any falsy value, including 0.

The various uses of the existential operator are examples of **idioms**. Until you know what they mean, you might find them cryptic; however, once learned they are quite powerful and reduce one's cognitive load in understanding code. They also are perfectly sensible. Consider `supervisor?.name`: if the supervisor is undefined (unknown), then so is the supervisor's name.

2.7 COFFEESCRIPT WRAP UP

In this chapter we were introduced to CoffeeScript and its concise syntax. We learned that:

- CoffeeScript is "just JavaScript." In fact, the language is defined by how its constructs are translated to, or *transpiled to*, JavaScript. What is truthy (or falsy) in one language is truthy (or falsy) in the other. CoffeeScript purposely transpiles only to a subset of JavaScript, intentionally avoiding some of JavaScript's messier parts.
- CoffeeScript does not avoid all of JavaScript's bad parts. For example, it inherits JavaScript's entire type system, and is therefore weakly-typed.
- CoffeeScript is a significant-whitespace language, using indentation to show structure. It provides excellent support for multiline strings.
- Thin arrow functions provide the value of the receiver (if any) to the `this`-expression, while fat arrow functions do not. The symbol `@` is an alias for `this`, and `@`x aliases `this.`x.
- In an extremely controversial design decision, CoffeeScript local variables do not shadow variables in outer scopes. To avoid potential bugs, programmers should adopt good practices such as keeping modules short, and using different naming conventions for top-level and local names. Only immediately-invoked function expressions (IIFEs) and the `do` keyword provide a means to force a variable to be "local."
- CoffeeScript constructors feature a slick syntax for initializing properties without assignment.

- Array comprehensions, destructuring assignment, and existential operators are among the features that make CoffeeScript a relatively concise language.

To continue your study of CoffeeScript beyond the introductory material of this chapter, you may wish to find and research the following:

- **Language features not covered in this chapter**. Symbols, regular expressions, property attributes, block comments, generators, `unless`, the operators `**`, `//`, and `%%`, `super`, the switch statement, and chained comparisons.
- **Open source projects using CoffeeScript**. Studying, and contributing to, open source projects is an excellent way to improve your proficiency in any language. You may enjoy the following projects written in CoffeeScript: atom (`https://github.com/atom/atom`), dynamics.js (`https://github.com/michaelvillar/dynamics.js`), Brunch (`https://github.com/brunch/brunch`), and Hubot (`https://github.com/github/hubot`).
- **Reference manuals, tutorials, and books**. CoffeeScript's home page, describing the entire language, and containing links to the implementation itself is at `http://coffeescript.org`. The page also contains a *Try CoffeeScript* section where you can type CoffeeScript and see the translation to JavaScript in real time. Two noteworthy books are Trevor Burnham's *CoffeeScript: Accelerated JavaScript Development* [14] and Alex McCaw's *The Little Book on CoffeeScript* [81].

EXERCISES

Now it's your turn. Continue exploring CoffeeScript with the activities and research questions below, and feel free to branch out on your own.

2.1 Read the CoffeeScript documentation page at `http://coffeescript.org`. Make a note of the language features not covered in this chapter.

2.2 What is the golden rule of CoffeeScript? What is the motivation for this rule?

2.3 Enter and execute several little scripts in the "Try CoffeeScript" window at the CoffeeScript home page. Pay attention to the JavaScript translations.

2.4 Install Node.js and the Node CoffeeScript module to your machine. Experiment with the CoffeeScript REPL. Execute the complete scripts from this chapter on the command line.

2.5 Research the `in` and `of` keywords, and demonstrate their use in a small script of your own. How do they differ from `in` and `of` in JavaScript?

2.6 Evaluate the CoffeeScript expressions `'#{2+2}'` and `"#{2+2}"` and explain the results.

2.7 JavaScript does not have a 100% free-form syntax, because of something known as *automatic semicolon insertion*, or ASI. This means that although many statements are required to end with semicolons, a language processor will insert them for you in certain cases where the language rules think you might have omitted them. Research the ASI rules. Have the ASI rules been considered a success? Give four examples of cases in which the ASI rules produce non-intuitive statement separation.

2.8 Find out what the expression `a[..]` (for some array *a*) means. Where might you use it?

2.9 CoffeeScript can iterate through key-value pairs of an object like so:

```
caps =
  'ME': 'Augusta'
  'VT': 'Montpelier'
  'NH': 'Concord'
  'MA': 'Boston'
  'RI': 'Providence'
  'CT': 'Hartford'

for state, capital of caps
  console.log "The capital of #{state} is #{capital}."
```

Why did we not apply this technique in our word count example near the beginning of this chapter?

2.10 In the median-of-three-program in this chapter, what would be output if the parentheses were removed in the return statement of the function? Why?

2.11 Argue for or against this claim: "Splats are simply a syntactic device and offer no real benefits in expressive power."

2.12 Research and then summarize the arguments for and against CoffeeScript's choice to reject shadowing. See if you can find a published anecdote of a local variable inadvertently becoming global in an actual software project.

2.13 Consider this short function:

```
distance = (p, q) ->
  dx = q.x - p.x
  dy = q.y - p.y
  Math.sqrt(dx * dx + dy * dy)
```

If used in a context in which `dx` and `dy` were global variables, calling this function would modify those globals. Rewrite the function so `dx` and `dy` are truly local.

2.14 Browse the code of a couple popular open source projects using CoffeeScript. Do the developers go out of their way to make local variables truly local? If not, do you find that they use naming conventions or short modules in order to mitigate the risk of naming collisions?

2.15 Create a little script, based on the animals script in this chapter, for shapes. The basic shape class should have a constructor for initializing the shape's color, and a method to produce a string stating the shape's area and perimeter. The actual shapes should be circles and rectangles with their own constructors and implementations of area and perimeter. You will have to research CoffeeScript's `super` keyword to do a nice job.

2.16 In the expression `highEarners = (e.name for e in employees when e.salary > 80000)` we evaluate a comprehension and assign the resulting array to the variable `highEarners`. Suppose the parentheses were omitted in this expression. What would the meaning of this modified expression be?

2.17 Write a CoffeeScript comprehension using two iteration variables that produces a "two-dimensional" array containing all pairs of numbers $[X, Y]$ in which each member of the pair is an integer between 1 and 10, inclusive.

2.18 What does the following CoffeeScript expression produce? Why?

```
[x*y for x in [1..12] for y in [1..12]]
```

2.19 In the word count example in the opening section of this chapter, we included, without fanfare, the line `{XRegExp} = require 'xregexp'`. Explain the meaning of this line. Hint: It is a destructuring assignment, but do explain exactly what is being "destructured."

2.20 For CoffeeScript practice, translate by hand the browser-based sketching program on page 24. Compare your hand-translation to the translation produced by the translator at `http://coffeescript.org`. Given that you should find some differences in translation, think about how human and machine translations are likely to be inherently different.

2.21 CoffeeScript does not have JavaScript's conditional operator. Yet the expression `x ? y : z` is syntactically legal in CoffeeScript. Explain in detail how this expression is parsed.

2.22 Explain how the expression `x? y : z` parsed in CoffeeScript. Note the lack of a space between the `x` and the question mark. (Hint: yes, the expression is quite different from the expression in the previous problem.)

2.23 Find out how CoffeeScript interprets text delimited with the backquote character (`` ` ``).

2.24 Research why CoffeeScript and JavaScript produce different values for the expression `-10 < -5 < -1`.

CHAPTER 3

Lua

Lua is a "powerful, fast, lightweight, embeddable scripting language." [77]

First appeared 1993
Creators Roberto Ierusalimschy, Waldemar Celes, Luiz Henrique de Figueiredo
Notable versions 5.0 (2003) • 5.1 (2006) • 5.2 (2011) • 5.3 (2015)
Recognized for Tables, Interoperability with C
Notable uses World of Warcraft, Angry Birds, Scripting
Tags Imperative, Prototypal, Scripting
Six words or less Lightweight, fast, powerful scripting language

Lua was born in 1993 at PUC-Rio, Pontifícia Universidade Católica do Rio de Janeiro. It has been evolving steadily, but is still quite small (in terms of the number of concepts and basic features) with lightweight and fast implementations. Its designers created a classic **scripting language**: typically an application's logic is written in Lua, with portions, such as time-critical or device-specific code, written in the host language.

The language features a small number of types and operators, a means for defining functions, and only *one* data structure—the **table**—that does double duty representing both traditional arrays as well as key-value objects. Lua's small size should not be confused with a lack of power; on the contrary, Lua is *extensible*. Lua tables can have metatables, providing the ability to customize lookup and override operators. Lua's functions are first-class objects, providing all the power of functional programming. It can interoperate with other languages, including C, C++, Java, Fortran, Smalltalk, and Erlang.

In this chapter, we will introduce Lua. Because the language is small, we will give pretty good coverage of its basic elements, including a full list of its operators and types. We will cover functions and scope in some detail. We will then see how Lua is able to work with only a single data structure, the table. We'll explain metatables, and how they provide a mechanism for user-defined types nearly identical to JavaScript's prototypes. We'll see how Lua allows its operators to take on new meanings under programmer control. We'll close with Lua's implementation of a popular mechanism for concurrency: coroutines.

3.1 HELLO LUA

Welcome to Lua. Our traditional first program, listing some right-triangles, is:

```
for c = 1, 40 do
  for b = 1, c-1 do
    for a = 1, b-1 do
      if a * a + b * b == c * c then
        print(string.format("%d, %d, %d", a, b, c))
      end
    end
  end
end
```

Here we've introduced the numeric `for`-loop (note that *both* bounds are inclusive) and `string.format` for nice output. Our second example introduces **tables**, used in Lua both for lists (indexed starting at 1) and for key-value pairs:

```
function generatePermutations(a, n)
  if n == 0 then
    print(utf8.char(table.unpack(a)))
  else
    for i = 1, n-1 do
      generatePermutations(a, n-1)
      local j = n % 2 == 0 and i or 1
      a[j], a[n] = a[n], a[j]
    end
    generatePermutations(a, n-1)
  end
end

if #arg ~= 1 then
  io.stderr:write('Exactly one argument required\n')
  os.exit(1)
end
word = {utf8.codepoint(arg[1], 1, utf8.len(arg[1]))}
generatePermutations(word, #word)
```

Here the `arg` table contains the command line arguments, `#` is the length operator, and `~=` is the not-equal operator. As in previous chapters, we write a message to standard error if we do not get exactly one command line argument. Strings are immutable, so we put our characters into a (mutable) table in order to generate the permutations by swapping.

Now let's get word counts from standard input:

```
counts = {}
for line in io.lines() do
  line:lower():gsub('[a-z\']+', function(word)
    counts[word] = (counts[word] or 0) + 1
  end)
end
```

```
report = {}
for word, count in pairs(counts) do
  table.insert(report, string.format('%s %d', word, count))
end
table.sort(report)
for _, line in ipairs(report) do
  print(line)
end
```

We need to do a little more work than in previous chapters, because Lua tables can only be sorted over their integer-indexed keys, and counting words (efficiently) requires a table keyed on words. Therefore we first collect the words (and their counts) into the table `counts`. The function `pairs` iterates over every pair in the table, regardless of its index type; we use it here to build a second, list-like, table (`report`) to store the output lines. This second table is indexed by integers, starting at 1, and `table.sort` sorts the elements by value. While this does seem like a lot of work, we can, at least, iterate through the lines of a file using the builtin `io.lines` function, without resorting to an external module as we needed to do in JavaScript and CoffeeScript.

3.2 THE BASICS

All values in Lua belong to one of eight types: nil, boolean, number, string, function, thread, userdata, and table. The first four are primitive types, and the latter four are reference types. *Primitive* and *reference* have the same meanings as in JavaScript: primitives are immutable; references allow access to objects through multiple variables simultaneously. Lua doesn't do as many implicit type conversions as JavaScript but it does do some: anything can be coerced to a boolean. Strings and numbers are coercible to each other[1], but `true` and `false` are not coerced to strings or numbers. For conversions to boolean, Lua departs significantly from JavaScript and CoffeeScript: the *only* things that are falsy in Lua are `nil` and `false`.

While referencing undeclared variables in JavaScript and CoffeeScript throws an error, the value of an undeclared variable in Lua is just `nil`. This means that both undeclared variables and nonexistent object properties are treated the same in Lua—you always get `nil`.

Strings are sequences of 8-bit values (not 16), so the length of the string "café" as computed with the `#` operator, is 5, not 4, since it is counting the bytes in the UTF-8 encoding. Use `utf8.len` to count characters.

```
assert(x == nil)              -- Does not fail! x is nil

s = "café"
assert(#s == 5)               -- counts bytes
assert(utf8.len(s) == 4)      -- counts characters

function firstFewPrimes()
  return {2, 3, 5, 7, 9, 11, 13}
end
```

[1]However the attempt to coerce a non-numeric-looking string to a number generates an error, not `NaN`.

```
assert(type(4.66E-2) == "number")
assert(type(true and false) == "boolean")
assert(type('message') == "string")
assert(type(nil) == "nil")
assert(type(firstFewPrimes) == "function")
assert(type(firstFewPrimes()) == "table")
assert(type(coroutine.create(firstFewPrimes) == "thread"))

assert(0 and "" and 0/0)          -- all of these are truthy!
assert(not(false or nil))         -- only false and nil are falsy
```

Lua (as of version 5.3) has 25 operators across 12 precedence levels. From highest to lowest precedence, they are:

Operator(s)	Associativity	Description
`^`	R	exponentiation
`not # - ~`	L	logical negation, length, numeric negation, bit complement
`* / // %`	L	multiplication, division, floor division, modulo
`+ -`	L	addition, subtraction
`..`	R	concatenation
`<< >>`	L	left shift, right shift
`&`	L	bit and
`~`	L	bit xor
`|`	L	bit or
`< <= == ~= >= >`	L	comparison operators
`and`	L	(short-circuit) return first operand if falsy, else second
`or`	L	(short-circuit) return first operand if truthy, else second

There are a few differences from JavaScript here. Lua adds an exponentiation operator, a length operator, a floor division operator[2], and uses different operators for string concatenation and numeric addition. This means that Lua avoids the classic surprise in which `"The answer is " + x + y` reports an "answer" of 22 even when x and y both have a numeric value of 2. Lua does not have a conditional operator (e.g., `_?_:_`), nor any assignment operators! Assignment in Lua is a statement, not an expression. This may be a good thing: Assignment has side-effects, and its appearance in the middle of an expression may be confused (by some) with an equality test.

In an assignment statement such as `x, y = 0, 0` we have a *name list* on the left and an *expression list* on the right. An expression list isn't an actual list object, nor is it a tuple. It's simply a list of expressions that get evaluated and then assigned to the variables on the left side. "Extra" expressions on the right are ignored; extra names on the left become `nil`. Expression lists also appear in function return statements, since a function may return multiple values:

[2]Floor division applies the floor operator to the quotient: `5 // 2 == 2` but `5 / 2 == 2.5`.

```
function computeThreeThings()
  return 5, 6, 7
end

a, b = computeThreeThings()          -- extra results ignored
c, d, e, f = computeThreeThings()    -- extra vars get undefined
g, h, i = 4, computeThreeThings()    -- right-hand side is 4, 5, 6, 7

assert(a == 5 and b == 6)
assert(c == 5 and d == 6 and e == 7 and f == nil)
assert(g == 4 and h == 5 and i == 6)
```

Lua's error-handling mechanism does not use the ubiquitous `throw` and `try` statements, and there is no special exception type. However, errors, whether from adding booleans, calling `nil` as a function, or explicitly calling the `error` function, are **propagated**, together with some optional diagnostic information, to the caller of the failing operation. If you wish to trap an error, make a **protected call**. Invoking `pcall(f,x,y)` performs `f(x,y)` and returns two values: a success indicator (`true` or `false`), and an error message, if applicable. If an error is trapped, the first returned value will be `false`.

```
function add(x, y)
  return x + y
end

success, result = pcall(add, 5, 3)
assert(success == true and result == 8)
success, result = pcall(add, 5, false)
assert(success == false)
```

3.3 SCOPE

Lua's visibility rules are fairly straightforward: (1) Variables are global unless defined using `local`; (2) shadowing happens; and (3) quoting the reference manual, "The scope of a local variable begins at the first statement after its declaration and lasts until the last non-void statement of the innermost block that includes the declaration." [78] Blocks are the bodies of `do`, `if`, `while`, `repeat`, and `for` statements, as well as function bodies. Rule 3 means that we can use a name prior to its declaration in a block, and even in the initializing expression of its declaration; in this case, it will refer to the nonlocal entity currently bound to the name:

```
x = 1
do
  assert(x == 1)      -- global x because local not yet seen
  local x = x + 2     -- uses global x on right-hand side
  assert(x == 3)      -- now, FINALLY, we see the local x
end
assert(x == 1)        -- back in the global scope, local gone
```

There is another interesting consequence of rule 3: the following attempt to define a factorial function fails, because the helper function, f, is not recursive:

```
factorial = function (n)
  local f = function (n, a)                 -- WRONG !!!!
    return n==0 and a or f(n-1, a*n)        -- refers to GLOBAL f, crashes
  end                                       -- because nil is not callable
  return f(n, 1)                            -- calls local f
end

ok, reason = pcall(factorial, 10)           -- hoping for 3628800
assert(ok == false)
print(reason)                               -- your output will vary
```

Because the local f is not in scope until after its declaration is complete, we can make a function recursive by declaring it on one line, then assigning to it on the next:

```
local f
f = function (n, a) ... end
```

Because recursion is not uncommon, Lua provides an equivalent sugared form:

```
local function f (n, a) ... end
```

3.4 TABLES

Like JavaScript and CoffeeScript, Lua provides a single structure that can be used to hold both named and numbered properties. Lua's structure is called a **table**. Here is a contrived example to show how tables are created and iterated over. The functions `pairs` and `ipairs` produce **iterators**.

```
widget = {
  weight = 5.0,
  ['part number'] = 'C8122-X',
  'green',                              -- key is 1
  'round',                              -- key is 2
  [4] = 'magnetic',
  imported = false,
  'metal',                              -- key is 3
}

print('pairs iterates through ALL pairs in arbitrary order')
for key, value in pairs(widget) do
  print(key .. ' => ' .. tostring(value))
end

print('ipairs iterates integer-keyed pairs from 1 in order')
for key, value in ipairs(widget) do
  print(key .. ' => ' .. tostring(value))
end
```

Keys are generally enclosed in brackets, except when the key is a simple name (letters, digits, and underscores, not beginning with a digit). Unspecified keys are auto-generated as the integers 1, 2, 3, and so on.[3]

Keys are not restricted to strings and numbers, but they cannot be `nil`. Values can't be `nil`, either, since `nil` turns out to be the response to a request for a value at a nonexistent key. This treatment of `nil` is consistent with Lua's evaluation of an undeclared variable as `nil`. A table value of `nil` would be indistinguishable from the key not being present in the table. Therefore, to remove an element from a table, set the value at the desired key to `nil`.

```
colors = {'red', 'blue', 'green'}
dog = {name = 'Lisichka', breed = 'G-SHEP', age = 13}

-- Length operator counts number of integer keys only
assert(#colors == 3)
assert(colors[1] == 'red')

-- Need our own function to count all pairs!
function number_of_pairs(t)
  local count = 0
  for _, _ in pairs(t) do count = count + 1 end
  return count
end

-- Assignment of nil removes a key pair
assert(number_of_pairs(dog) == 3)
dog.age = nil
assert(number_of_pairs(dog) == 2)
```

Global variables are actually kept in a table called the *global environment.* This table contains

- A string at index `_VERSION`;
- Functions, including `assert`, `pairs`, `pairs`, `print`, `pcall`, `tonumber`, `tostring`, `require` among others; and
- Tables, including `coroutine`, `io`, `debug`, `string`, `utf8`, `table`, `math`, `os`, `package`, among others.

When you create your own global variables, they are added to this table. Hence, Lua's global environment table is analogous to JavaScript's global object.

3.5 METATABLES

Recall that in JavaScript and CoffeeScript, every object has a prototype (unless explicitly created, by the programmer, without one) to which it delegates the search for missing properties. Lua supports delegation, too: a table may have a **metatable**, though metatables do much more than simply extend the search for properties!

[3]Yes, they start at 1, not 0.

To attach metatable m to a table t, invoke `setmetatable(t,m)`.[4] The metatable will contain zero or more of the following properties, which will define how t behaves:

```
__add __sub __mul __div __mod __pow __unm __idiv __band __bor __bxor
__bnot __shl __shr __concat __len __eq __lt __le __index __newindex
__call __mode __tostring __metatable __gc
```

When evaluating `x+y`, Lua first checks if x has a metatable m with an `__add` entry, and if so, produces `m.__add(x,y)`. If not, y is checked in a similar fashion. The next 18 properties work in much the same way: `__sub` is used for subtraction, `__unm` for unary minus, `__concat` for `..`, `__len` for `#` and so on.[5]

The `__index` property allows delegation of property lookup: its value can be a function that returns the value for the missing property, or a table holding the missing property. Let's recast our JavaScript delegation example from Chapter 1 into Lua:

```
unitCircle = {x = 0, y = 0, radius = 1, color = "black"}
c = {x = 4, color = "green"}
setmetatable(c, {__index = unitCircle})
assert(c.x == 4 and c.radius == 1)
```

Let's use metatables to create a "type" of two-dimensional vectors useful in graphics applications. For simplicity, we'll keep our code light and implement only construction, vector addition, vector dot product, a magnitude function, and a conversion to string. Here is one implementation (illustrated in Figure 3.1):

```
Vector = (function (class, meta, prototype)
  class.new = function (i, j)
    return setmetatable({i = i, j = j}, meta)
  end

  prototype.magnitude = function (self)
    return math.sqrt(self.i * self.i + self.j * self.j)
  end

  meta.__index = prototype
  meta.__add = function (self, v)
      return class.new(self.i + v.i, self.j + v.j)
  end
  meta.__mul = function (self, v)
    return self.i * v.i + self.j * v.j
  end
  meta.__tostring = function (self)
    return string.format('<%g,%g>', self.i, self.j)
  end
  return class
end)({}, {}, {})
```

[4] Actually, in Lua every value, not just every table, can have a metatable. However, setting the metatable for non-tables requires the C API, which is beyond the scope of our overview.

[5] There are some restrictions here and there for some of the operator stand-ins; details are in the Lua Reference Manual.

```
u = Vector.new(3, 4)
v = Vector.new(-5, 10)
assert(tostring(u) == "<3,4>")
assert(tostring(v) == "<-5,10>")
assert(u.j == 4)
assert(u:magnitude() == 5.0)
assert(tostring(u + v) == "<-2,14>")
assert(u * v == 25)
```

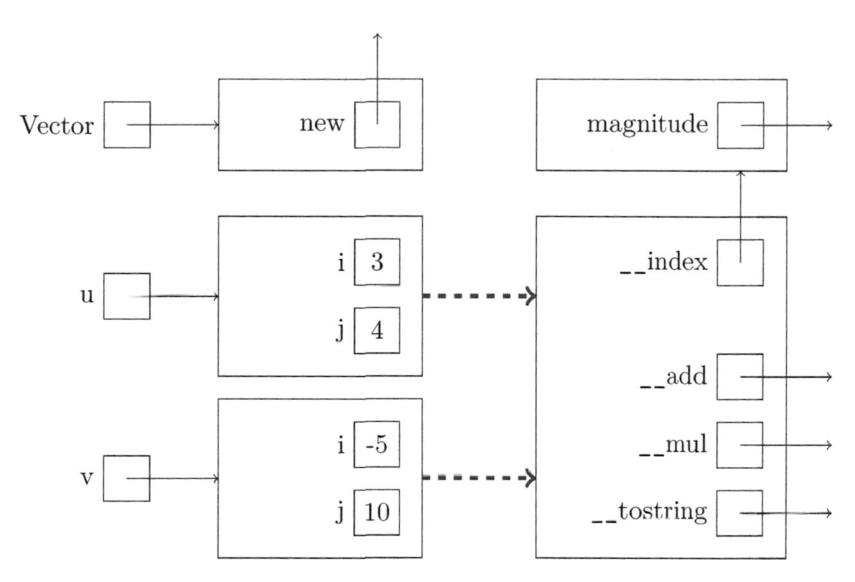

Figure 3.1 Vectors in Lua

Each call to `Vector.new` produces a vector instance, a table with `i` and `j` components, and a metatable containing operator implementations and a reference to a table with the shared `magnitude` property. Lua doesn't have the magic `this`-expression of JavaScript, so the magnitude function must explicitly be defined to take an argument. This means an invocation can be written:

```
v.magnitude(v)
```

However, Lua allows

```
v:magnitude()
```

as sugar for this expression, which is better because it evaluates v only once.

3.6 COROUTINES

A function runs until it terminates (via a normal return or an error), then returns to its caller. A **coroutine** can do the same, but also **yield**, in which case it can be *resumed* to continue where it left off. Use coroutines to build on-demand sequences:

```
nextSquare = coroutine.create(function ()
  for value = 1, 5 do
    coroutine.yield(value * value)
  end
  return "Thank you"
end)

for i = 1, 8 do
  local status = coroutine.status(nextSquare)
  local success, values = coroutine.resume(nextSquare)
  print(status, success, values)
end
```

Note that `coroutine.resume` returns a success flag (whether the coroutine has yielded or returned without error) followed by the values that were yielded or returned. The script above produces:

```
suspended   true    1
suspended   true    4
suspended   true    9
suspended   true    16
suspended   true    25
suspended   true    Thank you
dead        false   cannot resume dead coroutine
dead        false   cannot resume dead coroutine
```

A coroutine's status is one of

- **suspended**: if it has not yet started running or has yielded and waiting to be resumed,
- **running**: if it's running (and is the caller to `status`)
- **normal**: if it has resumed another coroutine and is waiting to be resumed itself, and
- **dead**: if it has terminated via a return or unprotected error.

Coroutines allow us to structure an application into concurrent activities, each written independently of each other. A game might contain coroutines for players, volcanos, tornados, waterfalls, and an event manager to listen for user input. Only one coroutine runs at a time (this is concurrency without parallelism), so each must contain explicit `yield` and `resume` calls, and not execute too long between these calls. This architecture is known as **cooperative multitasking**. Contrast this with **preemptive multitasking**, where the operating system can preempt a thread at anytime and give control to another thread.

3.7 LUA WRAP UP

In this chapter we were introduced to Lua, a high-performance, embeddable, lightweight, scripting language. We learned that:

- Lua is a relatively small language, with only a few statements, operators, types, and only one data structure.
- The only falsy values are `false` and `nil`.
- In Lua, unlike many other languages, assignment is a statement and not an expression.

- Lua has no conditional operator.
- Lua functions can return zero or more values. There is no tuple type in Lua; functions actually can return multiple values.
- Lua variables can have global scope or local scope. Local variables are introduced with the keyword `local`. Their scope begins on the statement after their declaration and runs until the end of the block. While this rule is simple, it does mean the programmer must take care in creating local recursive functions.
- Tables are used for both classic arrays (1-based sequences of numerically indexed values) and dictionaries (sets of key-value pairs). Both numeric and non-numeric indexed values can be mixed in a single table; it is up to the language implementation to keep the "array-part" access efficient.
- Like JavaScript, Lua's standard library is actually a collection of objects (e.g., `io`, `os`, `math`, etc.)
- A metatable can be attached to a table. The metatable can customize the manner in which table fields are read and written, and redefine the meaning of 19 of the operators. In addition, metatables can be used to build the same prototypal inheritance mechanism found in JavaScript.
- Lua directly supports coroutines: calling `coroutine.create` on a function object produces a coroutine (of type `thread`). Coroutines can yield and be resumed, and their status can be queried.

To continue your study of Lua beyond the introductory material of this chapter, you may wish to find and research the following:

- **Language features not covered in this chapter**. Chunks, environment variables, garbage collection metamethods, weak tables, the auxiliary library, the standard library, and interoperability with C (the API).
- **Open source projects using Lua**. Studying, and contributing to, open source projects is an excellent way to improve your proficiency in any language. Of the many projects using Lua, you may enjoy Luvit (`https://github.com/luvit/luvit`), CorsixTH (`https://github.com/CorsixTH/CorsixTH`), termtris (`https://github.com/tylerneylon/termtris`), and PacPac (`https://github.com/tylerneylon/pacpac`).
- **Reference manuals, tutorials, and books**. Lua's home page is `http://www.lua.org/`. The Reference Manual (as of Version 5.3) is freely available at `http://www.lua.org/manual/5.3/`. A good selection of Lua books can be found at `http://www.lua.org/docs.html#books`. We recommend the most recent version of *Programming in Lua* by Lua's creator, Roberto Ierusalimschy. You may also like one of the few books covering Lua's use in Game Programming.

EXERCISES

Now it's your turn. Continue exploring Lua with the activities and research questions below, and feel free to branch out on your own.

3.1 Find the Lua home page and the Lua Reference Manual online. Read Chapters 1–3 and 9 of the Reference Manual in full; skim other chapters according to your interests.

3.2 Practice with the Lua REPL.

3.3 What do you think about Lua's choice of allowing you read undeclared variables without an error? Do you find this error-prone, or useful? Why?

3.4 Contrast Lua's support for assignment with that of JavaScript and CoffeeScript. Consider assignments such as

```
{a: y, b: [_, c]} = {b: [10, true], a: 9, c: 5}
```

Does this work in JavaScript? CoffeeScript? Lua?

3.5 Write a Lua function that returns a random integer in the range 1 through 6 inclusive.

3.6 In a multiple assignment statement, how does Lua handle extra variables on the left-hand side? How does it handle extra expressions on the right?

3.7 To compensate for the lack of a conditional expression, some suggest simulating JavaScript's x?y:z with the Lua expression x `and` y `or` z. Is the simulation correct? If so, provide a justification for all boolean values of x, y, and z. If not, how exactly does it differ from the intent of the conditional?

3.8 What happens when a programmer intends to, but forgets, to mark a variable with `local`?

3.9 Contrast the behavior of the following Lua and C programs. How do you think C's scoping rules are defined?

```
x = 3
function main()
  local x = x
  print(x)
end
main()
```

```
#include <stdio.h>
int x = 3;
int main() {
    int x = x;
    printf("%d\n", x);
}
```

3.10 In Lua, `function f()...end` is sugar for `f = function()...end`. Knowing this fact, run and explain the output of the following script:

```
function outer()
  function inner()
    print(1)
  end
end

outer()
inner()
```

How would this script be different if the call to `outer` were removed? Why?

3.11 Write a Lua script to list the contents of the global environment. Print each of the items, one per line, to standard output as follows: first the key, then a tab character, then the type of the value, then a tab character, then the value.

3.12 What happens when you call `setmetatable` on a number? On a boolean? On `nil`? On a thread?

3.13 Research the metatable properties `call`, `__gc`, `__mode`, and `__metatable`. Build example Lua scripts that illustrate their behavior.

3.14 Rewrite the vector example in this chapter so that the metatable of each instantiated vector is `Vector` itself (Hint: it is a very small change). What are the advantages and disadvantages of this approach?

3.15 True or false: A function is just a coroutine that doesn't call `yield`. Explain.

3.16 Modify the word count program from the beginning of the chapter to use coroutines. Write one coroutine to fetch a line of text from standard input and break it up into words. Write another coroutine to count each word. The first coroutine should be able to send a signal to the second to tell it there are no more words. When the second coroutine has counted every word, it can call a regular function to produce, sort, and output the result table.

3.17 Lua is often used in applications that are partially written in C (or C++) and partially in Lua. Find an online tutorial or reference on embedding Lua in C++ applications. Write a small example program showing how a Lua function can be called from C++, and a C++ function called from Lua.

CHAPTER 4

Python

Python is "a programming language that lets you work quickly and integrate systems more effectively." [101]

First appeared 1991
Creator Guido van Rossum
Notable versions 2.2 (2001) • 3.0 (2008) • 2.7 (2010) • 3.5 (2015)
Recognized for Expressiveness
Notable uses Web servers, Scripting, Data science
Tags Imperative, Glue, Dynamic, Scripting
Six words or less Easy to learn and extremely productive

Guido van Rossum started working on Python in 1989, and has overseen much of this popular language's development over time. The Python community has given him the title Benevolent Dictator for Life, or BDFL.

Python is known for being easy to read and write. The syntax has a light and uncluttered feel. A large number of data types, including tuples, lists, sets, and dictionaries are built-in with convenient syntactic forms. Its design enables code to be written with far fewer characters than would be required in other popular languages. The language comes with a huge standard library and a massive repository of user-contributed packages, allowing Python applications to be built fairly rapidly. Python is particularly powerful as a scripting language, but it is a truly general-purpose language, having been used in domains from games and entertainment to web applications to statistics and data science. Though over twenty years old, it has evolved well over time through the Python Enhancement Process [125], and remains extremely popular.

Python is notable for its culture. Python aficionados, known as *Pythonistas*, take pride in writing elegant, expressive, *Pythonic* code. They learn and practice *The Zen of Python* [95], a collection of best practices and philosophies that help give Python applications a characteristic look. In fact, one of the principles from the Zen document is "There should be one—and preferably only one—obvious way to do it."

In 2008, Python made the jump from Version 2 to Version 3. In doing so, a large number of language quirks, outright problems, and aspects of the language that made it confusing to newcomers were completely changed without regards to backward compatibility. This made the cost of upgrading existing Python 2 code to Python 3 quite significant for some organizations, and indeed at the time of this writing, both versions of Python are widely used! This chapter, however, is all about Python 3. Python 3 is the preferred way to *learn* Python; if you later need to drop down to Python 2, you can learn the differences as needed. [19]

Our survey of Python will begin in the usual fashion, with our common three introductory scripts. We'll continue with a look at Python's type system, which differs a great deal from JavaScript's and Lua's (in particular, everything is an object, even types!) Python is the first language on our tour with a rich type system, with types organized into an ontology that supports something called **multiple inheritance**. We'll see how Python's handling of scope also differs from previous languages we've seen. We'll then take a look at many of the language's modern-ish features, including named parameter associations, default and keyword arguments, list comprehensions, its mechanism for custom operator definition (reminiscent of Lua's metamethods), and its support for iterators and generators. We'll close with a look at Python's decorators.

4.1 HELLO PYTHON

Let's get started in the usual way:

```
for c in range(1, 41):
    for b in range(1, c):
        for a in range(1, b):
            if a * a + b * b == c * c:
                print('{}, {}, {}'.format(a,b,c))
```

Python, like CoffeeScript, uses indentation to define structure. The `range` constructor builds a range object, inclusive at the bottom and exclusive at the top. Ranges produce their values on demand, so there is no need to worry that large lists are being created before iteration. String formatting uses curly braces within the format string; our empty braces indicate that we want the default formatting for our values. Had we wanted to display our values in three aligned columns, we could have written `{:3d}{:3d}{:3d}`.

Now let's write an anagram script directly implementing Heap's algorithm. (We'll see how to use a permutation generator from the standard library later in the chapter.)

```
import sys

def generatePermutations(a, n):
    if n == 0:
        print(''.join(a))
    else:
        for i in range(n):
            generatePermutations(a, n-1)
            j = 0 if n % 2 == 0 else i
            a[j], a[n] = a[n], a[j]
        generatePermutations(a, n-1)
```

```
if len(sys.argv) != 2:
    sys.stderr.write('Exactly one argument is required\n')
    sys.exit(1)
word = sys.argv[1]
generatePermutations(list(word), len(word)-1)
```

Command line arguments are stored in the list `argv` from the `sys` module, which we must explicitly *import*. The script name is contained in `argv` at index 0, so we check for a length of two to see that we have exactly one argument. We get the length of this list with the built-in `len` function.

Did you notice these two interesting language design choices?

- Python's conditional expression has the form y `if` x `else` z. Contrast with JavaScript's x `?` y `:` z and CoffeeScript's `if` x `then` y `else` z.
- To make a mutable list of characters from a string s—as required by the algorithm we are implementing—you need only write `list(`s`)` rather than invoking a `split` operation. Using a type name to "convert" a value from a different type is quite common.

Now let's read from standard input and count words:

```
import sys, re

counts = {}
for line in sys.stdin:
    for word in re.findall(r'[a-z\']+', line.lower()):
        counts[word] = counts.get(word, 0) + 1

for word, count in sorted(counts.items()):
    print(word, count)
```

The standard input stream (`sys.stdin`) is *iterable*, line by line, so it can be used in a `for` statement. The `findall` function from the `re` module produces the list of tokens on the line matching a pattern. We accumulate counts for each word in a **dictionary**. The dictionary lookup method `get` accepts a "default" value to return in case of a missing key (here 0), which makes counting quite elegant. And Python makes iterating the pairs of a dictionary in sorted order quite easy.

4.2 THE BASICS

If you've gotten comfortable with the type systems of JavaScript, CoffeeScript, and Lua, get ready for something completely different.

DIFFERENCE #1 *JavaScript, CoffeeScript, and Lua each have a small, fixed, number of types; Python has an unlimited number.* Lists, sets, dictionaries—and user-defined abstractions such as players, employees, blog posts, and calendar appointments—would all have type `Object` in JavaScript and type `table` in Lua. In Python, these are different types:

```
# Many built-in types have built-in names
assert type(5) == int
assert type(True) == bool
assert type(5.7) == float
assert type(9 + 5j) == complex
assert type((8, 'dog', False)) == tuple
assert type('hello') == str
assert type(b'hello') == bytes
assert type([1, '', False]) == list
assert type(range(1,10)) == range
assert type({1, 2, 3}) == set
assert type(frozenset([1, 2, 3])) == frozenset
assert type({'x': 1, 'y': 2}) == dict
assert type(slice([1, 2, 3])) == slice

# Built-in vs. User-defined functions
def plus_two(x):
    return x + 2
assert str(type(plus_two)) == "<class 'function'>"
assert str(type(max)) == "<class 'builtin_function_or_method'>"

# Even modules are types!
import math
assert str(type(math)) == "<class 'module'>"

# Many built-in modules define their own types
from datetime import date
assert type(date(1969,7,20)) == date
```

Difference #2: *Python types are objects.* In JavaScript, CoffeeScript, and Lua, asking for the type of the value 5 produces the *string* "number" in reply. In Python you get the object that *is* the type of integers.

```
assert type(int) == type
assert type(list) == type
assert type(type) == type  # Interesting...or obvious?
```

Difference #3: *You can create your own types in Python.* Python, like several other languages, uses the keyword class:

```
import math

class Circle:
    def __init__(self, radius):
        self.radius = radius
    def area(self):
        return math.pi * self.radius * self.radius

c = Circle(10)
```

```
assert type(c) == Circle                # Circle is a type!
assert type(Circle) == type             # It really is!

assert str(type(Circle.area)) == "<class 'function'>"
assert str(type(c.area)) == "<class 'method'>"

assert c.area() == math.pi * 100        # A method call...
assert Circle.area(c) == c.area()       # ...is just sugar
```

Here we've created a new type, `Circle`. Classes are **namespaces**: we did not define a global name `area`, we defined `Circle.area`. A class is also **callable**; the expression `Circle(10)` creates a new **instance** of type `Circle` and immediately passes it to `Circle.__init__` for initialization, here setting its `radius` **attribute**. Python provides syntactic sugar very similar to Lua's: writing `c.area` is shorthand for `Circle.area(c)`.[1] The sugared form gives you the feeling of a method invocation where *c* is the receiver; indeed, you may have noticed that `c.area` is a *method*, while `Circle.area` is a plain old function.

DIFFERENCE #4: *JavaScript, CoffeeScript, and Lua have primitives and objects, but Python has objects only.* There is no need for the complex distinction between primitives and objects. *All* values are objects. And every object has a unique identifier, accessible to the programmer:

```
a = 5                     # 5 is an object
b = {'x': 5, 'y': 3}      # dicts are objects
c = "hello"               # strings are objects too
d = c                     # two variables sharing an object
e = c.lower()             # should generate a new object
f = 8 * b['y'] - 19       # what happens here?

for obj in (a, b, b['x'], b['y'], c, d, e, f):
    print(id(obj))
```

One run of this program produced:

```
4329311472
4330887560
4329311472
4329311408
4330891000
4330891000
4332071152
4329311472
```

From this output, we can infer the objects are bound to variables as in Figure 4.1. Variables *c* and *d* reference the same object, since assignment copies references. Interestingly, we see there is only one copy of the integer 5 in the **object pool**; this is acceptable because integers are immutable. Python strings are immutable as well, so our implementation *could have* decided to share a single copy of the string `"hello"` among three variables, but chose not to.

[1] Note that Lua's sugared expression uses a colon (`c:area()`), while Python is able to simply use the dot (`c.area()`). Do you see why?

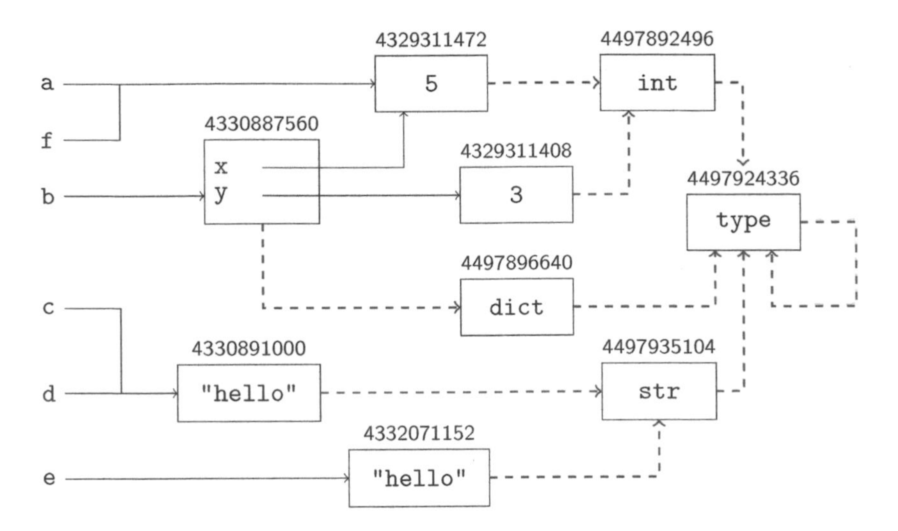

Figure 4.1 Python variables and objects

If you know something about computer architecture, you might worry that dealing with integers (and floats and booleans and similar things) as objects might be painfully slow. Are they dynamically allocated at runtime? Are they accessed via an indirect memory lookup? Maybe, but not necessarily so. Because these values are immutable, an implementation is free to implement them efficiently, as long as it returns a uniquely identifying value from the built-in `id` function. By treating *everything* as objects at the conceptual level, Python makes the programmer's mental model of values quite a bit simpler!

DIFFERENCE #5: *Python types have supertypes and subtypes.* JavaScript, CoffeeScript, and Lua require the programmer to set up prototype chains to simulate IS-A relationships between types. Python models this relationship directly; simply state the supertype(s) in the class definition:

```
class Animal:
    def __init__(self, name):
        self.name = name
    def speak(self):
        return '{} says {}'.format(self.name, self.sound())

class Cow(Animal):
    def sound(self):
        return 'moooo'

class Horse(Animal):
    def sound(self):
        return 'neigh'

class Sheep(Animal):
    def sound(self):
        return 'baaaa'
```

```
if __name__ == '__main__':
    s = Horse('CJ')
    assert s.speak() == 'CJ says neigh'
    c = Cow('Bessie')
    assert c.speak() == 'Bessie says moooo'
    assert Sheep('Little Lamb').speak() == 'Little Lamb says baaaa'
```

Here we've made `Animal` the **superclass** of its three **subclasses** `Horse`, `Cow`, and `Sheep`. `Animal` has a superclass too, namely `object`, the ultimate superclass of all classes in Python. Instances of subclasses **inherit** attributes of their superclass(es), so, for example, any instance of `Horse` we create not only has a `sound` attribute, but also `__init__` and `speak` attributes. We've illustrated this fact at the bottom of the script, where we create a horse, cow, and sheep and have them speak. (We've isolated our test code inside of an `if`-statement to execute only when the built-in variable `__name__` is set to `"__main__"`, which happens only when the script is run at the top-level, and not imported into another module. As a matter of fact, we will be importing this module in the next example, so our illustration of this common idiom is quite timely.)

Classes can have multiple direct superclasses, a condition known as **multiple inheritance**. Consider:

```
from subprocess import call
from animals import Cow

# Disclaimer: This only works on an O.S. with a say command.
class Vocalizer:
    def vocalize(self):
        call(['say', self.name + 'says' + self.sound()])

class VocalCow(Vocalizer, Cow):
    pass

c = VocalCow('Bessie')
print(c.speak())          # speak is inherited from Cow
c.vocalize()              # vocalize is inherited from Vocalizer
```

Nothing too surprising here: A `VocalCow` can both speak and vocalize (see Figure 4.2). But what if we had called the vocalizer method `speak` instead of `vocalize`? Would `VocalCow` have two `speak` attributes? Or would one be inherited and the other not? Or would the attempt to build the class with two superclasses fail because of the name clash?[2]

This question is ultimately one of how Python finds attributes of an object that the object's class does not directly define. In the languages we've seen previously, lookup proceeds along a single prototype chain, but in Python we need to know whether to look first in `Cow` *or* `Vocalizer` for attributes not in `VocalCow`. Python addresses this question by giving each class a **method resolution order**[3], or **MRO**. In our case, evaluating `VocalCow.mro()` produces:

```
[VocalCow, Vocalizer, Cow, Animal, object]
```

[2]That we can even ask this question has led some language designers to simply avoid multiple inheritance.
[3]Think of this as an attribute, rather than "just" a method, lookup order.

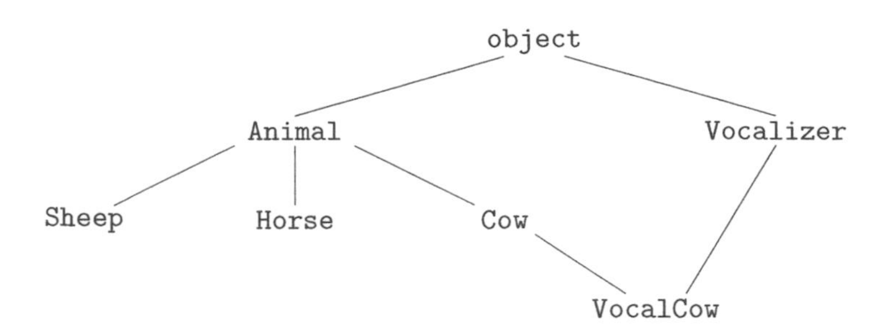

Figure 4.2 Python animal classes

This implies attributes in `Vocalizer` take precedence over those in `Cow`. We could, of course, modify the order to suit our needs. It can get tricky, but the mechanism is well-defined. See [110] for a detailed description of the algorithm used to compute the MRO.

Before moving on, let's see how Python's plethora of built-in types are related; see Figure 4.3. The common built-in types are all direct subtypes of `object`, with one notable exception: `bool` is a subclass of `int`. Interestingly, the various sequence types and numeric types do not share any common supertype (other than `object`); we've called out these informal groupings in the figure simply because programmers may find them useful.

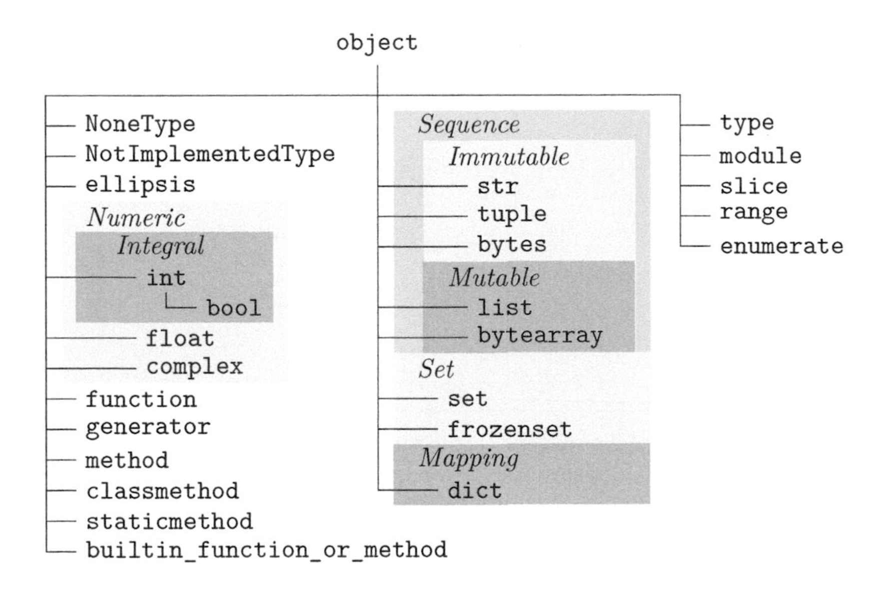

Figure 4.3 A few of the built-in types in Python

4.3 SCOPE

Python variables are either global or scoped to a function. There is no explicit variable declaration; assigning to a variable inside a function creates a local variable, unless marked `global` or `nonlocal`. Unlike CoffeeScript, Python locals **shadow** globals of the same name.

```
a, b, c = 1, 2, 3     # Three global variables

def f(x):
    d = 4             # Brand new local!
    a = 5             # Local to f, shadows global
    def g():
        x = 1         # Local to g, shadows f's x
        nonlocal a    # allows access to f's a
        a = 6         # updates the nonlocal
        global c
        c = 7
    g()
    assert x == 10    # g's x did not overwrite f's x
    assert a == 6     # assert that it changed
    assert b == 2     # globals are visible

f(10)
assert a == 1 and b == 2 and c == 7
```

Using a local before it is first written raises an `UnboundLocalError`; using a global before it is first assigned raises a `NameError`:

```
try:
    print(x)
except NameError as e:
    print('NameError raised as expected')

x = "I'm global"

def problem():
    print(x)          # Local but not bound yet
    x = "I'm Local"

try:
    problem()
except UnboundLocalError as e:
    print('UnboundLocalError raised as expected')
```

4.4 PARAMETER ASSOCIATION

Python passes arguments to parameters by passing each argument's id. This avoids the expense of copying large arguments and allows both the caller and callee to reference the same object. We call this mechanism **pass-by-sharing**. The language supports both default arguments and argument packing and unpacking. A single-starred parameter packs any additional arguments into a tuple. In a call, the single-star unpacks a sequence so that its constituent elements are passed to multiple parameters:

```
def f(x, y=1000, *z):
  print('x={} y={} z={}'.format(x,y,z))

f(0)                          # x=0, y=1000, z=()
f(0,1)                        # x=0, y=1, z=()
f(0,1,2)                      # x=0, y=1, z=(2,)
f(0,1,2,3)                    # x=0, y=1, z=(2,3)
f(0,1,2,3,4)                  # x=0, y=1, z=(2,3,4)
f(*[i for i in range(6)])     # x=0, y=1, z=(2,3,4,5)
f(*range(7))                  # x=0, y=1, z=(2,3,4,5,6)
```

But there is much more to parameter association in Python.

In his essay *Learnable Programming*, Bret Victor lays out nine design principles for building an environment and a programming language suitable for learning to program. One of his principles for a language is to MAKE MEANING TRANSPARENT. He motivates this principle by showing a snippet of code (not in Python) containing the lines

```
fill(161,219,114);
rect(105,20,60,60);
```

and suggests that a learner might ask “What do the numbers after **fill** mean?” and “What units are they in?” and “Why are there so many numbers?”

A simple way to help with the first question is to supply the parameter names in a call. And we do exactly that in idiomatic Python:

```
set_fill_color(red=161, green=219, blue=114)
draw_rectangle(corner=(105,20), other_corner=(60,60))
```

Python calls arguments passed in this fashion **keyword arguments**. In addition to making code easier to read, keyword arguments can be passed in any order (though they must follow non-keyword arguments). They provide a nice alternative to passing a dictionary. For a concrete example, let’s generate some Cascading Style Sheets text:

```
def write_rule(selector, **options):
    print(selector, '{')
    for prop, value in options.items():
        print('  {}: {};'.format(prop.replace('_','-'), value))
    print('}')

write_rule('h1', font_family='Helvetica', size='20px')
write_rule('p.error', color='red', margin='16px', padding='0')
```

4.5 SPECIAL METHODS

Python is a rich language with dozens of operators and built-in functions. It conveniently allows you to customize nearly all of them on your own classes. As in Lua, Python provides certain special functions that are implicitly invoked in certain contexts; Table 4.1 list some of the more commonly used ones.

Expression	What is Called	Expression	What is Called
+a	a.__pos__()	a > b	a.__gt__(b)
-a	a.__neg__()	a >= b	a.__ge__(b)
~a	a.__invert__()	a & b	a.__and__(b)
abs(a)	a.__abs__()	a | b	a.__or__(b)
a + b	a.__add__(b)	a ^ b	a.__xor__(b)
a - b	a.__sub__(b)	a << b	a.__lshift__(b)
a * b	a.__mul__(b)	a >> b	a.__rshift__(b)
a / b	a.__truediv__(b)	a *(in bool context)*	a.__bool__()
a // b	a.__floordiv__(b)	len(a)	a.__len__()
a % b	a.__mod__(b)	a in s	s.__contains__(a)
divmod(a,b)	a.__divmod__(b)	d[k]	d.__getitem__(k)
a ** b	a.__pow__(b)	d[k] = v	d.__setitem__(k, v)
a == b	a.__eq__(b)	iter(a)	a.__iter__()
a != b	a.__ne__(b)	next(a)	a.__next__()
a < b	a.__lt__(b)	str(a)	a.__str__()
a <= b	a.__lr__(b)	repr(a)	a.__repr__()

Table 4.1 Python special methods

Let's make a little class for two-dimensional vectors, translating as directly as possible our Lua program from the previous chapter:

```
import math

class Vector:
    def __init__(self, i, j):
        self.i = i
        self.j = j

    def magnitude(self):
        return math.sqrt(self.i * self.i + self.j * self.j)

    def __add__(self, v):
        return Vector(self.i + v.i, self.j + v.j)

    def __mul__(self, v):
        return self.i * v.i + self.j * v.j

    def __str__(self):
        return '<{},{}>'.format(self.i, self.j)

u = Vector(3, 4)
v = Vector(-5, 10)
assert u.i == 3
assert u.j == 4
assert u.magnitude() == 5.0
assert str(u + v) == '<-2,14>'
assert u * v == 25
```

There are a quite a few similarities with the Lua code (the use of special methods for one), but a few differences, too. Perhaps the most significant is that Python does not require explicit metatables or prototypes. An object's attributes come from its class and all of the class's **ancestors**. We say Python therefore features **classical inheritance**, while JavaScript, CoffeeScript, and Lua feature **prototypal inheritance**. We encourage you to compare and contrast the Python and Lua solutions in detail over as many dimensions as you can discover.

4.6 ITERATORS AND GENERATORS

An **iterable** is an object that produces a sequence of values. Lists, tuples, sets, dictionaries, and objects of a few other built-in types are iterables. Iterables can be used in `for` statements and a number of built-in functions, including `min`, `max`, `sum`, `all`, `any`, `filter`, `map`, `sorted`, and `zip`.

```
for x in (1,2,3): print(x)          # elements of a tuple
for x in [1,2,3]: print(x)          # elements of a list
for x in {1,2,3}: print(x)          # elements of a set
for c in 'hello': print(c)          # characters of a string

for k in {'x':1, 'y':2, 'z':3}:
    print(k)                        # keys of a dict

with open('colors') as f:
    for line in f:                  # lines of a file
        print(line.strip())
```

When Python needs to iterate an iterable, it applies the built-in `iter` function to the iterable, producing (via a call to the iterable's `__iter__` method—remember Table 4.1?) an **iterator**. The iterator's `__next__` method yields the next element of the sequence or raises a `StopIteration` exception when no more elements remain. This means you can write your *own* iterables:

```
class OneTwoThree:
    def __iter__(self):
        value = 0
        class OneTwoThreeIterator:
            def __next__(self):
                nonlocal value
                value += 1
                if value > 3:
                    raise StopIteration()
                return value
        return OneTwoThreeIterator()

x = OneTwoThree()
for i in x:
    print(i)          # prints 1 then 2 then 3
```

Writing iterables and iterators from scratch can be a bit tedious, so Python provides **generators** to simplify the creation of iterable sequences. A generator results from calling a function containing a **yield** statement. Executing such a function does not invoke the function's body, but rather returns a generator object. Here's a generator that produces successive powers of two, up to some limit:

```
def powers_of_two(limit):
    value = 1
    while value < limit:
        yield value
        value += value

# Use the generator
for i in powers_of_two(70):
    print(i)

# Explore the mechanism
g = powers_of_two(100)
assert str(type(powers_of_two)) == "<class 'function'>"
assert str(type(g)) == "<class 'generator'>"
assert g.__next__() == 1
assert g.__next__() == 2
assert next(g) == 4
assert next(g) == 8
```

Calling a generator function creates an object and assigns to it `__iter__` and `__next__` methods. It assembles, on the fly, the code to raise `StopIteration` at the proper time.

Interestingly, we can write an iterator that *never* raises a `StopIteration` exception—in other words, an infinite sequence.

```
def powers_of_two(limit):
    value = 1
    while True:
        yield value
        value += value
```

If the iteration scheme is sufficiently simple, you can create a generator with a **generator expression**, which looks like a list comprehension with parentheses instead of square brackets. Generator expressions have the advantage over list comprehensions of not having to compute and store the entire data set in memory.

```
# List comprehension.
# Computes all of its elements first.
# If large, it's too slow to produce and wastes memory.
bad = [x*x for x in range(10**9)]

# Generator expression.
# Produces values on demand, during iteration.
# Computed instantly and consumes almost no memory.
good = (x*x for x in range(10**9))
```

The module `itertools` from the standard library includes a number of functions for constructing and manipulating generators, including `permutations`, which builds a generator that produces permutations of a sequence as tuples. Let's use this to rewrite our anagrams script from the beginning of the chapter:

```
import sys
from itertools import permutations

if len(sys.argv) != 2:
    sys.stderr.write('Exactly one argument is required\n')
    sys.exit(1)

for word in (''.join(p) for p in permutations(sys.argv[1])):
    print(word)
```

If called with the string `"rat"`, the call to `permutations` produces a generator that delivers the tuples`('r','a','t'), ('r','t','a'), ('a','r','t')` and so on. We wrap *that* generator with a generator producing the (joined) string for each tuple, iterating over the new generator with a `for` statement to print each string.

4.7 DECORATORS

Time for some magic. We're going to change the behavior of a function *without ever rewriting its source code*. In Python, functions (and hence methods) can be **decorated**, allowing us to do some extra work when the function is called. Here is one of the simplest decorator examples; it prints the time the function is called as well as the duration taken by the call:

```
import datetime

def logged(f):
    def wrapper(*args, **kwargs):
        start = datetime.datetime.now()
        print('{} started at {}'.format(f.__name__, str(start)))
        f(*args, **kwargs)
        duration = datetime.datetime.now() - start
        print('{} took {}'.format(f.__name__, duration))
    return wrapper

@logged
def say_hello(name):
    print('Helllllloooooooooooo, {}'.format(name))

say_hello('Alice')
```

Prefixing the function definition of `say_hello` with the `@logged` decorator is equivalent to following its definition with `say_hello = logged(say_hello)`. So the decorator syntax is not essential, but it adds a great deal of expressiveness: all of the logging code is isolated from the mainline code, allowing us to see at a glance (without clutter) what our application should be doing, and knowing it will be logged.

Decorators give us a nice solution to **cross-cutting concerns** such as logging, transaction maintenance in databases, persistence, authentication, and so on. These concerns are called "cross-cutting" because one does not organize a system around authenticated objects vs. persistent objects; instead these concerns can apply almost anywhere. You will often see Python code with decorators such as `@persistent`, `@transactional`, and `@authenticated`. And here's another. It's well known that the naïve application of the definition

$$F(n) = \begin{cases} n & n \leq 1 \\ F(n-1) + F(n-2) & \text{otherwise} \end{cases}$$

leads to horrific performance: computing $F(40)$ results in 331,160,281 calls to the function. Over 40 trillion calls are needed to do $F(50)$, and over a septillion (10^{21}) calls for $F(100)$. The problem is that so many calls are repeated, as we can see from the evaluation:

$$\begin{aligned} F(40) &= F(39) + F(38) \\ &= (F(38) + F(37)) + F(38) \end{aligned}$$

As $F(38)$ is called twice, we should record its value after the first call and then simply reuse this value rather than making the second repeated call. This trick of reusing results, known as **memoization**, is applied also to $F(37)$, $F(36)$, and so on, reducing the number of calls from over 331 million down to 41.

Performing memorization within the `fib` function would violate the separation of concerns principle, so we write a decorator:

```
def memoized(f):
    cache = {}
    def wrapper(*args):
        if args in cache:
            return cache[args]
        cache[args] = f(*args)
        return cache[args]
    return wrapper

@memoized
def fib(n):
    return 1 if n <= 1 else fib(n-1) + fib(n-2)

print(fib(100))
```

Python decorators can also be parameterized, constructed with a class syntax (any object of a class with a `__call__` method becomes a decorator) and even applied to classes and methods. We'll leave to explore these additional forms on your own.

4.8 PYTHON WRAP UP

In this chapter we were introduced to Python. We learned that:

- Python is a mature, general-purpose language popular in scripting, web frameworks, and data science. It was created and is still overseen by Guido van Rossum.

- Python has a large number of built-in types, including many different numeric types and over a dozen collection types.
- Python has no primitives. All values are objects. All objects have an `id` and a `type`. Types are objects.
- New types are created by defining a `class`. (Contrast this with JavaScript and CoffeeScript, where the `class` keyword defines a function.)
- Classes, as well as all types in Python, are objects. (Contrast this with JavaScript, CoffeeScript, and Lua, where inquiring the type of an object produces the name of the type as a string.)
- Types in Python can have supertypes and subtypes. All paths through the chain of supertypes of an object will end at the type `object`. The supertype-subtype relationships are used in attribute lookup. Subtypes inherit attributes from their supertypes.
- Because types can have multiple supertypes, the language enforces a linear order in which all ancestor types must be searched to find inherited attributes. This is called the method resolution order, or MRO.
- Python local variables are introduced by assignment in a local scope. Non-local variables in enclosing lexical scopes can be read; to write to non-locals either `nonlocal` or `global` must be used.
- Many standard operators and functions can be overloaded by defining special methods in a class. These special methods have names such as `__add__`, `__eq__`, and `__contains__`.
- Python supports delayed evaluation through iterators and generators. Generators can be created by functions with a **yield** statement, or via generator expressions.
- Decorators enable ways to enhance the behavior of functions and classes, without actually adding any code directly inside the function or class.

To continue your study of Python beyond the introductory material of this chapter, you may wish to find and research the following:

- **Language features not covered in this chapter**. Exceptions, packages, chained comparisons, set and dictionary comprehensions, a large number of additional built-in functions, the full details of the sequence types, the `pass`, `del`, and `with` statements, interactive input, the incredibly extensive standard library, and the massive ecosystem of third-party libraries and frameworks.
- **Open source projects using Python**. Studying, and contributing to, open source projects is an excellent way to improve your proficiency in any language. You may enjoy the following projects written in Python: Requests (`https://github.com/kennethreitz/requests`), Flask (`https://github.com/mitsuhiko/flask`), Reddit (`https://github.com/reddit/reddit`), and Boto (`https://github.com/boto/boto`).
- **Reference manuals, tutorials, and books**. Python's home page is `https://www.python.org/`. The docs page, `https://docs.python.org/3/`, contains links to the official Tutorial, Language Reference, and Library Reference. A curated list of books can be found at `https://wiki.python.org/moin/PythonBooks`. Google has published an online Python course at `https://developers.google.com/edu/python/`.

EXERCISES

Now it's your turn. Continue exploring Python with the activities and research questions below, and feel free to branch out on your own.

4.1 Locate the Python home page and browse the official tutorial, language reference, and library reference.

4.2 Install Python and experiment with the REPL.

4.3 Read about the reasons for the development of Python 3. Describe three major language changes made in Python 3 that were viewed as correcting shortcomings or problems in Python 2.

4.4 In this chapter we've seen several uses of the `format` method of the `str` class, but we've only used default formatting. What kind of options are available for formatting numbers and strings?

4.5 What is the built-in type `enumerate` commonly used for? Construct a realistic example illustrating one such use.

4.6 Extend the `Circle` class from Section 4.2 with methods `circumference`, `str`, and `repr`. You will need to research the language conventions for the latter two.

4.7 Read *The Zen of Python.* [95]

4.8 We mentioned that Python classes are namespaces. But classes are not the only kind of namespace in the language. Research and describe, with examples, at least two other constructs in Python that serve as namespaces.

4.9 In your own words, describe the difference between a function and a method in Python.

4.10 Research the functions `isinstance` and `issubclass`. What do they do? See if you can find arguments for and against their usage in everyday code.

4.11 The term "multiple inheritance," introduced in this chapter, turns out to be an interesting topic because many have argued that programming languages are often better without it. What are the arguments *against* multiple inheritance?

4.12 Give a description of the Python MRO, in pseudocode.

4.13 Python locals shadow nonlocals and globals, but (as we saw in Chapter 2) CoffeeScript does not allow shadowing at all. Research the differences between these two approaches. What are the advantages and disadvantages of each?

4.14 JavaScript's `let` and Lua's `local` allow one to define a variable scoped to a block (e.g., the body of a `while`-statement). Is this even possible in Python? Why or why not?

4.15 Perform a detailed comparison between the `Vector` class implemented in this chapter and the `Vector` table written in Lua in Chapter 3.

4.16 If you have some experience with spatial mathematics, create Python classes for points, vectors, rays, and planes, "overloading" as many operators as is convenient. Your classes should include functionality, to among other things, subtract two points to produce a vector, and a point and a vector to get another point, multiply a vector by a number to change its magnitude, compute the dot and cross products of two vectors, and determine the normal vector of a plane.

4.17 Find out, how, in Python to (a) get all of the attributes of a class, (b) get all of the subclasses of a class, and (c) whether or not it is possible to add new attributes to an existing class.

4.18 Research *classmethods* and *staticmethods* in Python. What are they? How are they implemented?

4.19 How do you, programmatically, get the contents of a module?

4.20 The word count example at the beginning of the chapter used a plain dictionary to compute the counts. Rewrite this script to use a `Counter` object from the standard library.

4.21 In this chapter we saw an example of packing keyword arguments, but no examples of unpacking a dictionary in a call. Supply a realistic example.

4.22 Can a function have both single-starred and double-starred parameters? If so, how many of each are allowed? How might using both ever be considered useful?

4.23 The `OneTwoThree` example in this chapter used separate classes for the iterator and the iterable. Rewrite the example so there is only one class for both. Hint: the `__iter__` method will return `self`. What are the advantages and disadvantages relative to the solution presented in this chapter?

4.24 Is a Python `range` an `iterator`? If not, why might one be inclined to think it is?

4.25 Do a compare-and-contrast study between Lua coroutines and Python generators.

4.26 Use the `timeit` module to time both the unmemoized and memoized versions of the fibonacci function from this chapter when called with the argument 40.

4.27 A classic problem in computer science is the *0-1 knapsack problem*: Given a sack with capacity C and a set of items $\{(w_1, v_1)...(w_n, v_n)\}$ with weights w_i and values v_i, find the most valuable selection of items with total weight $< C$. There is a straightforward recursive solution, which, like the fibonacci example, ends up making multiple recursive calls of the same argument. Implement the straightforward solution. Generate a random set of items with random weights and values, and time your solution both with and without the `@memoized` decorator.

4.28 Consider the following slight modification to the Vocalizer class from this chapter.

```
import subprocess
class Vocalizer:
    def vocalize(self):
        subprocess.call(['say', str(self)])
```

Now simply adding `Vocalizer` to a classes supertype list should enable any object of that class to have its string representation vocalized. Try it. Can the same functionality be done through decorators? That is, can you decorate a class (perhaps with `@vocalized`) such that the decorator adds a vocalize method?

4.29 Research the class syntax for decorators and write a `@memoized` decorator using this syntax.

4.30 We did not cover Python's approaches to asynchronous computing. Research the use of `async` and `await` and illustrate their use in a small example application of your choice.

CHAPTER 5

Ruby

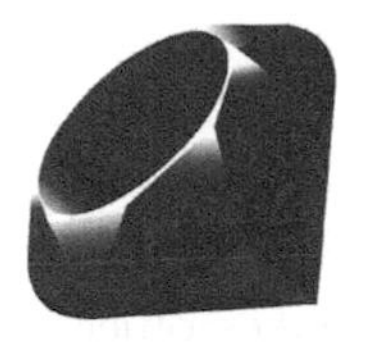

Ruby's designer expressed his language's purpose as follows: "I hope to see Ruby help every programmer in the world to be productive, and to enjoy programming, and to be happy. That is the primary purpose of Ruby language." [80]

First appeared 1995
Creator Yukihiro Matsumoto
Notable versions 1.8 (2003) • 1.9 (2007) • 2.0 (2013) • 2.3 (2015)
Recognized for OOP, Metaprogramming, DSL construction
Notable uses Web servers, Scripting
Tags Object-oriented, Expression-oriented, Dynamic
Six words or less "A Programmer's Best Friend"

Yukihiro Matsumoto (Matz) designed Ruby in the mid-1990s as a language in which one could be productive and have fun programming. Rather than building a language with a small, simple core (for which Lisp and Smalltalk are famous), Matz looked to be pragmatic, and aimed for human-friendliness. He once told an interviewer: "I want to emphasize the *how* part: how we feel while programming. That's Ruby's main difference from other language designs. I emphasize the feeling, in particular, how *I* feel using Ruby. I didn't work hard to make Ruby perfect for everyone, because you feel differently from me. No language can be perfect for everyone. I tried to make Ruby perfect for me, but maybe it's not perfect for you. The perfect language for Guido van Rossum is probably Python." [122]

But Ruby does have many, many fans, who find the language beautiful, powerful, and expressive. Of the many languages that influenced Ruby, two stand out. It's a scripting language like Perl, with operating system interfaces, string processing, regular expressions, and various syntactic niceties like flexible quoting syntaxes. Ruby's other major influence is Smalltalk. Ruby is object-oriented in the way Smalltalk is object-oriented, expressing nearly all computation with message passing. Ruby's blocks come from Smalltalk, too. And the metaprogramming facilities of Ruby are, as in Smalltalk, an area in which the language really shines.

Many programmers today know Ruby from the web frameworks built upon it, most notably Rails, but also Sinatra, Camping, Rack, Padrino, Pancake, Monk, and others. But Ruby is truly a *general-purpose* language, and you'll see it used in many popular applications. Its lineage as a scripting language makes it popular among operations teams; in fact, popular automation and testing tools like Puppet, Chef, and Vagrant are written in Ruby. Ruby's uncanny ability to implement domain-specific languages (DSLs) also fuels its popularity across domains.

Our Ruby tour will begin with our standard three example programs and a brief overview of the language. We will then introduce the object-oriented paradigm and show how Ruby makes object-oriented programming not only possible but natural. Ruby is the first, and only, language in this text that we will bother to label object-oriented, and we'll explain why. We'll then proceed to two important features of Ruby: blocks and mixins. These features will be followed by a section on access control, a feature Ruby makes relatively easy compared to the four languages we have seen up until now. We round out our coverage with Ruby's metaprogramming facilities.

5.1 HELLO RUBY

We begin with our familiar script to list small right triangle measurements:

```
1.upto(40) do |c|
  1.upto(c - 1) do |b|
    1.upto(b - 1) do |a|
      puts "#{a}, #{b}, #{c}" if a * a + b * b == c * c
    end
  end
end
```

Ruby sports a couple features we first saw in CoffeeScript—string interpolation with `#{}` and a trailing `if`-clause—but this little script is definitely not using a traditional `for`-statement. Instead, we start the script by sending the **message** `upto` to the **object** `1`. The message is sent with with argument `40` and a **block** (here delimited with `do` and `end`). The `upto` method executes its block for each value in the range from its receiver up to, and including, its argument.

Next, let's translate our running anagrams example—the one using Heap's Algorithm—into Ruby:

```
def generatePermutations(a, n)
  if n == 0
    puts a.join
  else
    0.upto(n-1) do |i|
      generatePermutations a, n-1
      j = n.even? ? 0 : i
      a[j], a[n] = a[n], a[j]
    end
    generatePermutations a, n-1
  end
end
```

```
if ARGV.length != 1
  STDERR.puts 'Exactly one argument is required'
  exit 1
end
generatePermutations ARGV[0].chars, ARGV[0].length-1
```

The constants `ARGV` for the command line argument array, and `STDERR` for the standard error stream, do not need to be explicitly imported or tagged with a module name. You might also notice that, like CoffeeScript, calls do not always require parentheses.

Like Python, Ruby already knows how to do permutations:

```
if ARGV.length != 1
  STDERR.puts 'Exactly one argument is required'
  exit 1
end
ARGV[0].chars.permutation.each{|s| puts s.join}
```

One line does all the work:

1. `ARGV[0]` is the first command line argument, a string.
2. `.chars` produces an array containing each character of this string.
3. `.permutation` produces an *enumerator* for all of the permutations of this array. An enumerator is not an actual array; think of it as a object that is ready to produce the next value whenever it is asked for.
4. The `each` method causes each permutation to be generated one-by-one. As each permutation is generated, it is passed to the block where the permuted array is joined up into a string and written on a line to standard output.

Now let's count words:

```
counts = Hash.new(0)
ARGF.each do |line|
  line.downcase.scan /[a-z']+/ do |word|
    counts[word] += 1
  end
end
counts.sort.each do |word, count|
  puts "#{word} #{count}"
end
```

This short, flowing script shows how writing Ruby is, for many, pleasant and productive. We store word-to-count mappings in a *hash*—Ruby's term for Lua's tables and Python's dicts. Lookup with a non-existent key normally produces `nil`, but here we tell Ruby to use 0 when the key is missing. Counting becomes elegant: `counts[word] += 1` always works, with no need for a `nil` check! `ARGF` was created with the scripter's happiness in mind. If no command line arguments are present, `ARGF` is standard input; if arguments are present, they're treated as filenames from which input will be read (in order). Our last statement is terse yet readable: calling the method `sort` on a hash produces an enumerator of key-value pairs sorted by key.

5.2 THE BASICS

Ruby has variables, assignments, the usual control structures, and even exceptions. As in Python, there are no primitives, only objects; you can find the unique identifier of any object x with x`.object_id`. Every object has a unique class, which can be found with x`.class`:

```
fail unless nil.class == NilClass
fail unless false.class == FalseClass
fail unless true.class == TrueClass
fail unless 3.class == Fixnum
fail unless (2**1000).class == Bignum
fail unless 2.0.class == Float
fail unless :dog.class == Symbol
fail unless "dog".class == String
fail unless (1..5).class == Range
fail unless [1,2,3,4,5].class == Array
fail unless {x: 1, y: 2}.class == Hash
fail unless {'x' => 1, 'y' => 2}.class == Hash
```

Ruby features a large number of built-in classes. A class can have at most one **superclass**. Superclasses allow us to express `IS-A` relationships: each of the expressions `5.is_a? Fixnum`, `5.is_a? Integer`, `5.is_a? Numeric`, `5.is_a? Object`, and `5.is_a? BasicObject` evaluate to `true`. Figure 5.1 shows part of this hierarchy.

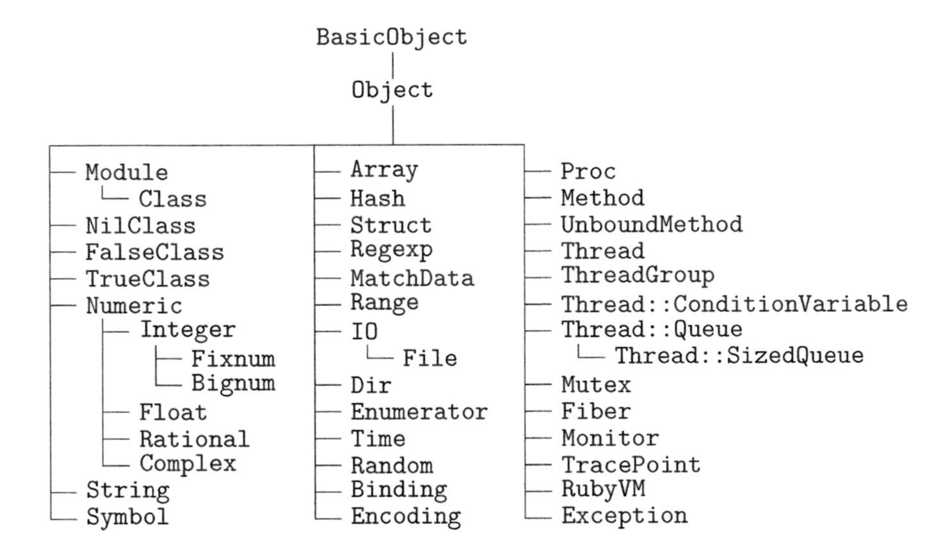

Figure 5.1 Part of the standard class hierarchy in Ruby

Like Python, Ruby classifies floating point numbers differently than integers, but *unlike* Python, it classifies smaller integers differently than bigger ones. In practice, the difference isn't too noticeable: raising the fixnum 2 to the power fixnum 100 happily results in the bignum 1267650600228229401496703205376. But not all types are so easily coerced. You can't, for example, add strings and numbers. But everything *can* be coerced to a boolean. Like Lua, the only expressions that are falsy are `false` and `nil`.

Ruby distinguishes strings and **symbols**. There is one—and only one—symbol `:dog` in the object pool, but there can be many strings with the three characters `"dog"`. Symbol names begin with a colon unless used as a hash key separated by its value with a colon, e.g., `{count: 0}`, which is equivalent to `{:count => 0}`. You can use non-symbols as hash keys under certain conditions (to be explored in an end-of-chapter exercise), but only with the hash rocket (`=>`) or `Hash.new`.

Ruby is designed with a number of convenient, and pragmatic, features that make programming fun. In addition to `while` and `if`, we have `until` and `unless`. And the language has not two, but *four* ways to disrupt a loop:

- `break`: Break out of the loop completely
- `next`: Immediately begin the next iteration of the loop
- `redo`: Restart the current iteration of the loop (without retesting a condition, if any)
- `retry`: Restart the entire loop from the first iteration

Names beginning with uppercase letters indicate the programmer intends them to be constants. If you try to change a constant, the world does not end—you just get a warning.

Programmers also get niceties such as named parameters, default parameters, and splats:

```
def positional(x, y, z)
  "#{x} #{y} #{z}"
end

def named(x:, y:, z:)
  "#{x} #{y} #{z}"
end

def splatted(x, *y)
  "#{x} #{y}"
end

def defaulted(x, y=100)
  "#{x} #{y}"
end

def defaulted_and_named(x, y:2)
  "#{x} #{y}"
end

fail unless positional(1, 2, 3) == "1 2 3"
fail unless named(z:3, x:1, y:2) == "1 2 3"
fail unless splatted(1) == "1 []"
fail unless splatted(1, 2, 3, 4) == "1 [2, 3, 4]"
fail unless positional(1, *[2, 3]) == "1 2 3"
fail unless defaulted(1) == "1 100"
fail unless defaulted(1, 2) == "1 2"
fail unless defaulted_and_named(1) == "1 2"
fail unless defaulted_and_named(1,y:3) == "1 3"
```

Here the method `positional` must be called with exactly three arguments, in order. `named` must be called with exactly three arguments as well. The names must be present in the call, but they can appear in any order. Callers must pass *at least one* argument to `splatted`; all arguments beyond the first are packed into the array bound to y. The method `defaulted` must be called with either one or two arguments, and ditto for `defaulted_and_named`. It's probably worth mentioning that positional arguments should not be called with names—the call will be interpreted as something very strange.

We really can't go much farther into our exploration of Ruby without covering the fact that the language is something we call object-oriented, in an essential way. It's time to see what that means.

5.3 OBJECT ORIENTATION

A **programming paradigm** is a style of expressing computations. The **functional paradigm** views computation as the composition of functions: the output(s) of one function are fed into another, and so on. The **imperative paradigm** views computation as a succession of actions, such as assignment statements and I/O operations, updating a global state. The **logic paradigm** views computation as the systematic satisfaction of constraints, specified by facts and inference rules. Wikipedians have described dozens of additional programming paradigms in the *Programming Paradigm* article. [136]

The popular **object-oriented paradigm** views computation as message passing among objects: an object responds to a message (when it can) by executing a **method**, possibly updating the object's internal state. They key insight of object-orientation is that a computer (or a complex software system) can be decomposed into smaller computers (objects) rather than into the dissimilar and less-powerful data structures or procedures.[1] There is a spectrum of opinions on what characterizes object-orientation, from (1) simply bundling code inside of data structures, to (2) requiring that objects have completely encapsulated (hidden) state, that objects be instances of classes, and that classes be grouped into inheritance hierarchies allowing for specialization of method behaviors. An excellent overview of object-orientation covering its goals, origins, mechanisms, models, and alternatives, is [129].

Ruby is an object-oriented programming language because it facilitates object-oriented programming. Let's see how.

5.3.1 Messaging

Like Python, all Ruby values are objects (there are no primitives), and with few exceptions, computation is carried out exclusively by sending messages to objects. Send a message with the expression *obj*`.send(`*message*`,`*args*`)`, denoting the message with an expression evaluating to a Ruby symbol.[2] If the message is known for sure, you can use the shorthand *obj*`.`*message*`(`*args*`)`.

[1] Alan Kay, who is said to have invented the term "object oriented," credits the genesis of this idea to his teacher, Bob Barton, who said "The basic principle of recursive design is making the parts as powerful as the whole." [71]

[2] You can pass a string for the message name if you wish; Ruby will convert the string to a symbol for you.

```
# Zero-argument messages
fail unless 5.send(:abs) == 5
fail unless 5.send('abs') == 5
fail unless 5.abs == 5

# A one-argument method
fail unless 5.send(:+, 3) == 8
fail unless 5.send('+', 3) == 8
fail unless 5.+(3) == 8
fail unless 5 + 3 == 8

# Messages can be stored in variables
operator = '-'
fail unless 5.send(operator, 2) == 3
```

5.3.2 Encapsulation

Like Python, Ruby's objects are instantiated from classes. Unlike Python, which exposes all of an object's attributes (both callable and non-callable), Ruby distinguishes **instance variables** (that hold state) from **methods** (which *respond* to messages), and does not allow code outside of the class to directly access the object's instance variables. *All* object interaction must occur through methods. The **state** of the object is therefore fully **encapsulated**.

Let's illustrate with a little circle class:

```
class Circle
  def initialize(x, y, r)
    @x = x
    @y = y
    @r = r
  end
  def center()
    [@x, @y]
  end
  def area()
    Math::PI * @r * @r
  end
  def to_s()
    "Circle at (#{@x}, #{@y}) with radius #{@r}"
  end
end

c = Circle.new(5, 4, 10)
fail unless c.center == [5, 4]
fail unless c.area == 100 * Math::PI

# to_s is automatically called when in string context
fail unless "#{c}" == "Circle at (5, 4) with radius 10"
```

The definition states a circle (1) is initialized with two center coordinates and a radius, (2) responds with an array of its center coordinates when queried,[3] (3) responds with its area when queried, and (4) can produce a string representation of itself. The code outside the class illustrates the creation of a circle by sending the message `new` to the class `Circle`. The variable *c* references the new object, to which we send `center` and `area` messages. There's a little magic here: Ruby automatically sends `initialize` upon object creation and `to_s` when an object is used in a string context.

An instance variable's name begins with `@` and is only accessible *within* the class. Were we to try to evaluate `c.@r` outside the class, a `SyntaxError` exception would be raised. And the expression `c.r` raises a `NoMethodError` since the dot-notation always refers to a method call. If we wanted to make the radius available to a query (e.g., `c.r`) we would have to supply a method to respond with it:

```
def r()
  @r
end
```

If you wish to write a method that does nothing but update an instance variable, Ruby lets you write:

```
def r=(new_radius)
  @r = new_radius
end
```

and this method will be invoked whenever `c.r` appears on the left of an assignment statement. Incidentally, you can let Ruby create these methods for you so you don't have to type them. Writing `attr_reader :r` creates the first method, `attr_writer :r` the second, and `attr_accessor :r` creates both.[4]

5.3.3 Inheritance

As mentioned earlier, simply bundling operations with encapsulated data counts as "object-oriented" in some circles. Cardelli and Wegner [16], however, define the term as having an additional ingredient, *inheritance* of instance variables and methods by subclasses. Here's our recurring example of inheritance seen in some previous chapters:

```
class Animal
  def initialize(name)
    @name = name
  end
  def speak()
    "#{@name} says #{sound()}"
  end
end
```

[3]The `return` keyword is optional. When omitted, the value of the last evaluated expression in a method body is returned.

[4]Just because Ruby allows you to easily write methods to directly read and write instance variables does not mean you should. The use of *getter* and *setter* methods is controversial because it violates encapsulation, exposing implementation details to callers, and encourages clients to pull data from an object to perform work which the object could (should?) have done itself. See [61] for details.

```
class Cow < Animal
  def sound()
    "moooo"
  end
end

class Horse < Animal
  def sound()
    "neigh"
  end
end

class Sheep < Animal
  def sound()
    "baaaa"
  end
end

if __FILE__ == $0
  s = Horse.new "CJ"
  fail unless s.speak == "CJ says neigh"
  c = Cow.new("Bessie")
  fail unless c.speak == "Bessie says moooo"
  fail unless Sheep.new("Little Lamb").speak == "Little Lamb says baaaa"
end
```

Cows, horses, and sheep inherit the `initialize` and `speak` methods from their superclass, as well as the instance variable `@name`. (We've also illustrated two of Ruby's global variables `__FILE__`, the currently executing file, and `$0`, the script being run. The code at the end of the file only executes when the file is run directly and not imported.)

5.3.4 Polymorphism

While all kinds of animals in our previous example have names and speak the same way, they all make different sounds. The expression `a.sound` will produce the proper sound of the animal referenced by the receiver *a*, even if we don't know the class of that animal until the script is run. This behavior is called **dynamic polymorphism**. The name `sound` is *polymorphic* (Greek: "of many forms") since it is bound to three entities, and the dispatch to the correct method is *dynamic* (happens at run-time).

5.3.5 Singleton Classes

In each of the four languages of the previous chapters, we have the ability to attach functions directly to objects, regardless of the object's class or prototype. But Ruby objects don't contain slots for methods—all methods live in classes. Creating object-specific methods would seem impossible; however, Ruby gives each object its very own **singleton class** (sometimes called an **eigenclass**) to hold object-specific methods. Let's create a dog class and two dogs. All dogs can bark, but we want Spike, and only Spike, to bite:

```
class Dog
  def bark()
    "woof"
  end
end

spot = Dog.new
spike = Dog.new

def spike.bite()
  "ouch"
end

spot.bark
spike.bite
```

Spike can bite because method lookup starts at an object's singleton class and then proceeds up the superclass chain. Figure 5.2 shows how these objects are related.

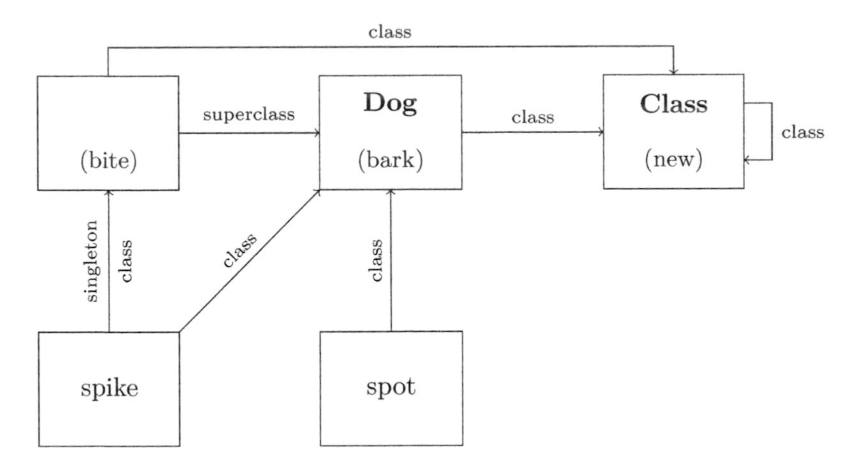

Figure 5.2 An object with a singleton class

5.4 BLOCKS

Every method in Ruby may be passed, in addition to its arguments, a **block**. We saw blocks in this chapter's opening example delimited with `do` and `end`, but curly braces work fine too:

```
3.times {puts 'Hello'}
```

Blocks may be parameterized (the parameters are placed between vertical bars at the beginning of the block):

```
mountains = 'Everest K2 Kangchenjunga Lhotse Makalu'.split
mountains.each_with_index {|name, i| puts "#{i+1}. #{name}"}
```

Block parameters have their own scope and shadow variables of the same name in enclosing scopes. Code within the block body *can* both read and mutate variables in the enclosing scope. In addition to parameters, blocks may have their own block-scoped variables. These appear after a semicolon in the block's "header," and also shadow variables in the enclosing scope:

```
def main
  x = 100                                 # x is a local var of main
  [1,2,3].map{|x| x * x}                  # a new block-local x
  fail unless x == 100                    # "outer" x unchanged at 100
  [1,2,3].each{|y| x = x + y}             # block can mutate outer x
  fail unless x == 106                    # x mutated to 106
  [1,2,3].each{|y; x| x = (x||0) + y}     # ; makes new x in block
  fail unless x == 106                    # x unchanged at 106
end

main
```

To have a method transfer control to its block, use `yield`. Let's see how this works. The following method generates powers of two up to some limit. What happens to those powers of two depends on the block supplied to the method call.

```
def powers_of_two(limit)
  x = 1
  while x <= limit
    yield x
    x += x
  end
end

powers_of_two 90 do |x|
  puts x
end
```

Blocks should remind you of anonymous functions. They have their own parameters and local variables, and can manipulate variables in the enclosing scopes. But to use blocks like first-class functions (assign them to variables and pass them around), we must create `Proc` objects from them. There are two kinds of `Procs`, regular procs and lambda procs; and a variety of ways to denote them:

```
plusTwo = proc {|x| x + 2}
plusThree = Proc.new {|x| x + 3}
plusFour = lambda {|x| x + 4}
plusFive = -> (x) {x + 5}

# How do we call them? What are they? Regular or lambda?
p [plusTwo.call(10), plusTwo.class, plusTwo.lambda?]
p [plusThree.call(10), plusThree.class, plusThree.lambda?]
p [plusFour.call(10), plusThree.class, plusFour.lambda?]
p [plusFive.call(10), plusThree.class, plusFive.lambda?]
```

```
plusTwo.call(100, 200, 300)     # Extra args? No problem!
begin
  plusFour.call(100, 200, 300)  # But it's a problem here!
rescue ArgumentError
  puts 'Rescued'                # Message will be output
end

a = [10, 20, 30, 40, 50]
p a.map(&plusTwo)               # All passed the same way
p a.map(&plusThree)
p a.map(&plusFour)
p a.map(&plusFive)
p a.map(&(->(x) {x * 10}))      # anonymous procs!
p a.map &->(x) {x * 10}         # Parens optional here too
```

The first form (`Proc.new` or `proc`) should be used sparingly, if at all—these kind of procs don't check the number of arguments passed to them, and a `return` statement actually returns from the enclosing method! The lambda forms behave as expected.

```
def conversation()
  puts 'Calling a lambda proc'
  (lambda {return}).call              # Returns from the proc
  puts 'Calling a non-lambda proc'
  (proc {return}).call                # Returns from conversation()
  puts 'This does not appear'
end

conversation
```

There's a neat little shorthand that you can use when the block contains a single expression that passes a message to the block parameter. The following two lines do the same thing:

```
[1,-2,3,-4,-5].map {|x| x.abs}
[1,-2,3,-4,-5].map &:abs
```

The unary ampersand (`&`) causes the `toProc` message to be sent to a symbol. The `Symbol` class defines a `to_proc` method to do the right thing. If you wish, you can create your own classes and define a `to_proc` method, enabling your own objects to work on methods expecting blocks.

5.5 MIXINS

In the previous chapter, we saw Python's multiple inheritance in action, with the vocal cow and its two supertypes, `Cow` and `Vocalizer`. Ruby classes may have only one superclass, but they may include any number of **modules**. A Ruby module is a container for methods; a class is a module that is allowed to have instances. Its methods are **mixed in** while defining a class. Here's the vocal cow example from the Python chapter, translated to Ruby:

```
require './animals.rb'

# The original Animal class did not define a name method.
# But in Ruby, we can *add* one! Classes are *open*.
class Animal
  attr_reader :name
end

module Vocalizer
  # Disclaimer: This only works on an O.S. with a say command.
  def vocalize()
    # Multiple arguments required to prevent shell injection attack.
    system 'say', self.name, ' says ', self.sound
  end
end

# A VocalCow will have all the methods of Cow and Vocalizer.
class VocalCow < Cow
  include Vocalizer
end

c = VocalCow.new 'Bessie'
puts c.speak
c.vocalize
```

Ruby has a number of built-in modules. One of the most useful is `Enumerable`, with methods `map`, `select`, `reduce`, `count`, `find`, `detect`, `first`, `min`, `max`, `sort`, `include?`, `all?`, `any?`, `one?`, `take`, `drop`, `cycle`, `find_all`, `group_by`, and many others. The following built-in classes include `Enumerable`:

`Array Hash Struct Range IO Dir Enumerator`

For example, to find all of the array elements, or hash keys, or lines in a file (`File` is a subclass of `IO`, the class of input-output streams), or files in an operating system folder (called a `Dir`), that begin with the letter `'A'`, you need only say:

```
obj.filter {|s| s.start_with? 'A'}
```

where *obj* is an instance of any one of those classes. How does this work? Each of the methods in `Enumerable` rely on the object providing an implementation of the method `each`. We can take advantage of this in classes we write ourselves. As long as we define `each` and include `Enumerable`, our objects can be mapped, reduced, and so on.

```
class OneTwoThree
  include Enumerable
  def each
    yield 1
    yield 2
    yield 3
  end
end
```

```
t = OneTwoThree.new
fail unless t.first == 1
fail unless t.map {|x| x * 100} == [100, 200, 300]
fail unless t.select {|x| x.even?} == [2]
fail unless t.all? {|x| x > 0} == true
```

Two other important predefined modules are:

- `Comparable`, which defines `<`, `<=`, `>`, `>=`, `==`, and `between`. Each of these are implemented to invoke the receiver's `<=>` method, which is intended to work as follows: `a <=> b` should produce -1, 0, $+1$, or `nil` when a is less than, equal to, greater than, or not comparable to, respectively, b. It is mixed in to the numeric, string, and time classes.
- `Kernel`, which defines dozens of useful methods, including `object_id`, `==`, `send`, `is_a?`,[5] `chop`, `rand`, `exit`, `abort`, `puts`, `system`, `require`, `sub`, and `lambda`. This module is mixed into `Object`, which is an ancestor class of every class (except `BasicObject`), so these methods are universally available.

Modules become "parents" of a class, just like the superclass, for the purpose of method lookup. So Ruby, like Python, must deal with a potentially complex ordering of its ancestors when looking up methods. Ruby looks up methods in this order:

1. The object's singleton class.
2. The modules mixed into the singleton class. Later `include` statements take precedence over earlier ones, but in the construct `include A,B`, A takes precedence over B.
3. The object's class.
4. The modules mixed into the class.
5. The class's superclass.

These rules are applied recursively.

5.6 ACCESS CONTROL

With the exception of various control structures (e.g., `while`, `break`, `retry`) and certain special operators (`&&`, `||`, `..`, `...`, `?:`, `and`, `or`, `not`, `defined?`, and the assignment operators), all computation in Ruby is carried out by sending messages to objects.

When applications become complex, we find it useful to have control over when and how methods can be invoked. In Ruby, **public methods** (the default visibility) can be invoked on an object from anywhere. **Protected methods** can only be invoked from within methods of the object's class or its subclasses. **Private methods** can only be invoked without an explicit receiver. We'll explain.

Let's consider a simple account class that logs deposits and withdrawls. (For simplicity, we'll stick to deposits only and log with `puts`.) We can send a `deposit` message to an account, which adjusts the balance and sends a `log` message to *itself*. The expression `self` refers to the object currently executing a method, so here's our first pass:

[5] By now, you've seen quite a few method names ending in a question mark. Generally, methods returning `true` or `false` will be named this way; similarly, methods that mutate a receiver will often have names ending in an exclamation mark. Alas, these are only conventions, and not enforced by the language.

```
# A demonstration of what not to do. All methods are public.
class Account
  def initialize; @balance = 0; end
  def deposit(amount); @balance += amount; self.log(amount); end
  def log(amount); puts "added: #{amount}, balance: #@balance"; end
end

a = Account.new
a.deposit(100)       # This is fine, we want to make deposits.
a.log(-999)          # Noooo! Shouldn't log from outside the class!
```

Whoops. We didn't want the `log` method to be part of the class's public interface. Let's make it protected:

```
# An attempt to fix the account. But it still can be subverted.
class Account
  def initialize; @balance = 0; end
  def deposit(amount); @balance += amount; self.log(amount); end
protected
  def log(amount); puts "added: #{amount}, balance: #@balance"; end
end

a = Account.new
a.deposit(100)
a.log(-999) rescue puts "Whew: Can't call protected method"
```

That helped a little, but we still have a problem. While we cannot call protected methods from outside the class, we *can* make calls from within the class *or in subclass methods.* So someone can construct a malicious subclass to invoke the `log` method on our existing account:

```
# But protected does not help you from subclasses
class BadAccount < Account
  def mess_up_log(otherAccount); otherAccount.log(-999); end
end

BadAccount.new.mess_up_log(a)    # Oh no!
```

It's time for `private`. Private methods can only be sent to `self`. Ruby enforces this by making us write *method(args)* instead of *object.method(args)*[6] and defines the *implicit* receiver to always be `self`.

```
# Making the log method private prevents the previous attack
class Account
  def initialize; @balance = 0; end
  def deposit(amount); @balance += amount; log(amount); end
private
  def log(amount); puts "added: #{amount}, balance: #@balance"; end
end
```

[6]There is one exception to this rule: methods whose names end in =.

```
a = Account.new
a.deposit(100)
a.log(-999) rescue puts "Whew: Can't call private method"
BadAccount.new.mess_up_log(a) rescue puts 'SAFE!!'
```

To be honest, we didn't really make our accounts perfectly secure; there are ways to circumvent access controls.[7] Access controls can even be changed at runtime. But we did show how to use the language to express our intent that `log` was part of an account object's internal implementation and not part of an external interface.

The ability to call methods without a receiver is the key to making Ruby look like it has functions when in reality it has only methods. Whenever we define a method outside of any class (like we have done on more than one occasion throughout this chapter!), Ruby makes it private method of the class `Object`. Let's verify this:

```
def hello
  "hello"
end

fail unless Object.private_instance_methods.include? :hello
```

As the class `Object` is an ancestor of every class, except those explicitly derived from `BasicObject` (see Figure 5.1), its private methods can be used almost anywhere. We get the appearance of top-level functions, but internally everything is object-oriented message passing.

5.7 METAPROGRAMMING

Ruby lets us create classes to model the things we care about. In a typical commerce application, we define customers, items, orders, and purchases. In a calendar application, we create contacts and appointments. We write methods to ask questions such as "How many customers have used a coupon in the last 30 days?" and perform operations such as "Reduce the price of all volleyball shoes by 20 percent." But since classes, modules, methods, and procs are objects, we can query and manipulate them too. Querying and manipulating the program itself is called **metaprogramming**.

The simplest kind of metaprogramming, querying the program elements, is called **introspection**. Ruby provides a large number of introspective methods. Among them, for class c, class or module m, and object o, are:

- c`.superclass`
- m`.name`
- m`.included_modules`
- m`.instance_methods`
- m`.public_instance_methods`
- m`.protected_instance_methods`
- m`.private_instance_methods`
- m`.constants`

[7]One way to do this is simply write, `a.send(:log, -999)` since `send` does not respect access controls.

- *o*.class
- *o*.singleton_class
- *o*.is_a?
- *o*.respond_to?
- *o*.methods
- local_variables
- global_variables

The first few methods tell us about *classes*—the class of an object, the superclass of a class, and the methods defined by a module or class. The `respond_to?` and `methods` methods are about *type*. Generally speaking, the type of an expression constrains its behavior; applying operations to the wrong kind of operands, like multiplying symbols with hashes, are type errors, not class errors. Since nearly all Ruby "operations" are essentially method invocations, asking an object whether it can respond to a given message, or asking which methods it does respond to, is a type query, not a class query. Consider:

```
def increment(x)
  x + 1
end
```

You may ask "what types of arguments are allowed for this method?" The answer is that we can pass any object that responds to the `:+` message with the argument `1`.[8] The specific class of the object does not matter; all that matters is the object understands the message.[9]

The last two methods on this list are quite interesting, as Ruby's introspective capabilities allow you to find out which variables—local and global—are currently in scope. Global variable names always start with a `$`. Ruby comes with several dozen predefined globals, including `$;`, `$!`, `$~`, `$0`, `$FILENAME`, `$.`, `` $` ``, `$&`, `$'`, `$PROGRAM_NAME`, `$*`, `$-a` and many more. You can create new globals by simply assigning to them.[10] Let's explore:

```
$x, $y = 1, 2                # globals
x, z = 10, 20                # private variables of main object

def example()
  x, y = 100, 200            # local variables
  p $x, x, $y, y             # globals and locals do not conflict
  p z rescue p 'No z'        # unlike blocks, methods don't capture
  p local_variables          # [:x, :y]
  p global_variables         # long list, including :$x and :$y
end

example
```

The second kind of metaprogramming involves (programmatically) changing the program itself. We can write code that adds or removes classes, adds or removes methods, and adds or removes instance variables. In fact, class definition and even method definition are actually done with message passing:

---

[8]That's because `x + 1` is sugar for `x.+(1)`.

[9]To fully understand why the class does not matter, remember that in Ruby you can add methods to a specific object, regardless of its class.

[10]But don't. There are almost always better ways to solve problems than to bring in global variables.

```
C = Class.new do
  define_method (:f) {|x| x + 1}
end

puts C.new.f(8)
```

This is pretty interesting, so let's break it down.

- We create a class by sending `new` to the class `Class` together with a block.
- We send `define_method` (it's a private method, so we don't mention the receiver) with a symbol for the method name and a block with the method's implementation.

The example shows that message passing is truly fundamental, and Ruby's use of the `class` expression, the use of `def`, and other fancy (but useful!) constructs are, more or less "just syntax."[11] Entire object-oriented languages can be built with hardly any syntax at all if desired—variables, messages, and parameters should suffice. If you find this approach to language design interesting, take a look at Io [24].

5.8 RUBY WRAP UP

In this chapter we were introduced to a genuine, but pragmatic, object-oriented language, Ruby. We learned that:

- Ruby was designed for programmers to be productive and have fun.
- Almost all computation is done via message-passing and an object's state is encapsulated, allowing Ruby to be called an object-oriented programming language.
- All objects in Ruby are instances of some class. Classes can have a single superclass and can have multiple subclasses. Method dispatch is dynamic.
- In addition to an object's class, an object may have a singleton class, in which to store object-specific behavior.
- Methods may be passed blocks in addition to their regular arguments. A method executes its block via the `yield` expression. Blocks provide the foundation for functional programming in Ruby.
- Blocks are closely related to procs. Procs have both lambda and non-lambda forms.
- While a class can have only one superclass, it may mix in many modules. (A class is just a module that can have instances.) This means classes can inherit methods from multiple ancestors, and a method resolution algorithm has been defined.
- The two most useful standard modules are `Enumerable`, which provides a means to iterate through arrays, hashes, directories, files, and the like, and `Kernel`, which provides dozens of methods applicable to any object. `Kernel` is mixed in to `Object`, which is the ancestor of (nearly) every other class.
- Ruby methods can be public, protected, or private. Private methods (with one tiny exception) cannot be called with an explicit receiver, enabling code to read as if it had global functions. Access controls can be subverted.

[11]This isn't exactly true, because the scoping rules kick in. Blocks can capture variables in their environments whereas class expressions do not.

- Ruby has a wealth of metaprogramming facilities. Objects, classes, modules and other entities can be introspected. Classes and even methods can be created programmatically at runtime.

To continue your study of Ruby beyond the introductory material of this chapter, you may wish to find and research the following:

- **Language features not covered in this chapter**. BEGIN and END blocks, Heredocs, alternate quotes (`%q`, `%Q`, `%w`, etc.), embedded documentation, regular expressions, class variables, class methods, predefined variables, the standard library.
- **Open source projects using Ruby**. Studying, and contributing to, open source projects is an excellent way to improve your proficiency in any language. You may enjoy the following projects written in Ruby: Rails (`https://github.com/rails/rails`), Homebrew (`https://github.com/Homebrew/homebrew`), Jekyll (`https://github.com/jekyll/jekyll`), Sass (`https://github.com/sass/sass`), and Vagrant (`https://github.com/mitchellh/vagrant`).
- **Reference manuals, tutorials, and books**. Ruby's home page is `https://www.ruby-lang.org/`. The docs page, `https://www.ruby-lang.org/en/documentation/`, contains links to the Getting Started Guide, the User's Guide, the Language Reference, and the Library Reference. Online resources for learning the language and exploring advanced features are plentiful; the Ruby Monk courses at `https://rubymonk.com/` cover a wide range of skill sets. Popular books include *Learn Ruby the Hard Way* [108] and *Programming Ruby 1.9 & 2.0* [118]. Ruby, like most languages, is evolving, so keep an eye out for the latest editions.

EXERCISES

Now it's your turn. Continue exploring Ruby with the activities and research questions below, and feel free to branch out on your own.

5.1 Experiment with the Ruby REPL.

5.2 Compare the manner in which Python and Ruby provide a "default value" when looking up values with nonexistent keys. (The major difference can be seen in the word count examples at the beginning of their respective chapters.)

5.3 The class hierarchy shown in Figure 5.1 is incomplete. The class `Exception`, for example, is the root of a sizable hierarchy of classes. What does this hierarchy look like? Are any other classes missing?

5.4 Familiarize yourself with the Ruby documentation. Find information on the following methods: `to_s`, `to_i`, `to_a`, and `to_h`. In which class or module are each of these defined? What do they do?

5.5 In Python, we saw that `7/2==3.5`, `7//2==3`, and `7.0//2==3.0`. The latter two expressions use the floor division operator. Does Ruby have a floor division operator? If so, what is it? If not, how we do simulate it?

5.6 Research the following operators: `===`, `<=>`, `**`, `&.`, and `=~`. Write a short script (or scripts) that use each in a meaningful way. What are the Python equivalents to each, if any?

5.7 Perform the following experiment in the Ruby REPL. Generate, by any means you can discover, two string objects that have the value "dog" but have different object ids. Send the `to_sym` method to each. Compare the object ids of each symbol. What does the result of this comparison tell you?

5.8 Can you make a hash whose keys are numeric? Strings? Arrays? Can you make a single hash with both numeric and symbol keys? Research and describe the exact conditions under which an object can be used as a hash key.

5.9 Suppose a Ruby method is defined with a single positional parameter, for example `def f(x); x; end`. What actually happens when the method is called as one would call a named parameter, that is, `f(x:3)`?

5.10 Write a Ruby function that takes a string of the form one-or-more digits followed by an operator from the set $\{+, -, *, /\}$ followed by one or more digits, and returns the evaluation of that string. Evaluate the string using `.send`.

5.11 Read Allen Holub's *Why getter and setter methods are evil* [61]. What do you think about the claim that getters and setters are often a poor programming choice? Find some articles or papers that rebut Holub's position (if any) and summarize their main arguments.

5.12 Finding out all of the technical details of a language's scope rules requires a look at the language's published documentation, or ideally, its formal semantics if one exists. However, we can always conduct experiments. Here is a script we've written to explore some scope questions.

```
def one
  x = 1
  [1,2,3].each {|x| y = 1}
  y
end

def two
  x, y = 1, 2
  [1,2,3].each {|x| y = 1}
  y
end

def three
  x, y = 1, 2
  [1,2,3].each {|x; y| y = 1}
  y
end
```

What does this script output? What questions do you think the script author is trying to discover? What can be learned from this experiment?

5.13 The following script

```
class C; def a; end; end
class C; def b; end; end

puts C.public_instance_methods(false)
```

features the definition of a class followed by... what? Is the second line a *re*-definition? Or an extension? Run the script to find out the answer to this question. Then run the equivalent "experiment" in Python. Did you notice a difference?

5.14 How are Ruby blocks like Lua coroutines? How are they different?

5.15 Ruby has both a `Method` and an `UnboundMethod` class. What is the difference between them?

5.16 In this chapter, we did not mention Ruby's class variables and class instance variables. Find out about these features. Explain them in your own words, and show how they are different from each other. What kind of problem solutions do they make more expressive than equivalent code without them?

5.17 The entity known as a "singleton class" in Ruby has been called an "eigenclass" or a "metaclass" or a "virtual class". The term "eigenclass" is not widely used, but the latter two terms are downright confusing because they have different meanings in different contexts. For example, "virtual class" is a common term in C++, and "metaclass" has a completely different meaning in the Smalltalk community. How exactly is the term used in the Smalltalk world, and how is such usage similar to, or different from, the Ruby singleton class?

5.18 Python has metaclasses, too (which we didn't cover). Compare and contrast the metaclass concept in Python with Ruby's singleton class.

5.19 Write a script that when given an object, produces the array of singleton classes, classes, and modules that are searched (in order) when resolving a message sent to the object.

5.20 Find the complete list of predefined global variables in Ruby. Experiment with them in the REPL.

5.21 One of the built-in classes in Ruby is `RubyVM`. Read the documentation for this class, and experiment with `RubyVM::InstructionSequence`. Would the use of this class be considered metaprogramming?

5.22 For each of the following classes and modules: `BasicObject`, `Object`, `Class`, `Module`, `Kernel`, `Numeric`, `Float`, and `Integer`, give its singleton class, its class, its superclass, and its list of included modules. Try your hand at a diagram of the relationship between these objects.

CHAPTER 6

Julia

Julia is a modern language for scientific and technical computing.

First appeared 2012
Creator Jeff Bezanson, Stefan Karpinski, Viral B. Shah, Alan Edelman
Notable versions 0.4 (2015) • 0.5 (2016)
Recognized for Multi-dispatch, Macros, Multicore support
Notable uses Data Science, Statistics, Numerical Analysis
Tags High-performance, Dynamic, Multi-dispatch
Six words or less A dynamic language for technical computing

Julia is targeted to high-performance numeric and scientific computing, where it competes with Matlab, Python, R, and Fortran—languages with well-established and mature libraries. While relatively new, Julia brings an impressive set of features not only of interest to the statistics and data science communities (such as its math-friendly syntax, powerful arrays and matrices), but that make it an efficient and powerful general purpose language. Language theorists will find its metaprogramming facilities of particular interest.

In this chapter, we'll tour this new language, beginning with our standard three introductory programs, then dive right into Julia's type system. We'll be seeing several topics for the first time in this book: **abstract types**, which do not allow instances, and **parametric types**, in which a type is dependent on certain values or other types. We'll see that Julia allows subtyping, but with an interesting restriction not shared by Ruby or Python. We'll also use Julia to introduce the type-theoretical concepts of sum and product types, covariance, contravariance, and invariance.

Next up in our tour is **multiple dispatch**. Each of the languages we've seen so far selects the appropriate method to execute based on the type of a single *receiver* at run time, but Julia goes farther, selecting based on the types of all of the method's *arguments*. We'll then look at metaprogramming facilities, which go well beyond what we've seen in the previous chapters. Julia can introspect not only objects and types, but also expressions and the internal structure of the program itself. We'll see **hygienic macros** for the first time. We'll close with a look at language support for parallel and distributed computing.

6.1 HELLO JULIA

We'll start with the usual right triangle script:

```
for c = 1:40, b = 1:c-1, a = 1:b-1
  if a * a + b * b == c * c
    println("$a, $b, $c")
  end
end
```

Julia's *ranges* (of the form `x:y`) have inclusive bounds, and string interpolation uses `$`. The `for`-statement has a neat compact form that avoids the usual requirement of nested loops. Heap's algorithm follows:

```
function generatePermutations(a, n)
  if n == 1
    println(join(a, ""))
  else
    for i = 1:n-1
      generatePermutations(a, n-1)
      local j = iseven(n) ? i : 1
      a[j], a[n] = a[n], a[j]
    end
    generatePermutations(a, n-1)
  end
end

if length(ARGS) != 1
  println(STDERR, "Exactly one argument is required")
  exit(1)
end
generatePermutations(split(ARGS[1], ""), length(ARGS[1]))
```

Julia uses expressions like `length(a)` rather than `a.length`; even `join` and `split` look more like top-level functions than methods on arrays and strings—though they really *are* methods as we will soon see. Writing to standard error is very easy, as is calling `exit`. The familiar ternary conditional operator is present, and swaps are easy to write. And, like Lua but unlike most other languages, the lower bound of arrays is 1.

Next, we count words from standard input:

```
counts = Dict{AbstractString, UInt64}()

for line in eachline(STDIN)
  for word in matchall(r"[a-z\']+", lowercase(line))
    counts[word] = get(counts, word, 0) + 1
  end
end
for (word, count) in sort(collect(counts))
  println("$word $count")
end
```

Here we first define a dictionary for storing the counts for each word. We constrain the dictionary so that keys must be strings, and values must be 64-bit unsigned integers.

The expression `eachline(STDIN)` gives us a line-by-line iterator over standard input. We lowercase and break each line into contiguous words with lowercase a-z (for simplicity) and the apostrophe. For each word, we bump its count in the dictionary if it's already there, and make a new entry with count 1 if not. The lookup operation in a map—`get(counts, word, 0)`—features a convenient third argument for the value to return if the key is not present. Zero is the perfect choice here.

The `collect` function turns our (unordered) dictionary into an array of key-value pairs, which we can sort by key, exactly what our problem specification requires. The destructuring iterator and the string interpolation make this whole script very short and clean.

6.2 THE BASICS

From our introductory programs, you may have noticed Julia looks a little like Lua. Structured statements are keyword terminated (e.g., `while...end`, `for...end`, `if...elseif...end`, `try...catch...end`), indexes start at 1 instead of 0, local variables are introduced with the `local` keyword, and there is no `class` keyword. Julia's coroutines are similar to Lua's, too. But Julia adds some nice syntactic forms. In addition to `function...end`, we have two additional ways to specify functions:

```
f(x, y) = 2x + y                  # call-like form on left-hand side
g = x -> x * x                    # arrow notation
@assert g(f(2, 9)) == 169
```

Functions can have optional and keyword parameters. You can omit the keyword `return` when returning the last expression in a function (as in CoffeeScript and Ruby):

```
function f(a, b=1 ; c=2, d=3)
  [a, b, c, d]
end

@assert f(100) == [100, 1, 2, 3]
@assert f(100, 200) == [100, 200, 2, 3]
@assert f(100, d=10) == [100, 1, 2, 10]
@assert f(100, 200, d=400, c=300) == [100, 200, 300, 400]
```

A semicolon separates the positional and keyword parameters, and keyword parameters must be given default values (in other words, all keyword parameters are also optional). Julia uses three dots (`...`) for packing and unpacking. Positional parameters pack and unpack into iterable collections of values; keyword parameters pack and unpack into arrays of $(symbol, value)$ pairs:

```
f(x, y...) = y
@assert f(1, 2, 3) == (2, 3)

g(x; y...) = y
@assert g(1, a=2, b=5) == [(:a,2),(:b,5)]
```

Here's the Julia version of a script we saw a while ago in Python:

```
function writerule(selector; options...)
  println("$selector {")
  for (prop, value) in options
    println("  $(replace(string(prop), "_", "-")): $value;")
  end
  println("}")
end

writerule("h1", font_family="Helvetica", size="20px")
writerule("p.error", color="red", margin="16px", padding="0")
```

Rather than having only global or function scopes, Julia's scoping regions coincide with the following constructs: function bodies, `while` and `for` loops, and blocks introduced with `try`, `catch`, `finally`, `let`, and `type`. Variables introduced in a scope are visible to nested scopes, and shadowing is permitted. We introduce new variables into the current scope with `local` or `const`; function parameters are automatically made part of the function body scope. If you assign to a variable without marking it local, a new local is introduced in the current scope *unless* it has been marked `local` or `global` in an enclosing scope. Marking a variable `global` is required to allow writing to a global variable in an "inner" (non-global) scope.

```
a, b, c = 1, 2, 3           # three globals
(function ()
  a = 10                    # introduces local and shadows
  global b = 20             # overwrites global
  d = 10c + 10              # new local, reading global
  local e = 5
  while true
    e = 50                  # outer e because marked local
    f = 60                  # local to while loop!
    break
  end
  @assert (a,b,c,d,e) == (10,20,3,40,50)
  @assert (try f catch -1 end) == -1
end)()
@assert (a,b,c) == (1,20,3)
```

Finally, Julia has a special rule for variables in comprehensions: new variables are created for each iteration. Python, in contrast, iterates with existing variables! Let's compare:

```
# Julia
julia> x = 0
julia> [x^2 for x in 0:9]
10-element Array{Int64,1}
julia> x
0
```

```
# Python
>>> x = 0
>>> [x**2 for x in range(10)]
[0, 1, 2, 3, 4, 5, 6, 7, 8, 9]
>>> x
9
```

That's it for the basics. Julia's type system goes well beyond anything we've seen up to this point, so we'll devote the next section to it.

6.3 TYPES

Julia's type system will remind you more of Python's and Ruby's than JavaScript's: there's no distinction between primitive and reference types, types are themselves objects (so there's a type called `Type`), types are organized into a hierarchy (though single-inheritance only), there are separate types for integers and floats, and you can define your own types. Julia's basic types go a bit further, distinguishing signed and unsigned integers, providing types for numbers of various bit lengths, and providing a rational number type.[1]

We show a part of the hierarchy of Julia's built-in types in Figure 6.1. Julia adds several typing concepts we've not yet seen: **abstract types**[2], **parametric types**, and **union types**. These are new and interesting concepts for us, so we'll look at each in detail.

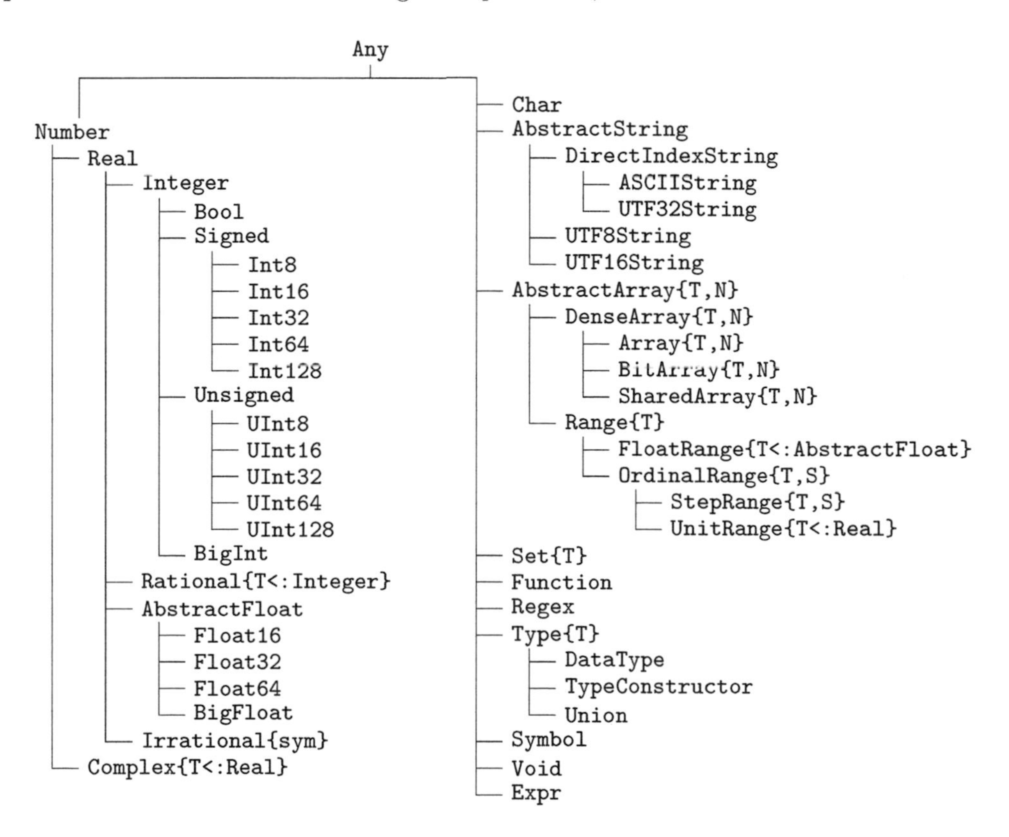

Figure 6.1 A few of the built-in Julia types

6.3.1 Abstract Types

Every type in Julia is either **abstract** or **concrete**. Every value has a single concrete type; for example, `2.54` has the type `Float64`. Let's explore a few more:

[1]You can find details of these various numeric types in Appendix A.

[2]Python does "support" the notion of an abstract class through operations in its `abc` library module, but Julia offers support for the concept directly in the language itself.

```
@assert typeof(3) == Int64
@assert typeof(0x22) == UInt8
@assert typeof(0xFA31) == UInt16
@assert typeof(false) == Bool
@assert typeof('é') == Char
@assert typeof("Hello") == ASCIIString
@assert typeof("Chloë") == UTF8String
@assert typeof(:hello) == Symbol
@assert typeof(r"\d+(\.\d+)") == Regex
@assert typeof(Float64) == DataType
@assert typeof(DataType) == DataType
```

Abstract types do nothing more than *generalize* other types; for example, the abstract type `AbstractFloat` generalizes the four concrete types `Float16`, `Float32`, `Float64`, and `BigFloat`. Calling `typeof` will never produce an abstract type, but you can use `isa` to check "membership" in an abstract type:

```
@assert typeof(88) == Int64
for t in [Int64, Integer, Signed, Real, Number, Any]
  @assert isa(88, t)
end
```

In Julia, *only* abstract types can have subtypes; in fact, in Figure 6.1, all leaf types are concrete and all non-leaf types are abstract.[3] We can explore the type system with the built-in methods `super` and `subtypes` (which do what you'd expect), and the `<:` operator, which determines whether one type is a descendant of another.

```
@assert super(Int32) == Signed
@assert super(Signed) == Integer
@assert super(Integer) == Real
@assert super(Symbol) == Any
@assert Set(subtypes(Type)) == Set([DataType, TypeConstructor, Union])

@assert Float64 <: Real
@assert isa(subtypes(Type), Array)
@assert isa(Array, Type)
```

Let's implement our own abstract and concrete types. We'll do so by directly translating our little animals script from Python and Ruby. We create an abstract type `Animal`, and three concrete subtypes. As before, all animals `speak` the same way, but each makes its own `sound`:

```
abstract Animal
speak(a::Animal) = "$(a.name) says $(sound(a))"

type Horse <: Animal
  name
end
sound(h::Horse) = "neigh"
```

[3]This is *not* the case in every language. Plenty of other languages allow concrete types to have subtypes.

```
type Cow <: Animal
  name
end
sound(c::Cow) = "moooo"

type Sheep <: Animal
  name
end
sound(s::Sheep) = "baaaa"

s = Horse("CJ")
@assert speak(s) == "CJ says neigh"
c = Cow("Bessie")
@assert speak(c) == "Bessie says moooo"
@assert speak(Sheep("Little Lamb")) == "Little Lamb says baaaa"
```

The Julia version looked much different than before! We did not create classes, nor did we embed `speak` and `sound` inside the type declarations. We did, however, write three distinct `sound` implementations. When the expression `sound(a)` is executed, Julia will **dispatch** to the correct implementation *based on the type of* `a`. In Julia's terminology, `sound` is a **generic function** with three separate **methods**. The function `methods` reports the methods for a generic function, as we show here in the REPL:

```
julia> sound
sound (generic function with 3 methods)

julia> methods(sound)
# 3 methods for generic function "sound":
sound(h::Horse)
sound(c::Cow)
sound(s::Sheep)
```

6.3.2 Parametric Types

Julia types can be parameterized by other types and certain other values. We can get a feel for these types by studying a few examples:

```
@assert typeof(1:10) == UnitRange{Int64}
@assert typeof(0x5 // 0x22) == Rational{UInt8}
@assert typeof(5 // 34) == Rational{Int64}
@assert typeof(8.75im) == Complex{Float64}
@assert typeof(e) == Irrational{:e}
@assert typeof([5,3]) == Array{Int64,1}
@assert typeof([3, "abc"]) == Array{Any, 1}
@assert typeof([]) == Array{Any, 1}
@assert typeof([1 0; 0 1]) == Array{Int64, 2}
@assert typeof(Set(4)) == Set{Int64}
@assert typeof(Set(['3', '$'])) == Set{Char}
```

To distinguish sets of integers from sets of characters, Julia defines a **parametric type** called `Set{T}` whose parameter T can be any type. Parameters can have restrictions as well. Rational numbers, for example, can have `UInt8` or `Int64` components but not string components, so Julia provides the type `Rational{T<:Integer}`, restricting T to a descendant type of `Integer`. You'll notice several families of parametric types in Julia, including those for sets, arrays, ranges, and complex numbers.

Julia's array type is parameterized not only by the type of its elements but by its **dimension**. So the type of one-dimensional integer arrays differs from the type of two-dimensional integer arrays. Here are some examples of arrays:

$$a = \begin{bmatrix} 10 \\ 20 \\ 30 \\ 40 \end{bmatrix} \quad b = \begin{bmatrix} 5 & 3 & 7 \end{bmatrix} \quad c = \begin{bmatrix} 1 & 0 & 9 \\ 0 & 1 & 6 \end{bmatrix}$$

Here a is written in Julia as `[10,20,30,40]` and is a *one*-dimensional array, or **vector**, of type `Array{Int64,1}`. Array b is written `[5 3 7]` (note spaces instead of commas), and is a 1×3 (one row, three columns) *two*-dimensional array, or **matrix**, of type `Array{Int64,2}`. Array c is a 2×3 array, also with type `Array{Int64,2}`, and is written `[1 0 9; 0 1 6]`. Higher dimensional arrays have no special syntax, though they can be constructed with comprehensions, e.g., `[i+j+k for i=1:4, j=1:3, k=1:5]`.

In an n-dimensional array, the expression $a[e_1, e_2, ...e_n]$ will select a single element from the array when each of the indexes e_i are integers. However, the index expressions can also be ranges or vectors, allowing some very interesting **slices** to be computed:

```
a = [11 12 13 14; 21 22 23 24; 31 32 33 34]

@assert a[2,3] == 23                         # element in row 2, col 3
@assert a[2,2:4] == [22 23 24]               # row 2, cols 2 through 4
@assert a[1:3,3] == [13; 23; 33]             # rows 1 through 3 of col 3
@assert a[3,1:end] == [31 32 33 34]          # row 3, all elements
@assert a[3,:] == [31 32 33 34]              # row 3, all elements
@assert a[2,[1;3;4]] == [21 23 24]           # row 2, cols 1, 3, 4
@assert a[[1;3],[1;4]] == [11 14; 31 34]     # very disjointed subarray
@assert a[7] == 13                           # enumerates by columns
```

Julia has *hundreds* of built-in methods for processing arrays, including `length(A)` for the number of elements in A, `ndims(A)` for the number of dimensions, and `size(A)` for a tuple of its dimensions. Some construct new arrays:

```
@assert zeros(Int64, 3) == [0, 0, 0]
@assert zeros(Int64, 2, 2) == [0 0; 0 0]
@assert ones(Int64, 3, 2) == [1 1; 1 1; 1 1]
@assert eye(Int32, 3) == [1 0 0; 0 1 0; 0 0 1]
@assert fill(5, 1, 4) == [5 5 5 5]
@assert transpose([1 3; 2 4]) == [1 2; 3 4]
@assert [1 3; 2 4]' == [1 2; 3 4]              # postfix ' transposes
@assert fill(10, 3, 1) == [10 10 10]'          # 2d column array!
@assert [2x for x in 1:5] == [2,4,6,8,10]
```

Many arithmetic operations extend to arrays. Array addition and subtraction (and a few other operations) are defined element-by-element, but multiplication, division, and exponentiation (and a few others) are not. Where an operation does not work element-by-element, a "dotted" version of the operator does work that way:

```
a = [1 2 3; 4 5 6; 7 8 9]
b = [1 0 0; 0 2 0; 9 9 9]

@assert a + b == [2 2 3; 4 7 6; 16 17 18]
@assert a * 2 == [2 4 6; 8 10 12; 14 16 18]
@assert a * b == [28 31 27; 58 64 54; 88 97 81]   # Matrix multiply
@assert a .* b == [1 0 0; 0 10 0; 63 72 81]       # Elementwise mul

@assert exp2(a) == [2.0 4.0 8.0; 16.0 32.0 64.0; 128.0 256.0 512.0]
```

Julia has a vast number of *built-in* functions useful in the field of linear algebra, including cross product; eigenvalue computation; Hessenberg and singular value decompositions; Givens rotations; computations of transpositions, determinants, triangles and diagonalizations; and a large number of factorizations. Dozens more methods are included in the standard library packages `Base.LinAlg.BLAS` and `Base.LinAlg.LAPACK`.

6.3.3 Sum and Product Types

Julia allows us to combine types T_1 and T_2 in an "algebraic" fashion to produce two new types, $T_1 + T_2$ and $T_1 \times T_2$. The **sum type** contains all the values from both types, while the **product type** contains pairs whose first element is from T_1 and the second from T_2. Let's build up these new types from `UInt8`s and `Bool`s:

	Sum	**Product**
Julia Type	`Union{UInt8, Bool}`	`Tuple{UInt8, Bool}`
Values	`0, 1, ... , 254, 255, false, true`	`(0,false), (0,true), (1,false), (1,true), ... (255,false), (255,true)`

Here are union types in action:

```
u = Union{UInt8, Bool}            # A new type
@assert typeof(u) == Union
@assert isa(u, Type)              # A union type is a type

@assert isa(0x08, u)              # UInt8 values belong
@assert isa(false, u)             # Boolean values belong
@assert isa(true, u)
@assert !isa(256, u)              # Not a UInt8, does not belong
```

and tuples:

```
t = Tuple{UInt8, Bool}              # a tuple type
@assert typeof(t) == DataType       # tuple types aren't special
@assert isa(t, Type)                # tuple types are types

@assert isa((0x08, false), t)       # tuples of the right type belong
@assert isa((0x7A, true), t)
@assert !isa((3, "wrong"), t)       # tuples of the wrong type do not
@assert !isa(0x25, t)               # non-tuples do not belong
```

You can remember the difference between sum types and product types by noting the following. There are 256 values in `UInt8` and 2 in `Bool`. The union (sum) type has $256+2=258$ values while the tuple (product) type has $256 \times 2 = 512$.

Sums and products are not limited to two constituent types:

```
# 3 component types
@assert typeof((4, false, [])) == Tuple{Int64, Bool, Array{Any, 1}}
@assert isa(false, Union{Int64, Bool, Array{Any, 1}})

# 1 component type
@assert typeof((5,)) == Tuple{Int64}
@assert isa(false, Union{Bool})
@assert Bool != Tuple{Bool}              # Because false != (false,)
@assert Bool == Union{Bool}              # Do you see why?

# No component types
@assert typeof(()) == Tuple{}
```

The type `Union{}` has no values (there are no underlying values to collect) while `Tuple{}` has exactly one, namely `()`, the zero-element tuple. If you've encountered abstract algebra, you'll recognize the analogy to 0 being the identity element of sum and 1 of product.

6.3.4 Type Annotations

Julia provides the ability to attach a type annotation to a variable, using the notation v`::`T for variable v and type T in local variable declarations and assignments. All assignments of a value to a variable with a type annotation will be type-converted to the annotated type if possible. (In Julia, a value v is converted to type T by calling `convert(T,v)`.)

```
(function()
  local x::Int64 = 0x02
  local y = 0x02
  @assert typeof(x) == Int64  # The UInt8 was converted
  @assert typeof(y) == UInt8  # No annotations, no coversion

  # There's no conversion from ASCIIString to Int64
  @assert isa(try x = "Oh no" catch (e) e end, Exception)
end)()
```

Type annotations may increase performance: a compiler can use knowledge that a variable has a given type to create efficient memory layout and run-time access code. Type annotations are completely optional in Julia. Some languages (e.g. CoffeeScript) have no notion of type annotations, while some languages (e.g., Java, coming up in the next chapter) require them for all variables.

6.3.5 Covariance, Contravariance, and Invariance

We've seen subtype-supertype relationships among simple, nonparameterized types, for example `Int64 <: Number` (i.e., a 64-bit integer is a number). But what about parameterized, union, and tuple types? Can we say anything about their subtype-supertype status based on the types of their component types? For example, how are `Set{Int64}` and `Set{Number}` related, if at all? First, some terminology:

- If `Set{Int64} <: Set{Number}`, sets would be **covariant**
- If `Set{Number} <: Set{Int64}`, sets would be **contravariant**
- If neither holds, sets would be **invariant**
- If both hold, sets would be **bivariant**

It turns out that in Julia, sets are invariant. Why are they not covariant? If they were, we would be able to declare a variable a with declared type "set of animals" and assign to it a value of type "set of dogs." Since a is declared as a set of animals, it would *appear* reasonable to add a cat to the set. But the underlying value of the variable is a set of dogs, which cannot hold a cat. Invariance avoids this pitfall.

Tuples, on the other hand, *are* covariant:

```
t = (3, "hello")
@assert typeof(t) == Tuple{Int64, ASCIIString}
@assert isa(t, Tuple{Real, ASCIIString})
@assert isa(t, Tuple{Integer, AbstractString})
@assert isa(t, Tuple{Number, Any})

@assert Tuple{Symbol, Float16, Union{}} <: Tuple{Any, Real, Any}
```

The previous problem with sets does not arise because tuples are immutable.

6.4 MULTIPLE DISPATCH

Let's turn now to functions and consider the following problem: how can we write a function whose behavior depends on the *types* of its arguments? For example, consider a function called `times` such that:

- If x and y are numbers, the function multiplies, e.g., `times(8,3)` ⇒ `24`
- If x is a string and y is an unsigned integer, we have repetition, e.g., `times("ho",3)` ⇒ `"hohoho"`
- If x is a vector and y is a number, we have element-wise multiplication, e.g., `times([1,2,3],3)` ⇒ `[3,6,9]`
- If x is a number and y is a vector, we also have element-wise multiplication, e.g., `times(5,[1,2,3])` ⇒ `[5,10,15]`

We *could* write a single function that checks the types of its arguments in a multi-way `if`-statement, but the clutter, the mixing of disparate concerns, and the impossibility of extensibility without modifying existing source code demands that we write a separate function for each case. We could write four functions with distinct names, or, as the previous languages in this book allow, attach methods directly to an object, or define methods in a class, prototype, or metatable. JavaScript, CoffeeScript, Lua, Python, and Ruby, however, won't permit *all four* methods to have the same name: *num*.`times`(*vec*) and *num*.`times`(*num*) (as methods of a number class or prototype) require different names or run-time type checks.

Julia can do better. Rather than attaching methods to a single receiver, it takes into account the types of *all* of its arguments. So we can write:

```
times(x::Number, y::Number) = x * y
times{T<:Number}(x::AbstractArray{T,1}, n::Number) = x .* n
times{T<:Number}(n::Number, x::AbstractArray{T,1}) = x .* n
times(s::AbstractString, n::Unsigned) = repeat(s, n)

@assert times(5, -2.5) == -12.5
@assert times([1.0, 5, 2//3], 3) == [3.0, 15, 2]
@assert times(3, [1.0, 5, 0]) == [3.0, 15, 0]
@assert times("ho", UInt64(3)) == "hohoho"
```

Here `times` is a **generic function** with four **methods**. Considering the types of multiple arguments when determining which method to call is known as **multiple dispatch**.

The first call in our example dispatches to the first method, because the argument types `(Int64, Float64)` can be passed to the parameter tuple `(Number, Number)`. But notice the definition of the second method. It is tempting to define its first parameter as an `AbstractArray{Number, 1}`, but this will not work! Passing an array of `Float64` objects to an array of numbers fails because arrays are not covariant on their element type. Julia's **parametric methods** rescue us here. We introduce a type parameter T and define the method's first parameter as an array of elements whose type is `Number` or any descendant type of `Number`. We'll encounter this technique of parameterizing methods by type to get around the invariance of collection types in the next chapter (on Java) as well.

6.5 METAPROGRAMMING

We've seen a few introspective capabilities in previous chapters, such as asking a Python or Ruby class for a list of its methods, or asking a JavaScript function for the number of arguments it accepts. Julia has dozens of such functions, including `super`, `subtypes`, `typeof`, and `isa` (seen before), as well as `fieldnames` (of a datatype), `names` (of a module), and `methods` (to list the methods of a generic function). The surprising functions `code_llvm` and `code_native` will show you the intermediate and target assembly language code for a method.

But Julia goes even farther: individual *expressions*, in their *unevaluated* form, can be accessed and manipulated as objects within the program itself. There are several ways to create expression objects:

```
e1 = parse("x * sqrt(9) - 5")
e2 = :(x * sqrt(9) - 5)
e3 = Expr(:call, :-, Expr(:call, :*, :x, Expr(:call, :sqrt, 9)), 5)

@assert e1 == e2 && e2 == e3 && typeof(e1) == Expr
```

Objects of type `Expr` have three fields: `head`, `args`, and `typ`. The internal structure of the expression in the previous example (without the `typ` fields) is shown in Figure 6.2.

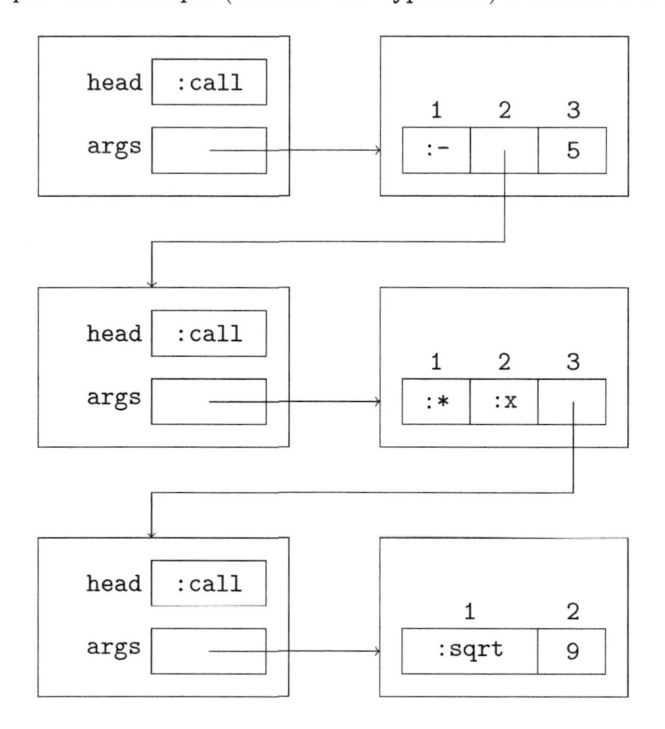

Figure 6.2 Internal structure of the Julia Expression `:(x*sqrt(9)-5)`

Unevaluated expressions (`Expr` objects) are also known as **quoted** expressions, named after the way we use quotes in prose to distinguish a *mention* of a word from its *use*: we say that ice is frozen water but "ice" has three letters. To make things interesting, Julia allows evaluation during the construction of a quoted expression, via **splicing** (intentionally using the same syntax for string interpolation we saw earlier):

```
y = 10
e = :(x + $(y / 2))     # y/2 evaluated here
@assert e == :(x + 5)
```

Access to unevaluated expressions allows us to rewrite code before evaluation, sometimes to optimize, sometimes to enhance (by adding, for example, timing, logging, or persistence), and sometimes to extend the language with new syntactic forms. While it is possible in theory to write functions that accept expressions as arguments and return expressions, Julia provides **macros** to more elegantly manipulate expressions. Macros are evaluated at the time the script is parsed, and return expressions that are then compiled into the context in which they are called. For example, the following script

```
macro tenTimesX()
  println("hello");         # executed at parse time
  :(x * 10)                 # expression returned at parse time
end

function f()
  x = 2;
  @tenTimesX()              # replaced with x*10 at parse time
end                         # when called, f executes x=2; x*10
```

prints a message, even though the function f is never called! While parsing the script, Julia executes the macro call within f (printing a message). The returned expression, `:(x * 10)`, can then be compiled into the body of f. This rewrites the function to:

```
function f()
  x = 2
  x * 10
end
```

Julia takes care that writable variables in a macro's result expression do not modify variables of the same name in the caller's context. In technical terms, Julia macros are **hygienic**. In the following example, the macro's own local y does *not* overwrite the caller's y:

```
macro tenTimesX()
  :(y = 10; x * y)
end

(function ()
  x = 2;
  y = 3;                    # Will this y be changed?
  z = @tenTimesX()          # Changed to 10 perhaps?
  @assert y == 3 && z == 20 # No, y will not change
end)()
```

We can use the function `macroexpand` in the REPL to see why no change occurs:

```
julia> macro tenTimesX()
         :(y = 10; x * y)
       end

julia> macroexpand(:((x,y) = (3,2); @tenTimesX()))
quote
    (x,y) = (3,2)
    begin
        #1#y = 10
        x * #1#y
    end
end
```

During macroexpansion, Julia generated a new symbol, `#1#y`, to use in place of the variable y, avoiding a name clash. Julia will generate new names for result variables that are (1) explicitly marked local, (2) assigned to but not marked global, or (3) used a function argument. Other variables are resolved in the context of the macro definition's environment, so the macro's use of well-known globals will not be overwritten by names the caller might have inadvertently redefined.

There are times, however, when we *do* wish a macro to manipulate variables from the caller's context. This is convenient when making use of macros' ability to "extend" the language with what appear to be new syntactic forms—a benefit of macros' ability to take arguments as unevaluated expressions, which, for convenience, are never quoted. Let's create and test an `unless` statement. We want the body of our generated expression to evaluate variables in the caller's context, so we suppress the hygienic generation of new variables with `esc`:

```
macro unless(condition, body)
  :(if !$condition
    $(esc(body))
  end)
end

x, y = 0, 1
@unless (x == 5) begin
  y = 3                      # should execute
end
@unless (x == 0) begin
  y = 5                      # should not execute
end

@assert y == 3
```

This macro rewrites the code fragment

```
@unless (x < 5) begin y = 2 end
```

into

```
if !(x < 5); y = 2; end
```

Without `esc` the macro would have rewritten our code fragment into:

```
if !(x < 5); #2#y = 2; end
```

The `esc` function explicitly bypasses hygiene: Escaped expressions are compiled directly into the output without any new symbols being generated.

6.6 PARALLEL COMPUTING

To be successful as a language for scientific computing, a language must support efficient execution over large datasets; it should take advantage of processors with multiple cores or multiple computers in a cluster. Running related computations on multiple processing units simultaneously is called **parallel programming**. When multiple independent processes communicate exclusively via message passing (as opposed to shared memory), we have **distributed computing**. Julia provides both primitives and high-level constructs for computing across multiple independent processes.

A Julia **remote call** invokes a function asynchronously on a specific process, returning immediately with a **remote reference**, or **future**. Calling `fetch` on the remote reference will wait for the function to complete on the remote process, and produce the function's result. The following script creates three new processes,[4] and instructs process #4 to factor a large number:

```
addprocs(3)
@assert nprocs() == 4
ref = remotecall(4, factor, 21883298135690819)
@assert isa(ref, RemoteRef)
factors = fetch(ref)
@assert factors == Dict(234711901=>1,93234719=>1)
```

The `@spawn` macro is often used in place of `remotecall`; it not only takes advantage of a clearer syntax (since macros take unevaluated expressions as arguments), but it chooses a process for you:

```
addprocs(3)
ref = @spawn factor(21883298135690819)
factors = fetch(ref)
@assert factors == Dict(234711901=>1,93234719=>1)
```

Between the remote call and fetching the result, the calling process is free to do other work. When no intermediate work is necessary, the caller can combine the call and fetch with `remotecall_fetch` (in place of `remotecall` and `fetch`) or `@fetch` (in place of `@spawn` and `fetch`).

A common example of parallel programming across multiple cores computes an approximation of π by generating random points in the square $(0,0)...(1,1)$. Counting the number of random points within the inscribed arc (the points (x, y) for which $x^2+y^2 \leq 1$ in Figure 6.3) divided by the total number of points within the square approximates the value $\frac{\pi}{4}$. Let's generate a million random points in parallel on three cores:

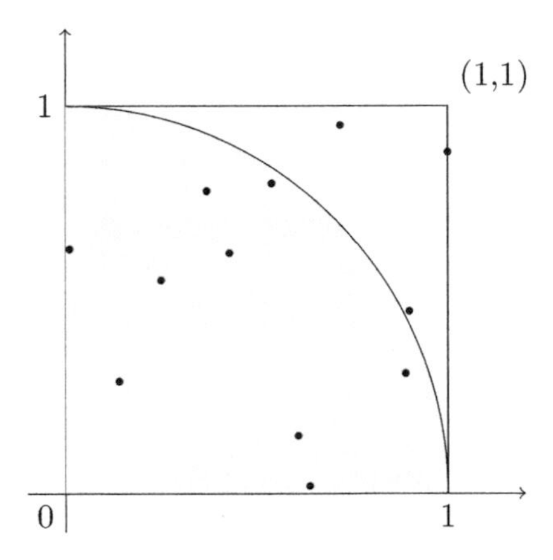

Figure 6.3 Approximation of $\frac{\pi}{4}$

[4]Calling `addprocs` with an integer argument creates processors on the local machine to take advantage of multiple cores. There are other variants of `addprocs` to add processors on remote machines either via user-host-port strings or cluster managers; we will not cover these alternatives in this text.

```
addprocs(3)

@everywhere function approximate_pi(trials::Int64)
  hits = 0
  for i in 1:trials
    if sum(rand(2).^2) < 1
      hits += 1
    end
  end
  hits/trials * 4
end

references = [@spawn approximate_pi(1000000) for i=1:3]
approximation = sum([fetch(ref) for ref in references])/3
@assert 3.1 < approximation < 3.2
```

Here the `@everywhere` macro ensures our function is available to each process. We spawn three executions on each of the three cores, then fetch and average the results. But we can do a bit better, letting Julia do most of the work with its `@parallel` macro:

```
addprocs(3)

trials = 4000000
hits = @parallel (+) for i = 1:trials
  Int(sum(rand(2).^2) < 1)
end

approximation = hits/trials * 4
@assert(3.1 < approximation < 3.2)
```

The parallel for-loop partitions the range over all of the available workers (all processes except the "main" process) and applies the reduction function (here we've used `+`) locally within each process. The calling process applies the function again to each process's result.

In addition to support for parallel and distributed execution, Julia offers other concurrent features that we won't cover here, including coroutines and channels. We've seen coroutines before, and we'll visit channels later in the book when we visit Erlang and Go.

6.7 JULIA WRAP UP

In this chapter we explored the Julia language. We learned that:

- Julia was designed for applications in data science, numerical analysis, and scientific computing.
- Julia features inclusive ranges and array bounds starting at 1, rather than 0.
- In function declarations, positional parameters come first; keyword parameters, if any, are separated from the positional parameters with a semicolon. Positional parameters *may* have defaults; keyword arguments *must* have defaults.

- Scopes coincide with function bodies, `while` and `for` loops, and blocks introduced with `try`, `catch`, `finally`, `let`, and `type`. The construct `begin...end` in particular, is not a scope. Variables introduced in a comprehension are scoped to each iteration.
- All Julia types reside in a single supertype-subtype hierarchy with `Any` at the top. Each type can have at most one supertype. Types can be abstract or concrete. Only abstract types can have subtypes. Types are first-class objects, and the type system is easy to introspect via methods such as `super`, `subtypes`, `isa`, `typeof`, and the operator `<:`.
- Types can be parameterized. Parameterized collection types in Julia are invariant.
- Julia features sum (`Union`) and product (`Tuple`) types. Tuples are covariant.
- Type annotations provide the foundation for multiple-dispatch, in which methods are selected based on the types of all of a function's arguments. Annotations are also frequently employed to improve performance.
- Julia expressions can be manipulated and modified before evaluation, enabling sophisticated forms of metaprogramming.
- Macros are executed when code is parsed, that is, before it is run. Macros return expressions which are compiled in to the context of the call. Macros take their arguments in unevaluated form, without the need to explicitly quote them.
- Julia provides many features for parallel and concurrent programming. Processes can be created to take advantage of multiple cores on a processor or multiple processors in a cluster.
- The low-level constructs for parallel programming are remote calls and remote references. Various macros, such as `@spawn`, `@fetch`, and `@parallel` provide a higher-level programming model.

To continue your study of Julia beyond the introductory material of this chapter, you may wish to find and research the following:

- **Language features not covered in this chapter**. Modules, constructors, networking, vararg tuple types, coroutines, channels, interfacing with C, Python, and Fortran, the standard library, and the powerful package system.
- **Open source projects using Julia**. Studying, and contributing to, open source projects is an excellent way to improve your proficiency in any language. You may enjoy the following projects written in Julia: Gadfly (`https://github.com/dcjones/Gadfly.jl`), Compose (`https://github.com/dcjones/Compose.jl`), Morsel (`https://github.com/JuliaWeb/Morsel.jl`), and of course, the language implementation itself (`https://github.com/JuliaLang/julia`).
- **Reference manuals, tutorials, and books**. Julia's home page is `http://julialang.org/`. The docs page, `http://docs.julialang.org/`, links to a Language Manual and the Standard Library Reference. Professional Julia programmers should not miss the *Performance Tips* section of the official documentation. Online resources for learning the language and exploring advanced features are plentiful; find curated learning resources at `http://julialang.org/learning/` and advanced publications at `http://julialang.org/publications/`. Leah Hanson in particular has a wealth of good online tutorials and presentations.

EXERCISES

Now it's your turn. Continue exploring Julia with the activities and research questions below, and feel free to branch out on your own.

6.1 Find the Julia home page. Download the latest distribution and experiment with the REPL.

6.2 Browse the standard library documentation at `http://julia-demo.readthedocs.org/en/latest/stdlib/base.html`.

6.3 Research and explain, via interesting examples, the differences between `==`, `===`, and `is`. Does the `==` operator perform a shallow or deep equality test?

6.4 What operator does Julia use for string concatenation?

6.5 Compare and contrast Python and Julia's approaches to positional, optional, and keyword parameters, and their corresponding arguments. Consider questions for each language such as: Can positional and keywords be optional? Can positional and keyword parameters and arguments appear in any order?

6.6 Define a function called `fib` to compute the nth Fibonacci number, using tail recursion and optional parameters. Calling `fib(7)` should return 13, via the following calls `fib(7)` ⇒ `fib(6,0,1)` ⇒ `fib(5,1,1)` ⇒ `fib(4,1,2)` ⇒ `fib(3,2,3)` ⇒ `fib(2,3,5)` ⇒ `fib(1,5,8)` ⇒ `fib(0,8,13)` ⇒ `13`.

6.7 If you were to browse Julia source code in both the online documentation and in online public repositories, you may notice that the `local` keyword is somewhat rare. Why do you think this is? In what situations would you consider the `local` keyword to be necessary?

6.8 Do the Julia expressions `[1 2]` and `[1 2 3]` have the same type? If so, what is that type?

6.9 Given the two expressions `[1 [2 3]]` and `[1;[2;3]]`, (a) what are the types of each? (b) draw pictures of the arrays (or matrices) that the expressions represent.

6.10 The expression `isa(e,Irrational) && isa(π,Irrational)` evaulates to true, yet `typeof(e)==typeof(π)` evaulates to false. Why, exactly?

6.11 Why is the Julia type `Type{T}` a parameterized type? What exactly, is the type `Type{Int64}`? When, if ever, would the types `Type` are `Type{Type}` be useful?

6.12 The type `Any` is sometimes called a "top" type. Find out what this means. Does Julia have a "bottom" type? If so, what is it?

6.13 Evaluate each of the following expressions and explain each result:

```
a. isa([0, 1], Array)
b. isa([0, 1], Array{Integer})
c. isa([0, 1], Array{Int64})
d. isa([0, 1], Array{Int64,1})
e. isa([0, 1], Array{Int64,2})
```

6.14 What is the type of the expression `[i+j+k/2 for i=1:4, j=1:3, k=1:5]`? How does Julia infer this type?

6.15 In your version of Julia, how many distinct methods exist for the following generic functions: `+`, `convert`, `length`, and `&`.

6.16 Does Julia use contravariance anywhere? If so, where? If not, find one other language in which it is used and give an example.

6.17 Research advantages and disadvantages of multiple dispatch over single dispatch. Why did Julia adopt multiple dispatch?

6.18 Are type annotations checked before the program is run or during? Write a program that proves your answer.

6.19 We covered three techniques for denoting `Expr` objects in this chapter. There is a fourth way, namely the `quote...end` construct. How does it differ from the other three? As it is more complicated, when is it ever advantageous to use it?

6.20 Are macros necessary? If you answer yes, give an example of a (useful) macro which cannot be easily simulated by regular "run-time" code that may, of course, use unevaluated expression objects. If you answer no, show how to duplicate the functionality of each of the macros defined in this chapter.

6.21 The variables generated during Julia's hygienic macro expressions start with the hash (`#`) character. Why is this not problematic, given that the hash character begins comments in Julia source code?

6.22 Design a macro for an "`until` statement," similar to the "`unless` statement" presented in this chapter.

6.23 We did not cover coroutines (called tasks in Julia) in this chapter, so here is your chance to learn about them. Create two tasks, one that produces a million random integers in the range 0..99, and another that consumes these values by histogramming them into buckets for the values 0–9, 10–19, 20–29, and so on.

6.24 What are the possible status values of a Julia task? How do these compare to the statuses of Lua's coroutines?

6.25 In the REPL, the return value of a remote call will display something like `RemoteRef{Channel{Any}}(4,1,13)`. Explain exactly what is being presented here.

6.26 If you are on a Unix or Unix-like system with multiple cores, run a π-approximation script from this chapter under the Unix `time` command. You should notice that the `real` time is less than the `user` time. Experiment with different values of arguments to `addprocs`. Which value on your system gives the most "speedup"?

6.27 Research Julia's shared arrays and create an example that illustrates their usefulness.

CHAPTER 7

Java

Java is an extremely popular enterprise programming language, designed with, among other things, thread-based concurrency and security in mind.

First appeared 1995
Designer James Gosling
Notable versions 1.1 (1997) • 5 (2004) • 8 (2014) • 9 (2017)
Recognized for Classes, Threads, Mature platform
Notable uses Enterprise computing, Android applications
Tags Architecture-neutral, Multithreaded, High-performance
Six words or less "Write once, run anywhere"

Engineers within Sun Microsystems began a project in 1991 for programming smart appliances. While the team originally considered C++ for the project's software platform, they found the language lacking in areas such as portability (implementations across machines could vary widely) and threading (as C++ had no built-in concurrency support at the time), and so a new language, Oak, was created. Sun renamed the language Java in 1994 and released it to the world the following year.

Java gained massive popularity in the early days of the World Wide Web as various browser manufacturers, notably Sun and Netscape Communications, added browser support for downloading and executing compiled Java code, known as applets. While Java did not become the most popular language for running apps in the browser (that title has gone to a competing—and remarkably different—language, JavaScript), Java ultimately gained huge popularity among developers, particularly for server-side (enterprise) applications, and as the primary language for application development in the Android ecosystem. Oracle Corporation, which acquired Sun in 2010 and owns the current implementation of the Java Platform, estimates that there are nine million Java developers worldwide today.

The power and influence of Java comes not only from the language itself, but also from its state-of-the-art virtual machine, the Java Virtual Machine (JVM). Compiled Java code can run on any JVM, regardless of computer architecture, allowing the developer to "write once, run anywhere." Though its name might lead you to believe otherwise, the Java Virtual Machine actually knows nothing of the Java programming language; it only knows of the binary `class` file format. Any language with a compiler capable of producing `.class` files can use the JVM. The popularity of the JVM gave rise to a massive ecosystem of both standard and third party libraries containing tens of thousands of class files for almost any computing task imaginable. Because of its wide availability, machine-independent platform, and performance, additional languages (Clojure, Groovy, Scala, Ceylon, Kotlin, JRuby, and Jython, to name a few) have been designed to target the JVM.

Our tour of Java in this chapter focuses on the language itself, beginning with our three introductory programs. We'll then look at the relatively simple-to-understand type system. Classes, interfaces, and static typing will take up the bulk of the chapter, after which we will turn to concurrency matters. Java's designers, we'll see, made the choice to have a small amount of in-language concurrency support while putting the bulk of functionality in a standard library.

7.1 HELLO JAVA

Let's begin our exploration of Java with our usual triangle program:

```
public class TripleApp {
    public static void main(String[] args) {
        for (int c = 1; c <= 40; c++) {
            for (int b = 1; b < c; b++) {
                for (int a = 1; a < b; a++) {
                    if (a * a + b * b == c * c) {
                        System.out.printf("%d, %d, %d\n", a, b, c);
                    }
                }
            }
        }
    }
}
```

Java is the most verbose language we've seen so far, but the verbosity is intentional. Java programs consist *entirely* of classes, optionally grouped into **packages**. All code must therefore appear within type declaration. The interpreter runs scripts by loading a class and passing command line arguments to a `static` function named `main` that accepts the arguments array as its sole parameter.

Our permutations example follows:

```
public class AnagramApp {
    public static void main(String[] args) {
        if (args.length != 1) {
            System.err.println("Exactly one argument is required");
            System.exit(1);
        }
        String word = args[0];
        generatePermutations(word.toCharArray(), word.length()-1);
    }
```

```
    private static void generatePermutations(char[] a, int n) {
        if (n == 0) {
            System.out.println(String.valueOf(a));
        } else {
            for (int i = 0; i < n; i++) {
                generatePermutations(a, n-1);
                swap(a, n % 2 == 0 ? 0 : i, n);
            }
            generatePermutations(a, n-1);
        }
    }

    private static void swap(char[] a, int i, int j) {
        char saved = a[i];
        a[i] = a[j];
        a[j] = saved;
    }
}
```

Now we start to see the primary reason for Java's verbosity: nearly everything is marked—variable types, method parameter types, return types, and **access levels** of classes and class members. These markings do come with some benefits. They provide additional documentation for the reader of the code, and they enable tools, such as those found in integrated development environments, to do a good job of **static analysis** [140], assisting programmers in detecting certain types of errors *before* running the code.

The script itself highlights little of interest, except perhaps that (1) the array of command line arguments does *not* contain the program name, and (2) strings are immutable, necessitating conversion to a mutable array for swapping.

Next, we turn to the word count application.

```
import java.util.Scanner;
import java.util.SortedMap;
import java.util.TreeMap;
import java.util.regex.Pattern;
import java.util.regex.Matcher;

public class TraditionalWordCountApp {
    public static void main(String[] args) {
        SortedMap<String, Integer> counts = new TreeMap<>();
        Pattern wordPattern = Pattern.compile("[a-z']+");
        Scanner scanner = new Scanner(System.in);
        while (scanner.hasNext()) {
            String line = scanner.nextLine().toLowerCase();
            Matcher matcher = wordPattern.matcher(line);
            while (matcher.find()) {
                String word = matcher.group();
                counts.put(word, counts.getOrDefault(word, 0) + 1);
            }
        }
```

```
        for (SortedMap.Entry<String, Integer> e : counts.entrySet()) {
            System.out.printf("%s %d\n", e.getKey(), e.getValue());
        }
    }
}
```

This script makes use of several classes from Java's standard library (e.g. `Pattern` from the package `java.util.regex` and `TreeMap` from the package `java.util`). Unlike Python, Java does not *require* `import` statements; importing a class simply lets us avoid specifying the full **qualified names** of these classes throughout our code. The class `String` comes from a special package `java.lang`, whose classes are imported by default.

Java's standard library requires a fair amount of code to iterate the lines of a file (we have to wrap standard input in a scanner and loop while lines are available), and matching regular expressions (we don't get an array of matches out of the box). Professional Java programmers often make use of third party libraries, such as Google Guava [41] or Apache Commons [2] to simplify the code. Still, our script is rather accessible: we read and lowercase each line, apply a **matcher** to it,[1] iterate through the matches filling a sorted map,[2] then iterate the map printing its contents.

Many Java programmers may prefer to write this script using **streams**:

```
import java.util.TreeMap;
import java.util.regex.Pattern;
import java.io.BufferedReader;
import java.io.InputStreamReader;
import java.util.function.Function;
import java.util.stream.Collectors;

public class WordCountApp {
    public static void main(String[] args) {
        Pattern nonWord = Pattern.compile("[^a-z']+");
        new BufferedReader(new InputStreamReader(System.in))
            .lines()
            .flatMap(line ->
                nonWord.splitAsStream(line.toLowerCase()))
            .filter(word -> !word.isEmpty())
            .collect(Collectors.groupingBy(Function.identity(),
                TreeMap::new, Collectors.counting()))
            .forEach((word, count) ->
                System.out.printf("%s %d\n", word, count));
    }
}
```

Our `InputStreamReader` transforms the byte sequence from standard input into a character sequence. This reader is then **decorated** by a `BufferedReader`, which buffers the characters for efficiency. The `lines` method produces a stream of lines, which `flatMap` not only splits into individual words but gathers the words across all lines into a single word stream. After

[1]Notice the empahsis on nouns: patterns, matchers, scanners, etc. Steve Yegge has written a humorous take on this part of Java programming culture in [141].

[2]Java's standard `TreeMap` is traditionally implemented with a Red-Black tree [137].

filtering the empty words produced by `split`, we collect and count each word, and print a report, automatically sorted thanks to Java's `TreeMap`.

7.2 THE BASICS

Like JavaScript and Lua, Java's types are subdivided into primitive types and reference types. Primitive types include `byte`, `short`, `int`, and `long` types, two's complement integers of 8, 16, 32, and 64 bits respectively; `float` and `double` types, which contain single- and double-precision IEEE 754 floating-point numbers respectively; a `char` type, the values of which represent UTF-16 **code units**;[3] and a `boolean` type. Classes and interfaces, together with arrays, comprise Java's reference types. The value of a reference type is either `null` or a reference to a class or array **instance**. Interfaces are not allowed any instances. The `==` operator on references *always* compares whether the references are the same; to compare the referent values, you must use the `equals` operator, if one is defined:

```
import java.time.LocalDate;
public class EqualsDemo {
    public static void main(String[] args) {
        LocalDate d1 = LocalDate.of(2020, 2, 29);
        LocalDate d2 = LocalDate.of(2020, 2, 29);
        assert !(d1 == d2) && d1.equals(d2);
    }
}
```

The value `null` belongs not to its own type, but to *every reference type*. This is considered by some to be an unfortunate design choice, but one that has a long history in programming language design. Its inventor, Sir Charles Anthony Richard Hoare, refers to this as his **billion dollar mistake** [60]. But don't be too hard on Sir Tony; his accomplishments include Quicksort, Communicating Sequential Processes (a highly influential process algebra), and Hoare logic (used in proving the correctness of programs with assignment statements). We'll have more to say about the billion dollar mistake when we discuss generics later in this chapter.

Java has no built-in syntactic support for lists, sets, and maps; you'll find them defined via interfaces and classes in various standard libraries. Interestingly, collections can only hold instances of reference types, so Java provides classes for each of the primitive data types that **wrap** the primitive in an object. If necessary, Java will automatically convert a primitive to (**autoboxing**) and from (**unboxing**) an instance of its corresponding wrapper class. If desired, the programmer may box and unbox manually.

Though Java is mostly strongly-typed, it will implicitly perform any type conversion that is considered to be a **widening conversion**. With primitive types, a widening conversion occurs when the target type is represented with more bits and thus a wider range of legal values (e.g. from 32-bit `float` to 64-bit `double`). With reference types, a widening conversion occurs when the type being converted to is a superclass of the original type (e.g. from `String` to `Object`, the ultimate ancestor of all classes). A **narrowing conversion** is the opposite of a widening conversion, and is never performed implicitly due to the chance of precision loss.

[3]Note that a code unit is *not* the same as a code point; in each encoding, code points are mapped to one or more code units.

Java has a bit in common with the class-based languages we've seen before, Python and Ruby, but there are some very strong differences, too. Here are a few:

- Unlike Python and Ruby, where all values are references to objects, Java has both reference values and primitive values.
- Like Ruby, and unlike Python, a class in Java can have at most one superclass. Java's ultimate superclass, which all classes (including arrays) either directly or indirectly inherit from, is the class `Object`.
- Like Python, and unlike Ruby, Java allows both methods and non-method properties (called **fields** in Java) to be members of a class.
- Unlike Ruby and Python, Java has explicit constructors that are not considered members of the class (and therefore are never inherited).
- Unlike Python, which makes all class members visible, and Ruby, which hides all of its instance variables, Java allows for selective visibility controls. Java's `public`, `private`, and `protected` keywords have completely different meanings from Ruby's.

Each compilation unit in Java defines one or more classes or interfaces, optionally as part of a **package**. If no package is defined, the classes will land in a **default package**.[4] Classes can contain fields, methods, constructors, static and instance initializers, and other classes. Classes and their contents (other than initializers) can have **modifiers**. Table 7.1 gives the complete list; the following circle class manages to use a few:[5]

```
class Circle {

    private double x;
    private double y;
    private double r;

    public Circle(double x, double y, double r) {
        this.x = x;
        this.y = y;
        this.r = r;
    }

    public static Circle create(double x, double y, double r) {
        return new Circle(x, y, r);
    }

    public double area() {
        return Math.PI * this.r * this.r;
    }

    public static double area(double r) {
        return Math.PI * r * r;
    }
}
```

[4]Java best practices dictate all code should go in a named package, but examples in this chapter will use the default package for brevity.

[5]You will explore initializers and nested classes in the end-of-chapter exercises.

Class Modifier	Description
no modifier	accessible only within its own package
`public`	accessible wherever its package is
`abstract`	cannot be instantiated (may have abstract methods)
`final`	cannot be extended (subclassed)
`static`	specifies a nested top-level class (See end-of-chapter exercises)
Member or Constructor Modifier	**Description**
no modifier	accessible only within its package
`private`	accessible only within its class
`protected`	accessible only within its package and subclasses
`public`	accessible wherever its class is
Field Modifier	**Description**
`final`	value may not be changed once assigned
`static`	only one instance of the field shared by all objects of the class
`transient`	not serialized
`volatile`	value may change asynchronously, so the compiler must avoid certain optimizations
Method Modifier	**Description**
`final`	may not be overridden
`static`	method does not apply to a particular instance
`abstract`	no body; to be implemented by a subclass
`synchronized`	requires locking the object before execution
`native`	implementation is not written in Java, but rather in some platform-dependent way

Table 7.1 Java Modifiers

```
public class CircleApp {
    public static void main(String[] args) {
        Circle c1 = new Circle(3, 5, 1);
        Circle c2 = Circle.create(2, 8, 10);
        assert c1.area() == Math.PI && c2.area() == 100*Math.PI;
        assert Circle.area(2) == 4*Math.PI;
    }
}
```

Note how static methods are called *as if* they were methods defined in the class `Class` (of which `Circle` is an instance), though technically, this is not the case. Both static methods and instance methods belong to `Circle`, though only instance methods have a receiver.

Like Julia, Java has both abstract and concrete classes. Java allows both abstract and concrete classes to have subclasses (unlike Julia, where only abstract types can have subtypes). Our recurring animals example illustrates inheritance in Java:

```
abstract class Animal {
    private String name;
    public Animal(String name) {
        this.name = name;
    }
    abstract String sound();
    public String speak() {
        return name + " says " + sound();
    }
}
```

```
class Horse extends Animal {
    public Horse(String name) {
        super(name);
    }
    @Override String sound() {
        return "neigh";
    }
}
```

```
class Cow extends Animal {
    public Cow(String name) {
        super(name);
    }
    @Override String sound() {
        return "moooo";
    }
}
```

```
class Sheep extends Animal {
    public Sheep(String name) {
        super(name);
    }
    @Override String sound() {
        return "baaaa";
    }
}
```

```
public class AnimalsApp {
    public static void main(String[] args) {
        Animal s = new Horse("CJ");
        assert s.speak().equals("CJ says neigh");
        Animal c = new Cow("Bessie");
        assert c.speak().equals("Bessie says moooo");
        assert(new Sheep("Little Lamb").speak().equals(
            "Little Lamb says baaaa"));
    }
}
```

The Java version of the animal hierarchy differs dramatically from those in previous chapters. A superclass expecting subclasses to implement a particular method—in our case `sound`—must explicitly declare that method an **abstract method**. A class extending an abstract class must either implement all abstract methods of its superclass or declare itself abstract. In none of the previous chapters was such a declaration necessary. Were we to call the `speak` method on a subclass in CoffeeScript, Python, Ruby, or Julia that did not implement the `sound` method, a *run-time error* would occur. In Java, such a subclass could not even exist—attempting to create a concrete subclass without overriding an abstract method generates a *compile-time error*.

Inheritance applies to fields and methods only. Subclasses do not inherit constructors from superclasses; they can, however, refer to the superclass constructor within their own constructors via the `super` keyword. A subclass that redefines a method from a concrete superclass, or provides an implementation of an abstract method from an abstract superclass, is said to **override** the superclass method. Programmers may use the optional **`@Override` annotation** to mark method overrides.[6]

7.3 INTERFACES

Java's **interfaces** define a set of related constants and methods that provide a contract, or application program interface (API), that implementing classes must conform to. An interface acts (very!) much like an abstract class; it defines a new reference type, it can contain abstract methods, and it cannot be instantiated. However, while abstract classes can both define (non-static) fields that will belong to instances of concrete subclasses, and define non-public methods, interfaces cannot. Interface fields are always public, static, and final (whether these modifiers are used or not!), and all interface methods are public (whether the modifier `public` appears or not). Methods not marked `default` or `static` are abstract.

The following interface defines two public abstract methods:

```
interface Drawable {
    void outline(Color color);
    void fill(Color color);
}
```

Originally, Java interfaces could *only* contain abstract methods. Adding a method to an interface would thus mean breaking existing classes implementing the interface.[7] Modern Java allows interfaces to contain **default methods** and **static methods**, both with implementations, so library maintainers can add these methods to existing interfaces. Let's add a default method to our previous interface:

[6]While the `@Override` annotation is optional, its use is considered a best practice. Even without the annotation, a method that can override will override; however, a typo in either the method name or signature will result in a *new* method in the subclass, perhaps leading to difficult-to-diagnose errors at run time. The presence of the annotation requires the compiler to ensure the overriding method actually exists in a superclass with an "override-equivalent" method signature.

[7]Until at least all implementations were tracked down and given their own new overriding method implementations.

```
interface Drawable {
    void outline(Color color);
    void fill(Color color);
    default void render(Color outlineColor, Color fillColor) {
        fill(fillColor);
        outline(outlineColor);
    }
}
```

Java restricts classes to extend at most a single superclass, though it allows classes to **implement** any number of interfaces. Restricting extension to a single superclass nicely prevents a class from inheriting two or more (instance) fields with the same name and type, or two or more methods with the same name and different signatures—a situation known as **multiple inheritance**. A wise move, perhaps, as multiple inheritance leads to questions. Are both instance fields inherited or only the first one? Are both methods inherited? Or only one? If so, which one? Is there a method resolution order, as we saw in Python?

But Java allows the implementation of multiple interfaces. Does this lead to multiple inheritance with all of its requirements for disambiguation? Let's see. Given:

```
interface A {
    int X = 1;              // implicitly public static final
    default void f() {}     // implicitly public
}

interface B {
    int X = 2;              // implicitly public static final
    default void f() {}     // implicitly public
```

Can we define the class

```
class C implements A, B {}
```

No, Java prohibits this class definition with the compile time error

```
class C inherits unrelated defaults for f() from types A and B
```

The error occurs only if you attempt to inherit *both* defaults; were class C to prove its own overridden implementation of f, all would be well. Now what about the differing definitions of (the field) X? No problem: within the class C, you may refer to both `A.X` and `B.X`, but you will get a compile-time error if you fail to qualify X with the originating interface name.

Java simply does not allow multiple inheritance at all.

An interface with exactly one abstract method is called a **functional interface**. Java provides nice syntactic sugar for instances of classes that implement a functional interface: you can define instances directly without ever specifying the implementing class!

```
@FunctionalInterface
interface Function {
    double apply(double x);
}
```

```
public class TwiceApp {
    private static double twice(Function f, double x) {
        return f.apply(f.apply(x));
    }

    public static void main(String[] args) {
        assert twice(x->x+2, 10) == 14;
        assert twice(Math::floor, -11.77) == -12;
    }
}
```

We've marked our `Function` interface with the `@FunctionalInterface` annotation; though optional, it's considered good practice. The sole abstract method within the interface denotes a function with a single `double` parameter and a `double` result. The expression `x->x*2` conforms to this type specification, and thus makes an acceptable first argument to our `twice` method. The same holds for `Math::floor`, a function object whose evaluation produces the value of the static `floor` method from the class `Math`.

7.4 STATIC TYPING

Java differs from the languages in the previous chapters in catching nearly all type errors *prior to execution*, i.e., **statically**. Multiplying an Animal by a boolean? Java never even lets your program run. In contrast, Python raises a `TypeError` at run time, and JavaScript coerces the operands into numbers. Java's compile-time type checking works even for expressions involving variables; for example, within the scope of the declaration `int x = 3;`, we can infer the expression `x + 5.0` to have type `double`. In the animals example from Page 125, the declaration `Animal s = new Horse("CJ");` implies the expression `s.speak()` has type `String` and the expression `s` itself has type `Animal`. Java requires type constraints *on the variables themselves*, restricting them to contain only values of a compatible type.

We often use the term **static typing** for pre-execution type checking and **dynamic typing** for runtime type checking. Granted, more formal definitions exist and can be found in the literature on types and programming languages, of which Pierce [97] is a classic reference. Static typing fits nicely with Pierce's definition of a type system as "a syntactic method for automatically checking the absence of certain erroneous behaviors by classifying program phrases according to the kinds of values they compute." Types tell us what we can or cannot even say. In Java, for example, associating the variable a with the type `Animal` limits us to using the variable in an expression only with `Animal` operations, even if it holds an object of class `Cow` with additional operations. Robert Harper even goes so far as to say "dynamic typing is but a special case of static typing." [48] Why? Compare these equivalent Java and Ruby fragments:

```
// Java
Object x;
if (Math.random() < 0.5) {
    x = "a string";
} else {
    x = 5;
}
System.out.println(x.getClass());
```

```
# Ruby
if rand() < 0.5
  x = "a string"
else
  x = 5
end
puts x.class
```

Harper points out, correctly, the difference between **type** and **class** (see Figure 7.1). This is immediately apparent in the Java side, with a variable constrained to type `Object`, which easily holds values of *classes* `String` or `Integer`. On the Ruby side, the same situation occurs, but without explicitly attaching a type to x. But x *does* have a (static type) and that type is the union of all Ruby classes. It's as if there's a single type, and indeed some authors, including Harper, refer to so-called "dynamically typed languages" as being **unityped**. [48] With only one static type, we cannot state certain invariants in a declarative fashion.

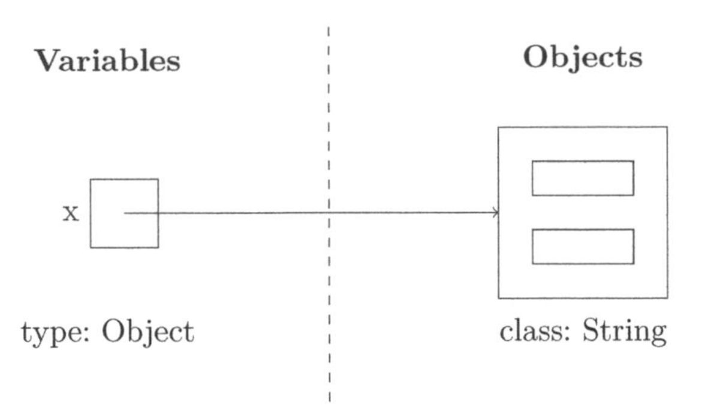

Figure 7.1 Types and Classes in Java

Note that the terms static and dynamic typing refer to mechanisms rather than languages themselves. While Java does most of its type checking at compile time, situations do arise where type checking occurs at runtime. For instance, (Java) arrays are covariant, so assigning an array of sheep to an animal array variable is A-OK. But an arbitrary animal array can contain cows, so the compiler will not stop you from adding a cow to an object attached to a variable of type `Animal[]`. If the associated object is a sheep array, you will see a type check exception at run time:

```
public class CovariantArrayDemo {

    public static void main(String[] args) {
        Animal[] animals = new Sheep[10];          // Covariance, sadly
        try {
            animals[0] = new Cow("Bessie");
        } catch (ArrayStoreException e) {
            System.out.println("Runtime type checking!");
        }
    }
}
```

While arrays are covariant, the other collections (lists, sets, maps, etc.) are not. In a statically typed language, it makes sense to use parameterized types to implement these collections. Java calls these generic types, and these are next up on our tour.

7.5 GENERICS

As Java tends to favor static typing, parameterized types, known as **generic types**, appear quite often. Lists and sets provide a typical illustration:

```
import java.util.Arrays;
import java.util.List;
import java.util.Set;
import java.util.HashSet;

public class ListAndSetExample {
    public static void main(String[] args) {
        List<String> words = Arrays.asList("do", "while", "if", "a");
        Set<Integer> sizes = new HashSet<>();
        for (String word: words) {
            sizes.add(word.length());
        }
        assert sizes.equals(new HashSet<Integer>(Arrays.asList(5,2,1)));
    }
}
```

We've created two variables, `words` of type list-of-strings, and `sizes`, of type set-of-integers. Both `List<E>` and `Set<E>` are interfaces in the package `java.util`; *E* is (as we've seen in Julia) a type parameter. We've initialized `sizes` with an object of class `HashSet<Integer>`[8] (note that the type argument `Integer` is inferred). The variable `words` is initialized to an object of *some* class implementing the list interface; we don't exactly know, nor care, which class this is—we access the object through the methods defined by the interface type.

Java collections are invariant, so a list of cows may *not* be assigned to a list of animals. In order to write a method that operates on a list of animals, while accepting as an argument a list of sheep or a list of horses, we may use a **wildcard** (?):

```
import java.util.List;
import java.util.Arrays;

public class AnimalChorusApp {
    public static void chorus(List<? extends Animal> animals) {
        for (Animal animal : animals) {
            System.out.println(animal.speak());
        }
    }

    public static void main(String[] args) {
        List<Cow> cows = Arrays.asList(new Cow("Bessie"));
        List<Sheep> sheep = Arrays.asList(new Sheep("Wooly"));
        chorus(cows);
        chorus(sheep);
    }
}
```

Java's `<? extends Animal>` is analogous to Julia's `{T<:Animal}`. The `extends` keyword signals a **upper bounded wildcard**, meaning the type parameter can be instantiated only with a type argument of class `Animal` or any Animal subclass (direct or indirect).

[8]`HashSet<E>` is one of eight implementations of the set interface in the Java standard library; others include `TreeSet<E>`, `EnumSet<E>`, and `ConcurrentSkipListSet<E>`.

Let's take a brief look now at **lower bounded wildcards**, with a trivial example. We can add sheep to sets of sheep, sets of animals, or even sets of objects:

```
import java.util.Set;
import java.util.HashSet;

public class SheepAdder {
    public static void main(String[] args) {
        Set<? super Sheep> s1 = new HashSet<Object>();
        Set<? super Sheep> s2 = new HashSet<Animal>();
        Set<? super Sheep> s3 = new HashSet<Sheep>();
        Sheep sheep = new Sheep("Rafe");
        s1.add(sheep);
        s2.add(sheep);
        s3.add(sheep);
    }
}
```

Wildcards cannot be used to instantiate objects; however, **bounded type parameters** exist for this purpose. Bounded type parameters can only have upper bounds. We'll illustrate their usage with an example:

```
import java.util.List;
import java.util.Arrays;

public class MinMax<C extends Comparable> {
    private C min;
    private C max;
    public MinMax(C x, C y) {
        min = x.compareTo(y) < 0 ? x : y;
        max = x.compareTo(y) > 0 ? x : y;
    }
    public C getMin() {return min;}
    public C getMax() {return max;}
}
```

Note that we've given both fields the type C, restricted to classes implementing interface `Comparable`.[9] Only by using the same type variable for both fields could we state that both fields had the same type. A wildcard cannot express this constraint.

To round out our tour of generics, we'll look at **optionals**, one approach to mitigating the previously mentioned "billion dollar mistake." The mistake, you'll recall, is the inclusion of the value `null` in every reference type. Given a variable s of type `String`, the expression `s.length()` may fail—throw a `NullPointerException`, or NPE—whenever s is `null`. Static typing doesn't help us here: the Java type system has no type of exclusively non-null strings. NPEs can happen any time, so the defensive programmer often resorts to surrounding method calls with if-not-null checks.

Why do `null` values exist? Null expresses the *absence* of a value. Consider modeling a person with a required name, but optional boss. We could type the name and boss fields as

[9] `Comparable`'s sole method `compareTo` should be implemented as follows: return a negative integer from x`.compareTo(`y`)` if $x < y$, 0 if $x = y$, and a positive integer if $x > y$.

`String` and `Person` respectively, and check the name field for `null` on construction, and let the boss be `null` if necessary. But expressions such as `p.getBoss().getName()` may fail at run-time. But we'd like the compiler to type check expressions, so we bring in `Optional<T>` from the package `java.util`. Think of an optional as a wrapper: an optional person either wraps a person, or is empty (see Figure 7.2).

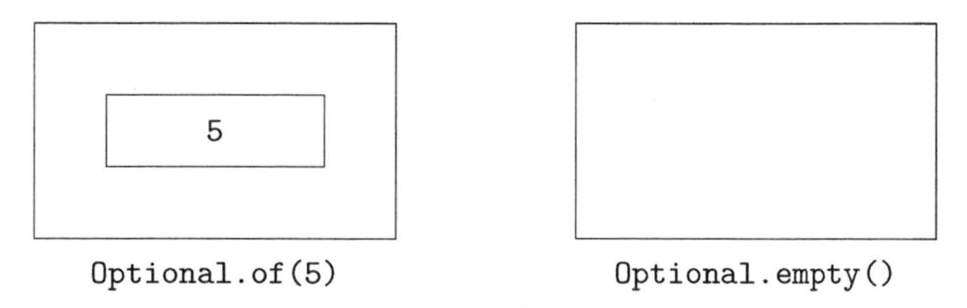

Figure 7.2 Optionals

```
import java.util.Optional;
import java.util.Objects;

class Person {
    private String name;
    private Optional<Person> boss;

    private Person(String name, Optional<Person> boss) {
        this.name = Objects.requireNonNull(name);
        this.boss = boss;
    }

    public Person(String name) {
        this(name, Optional.empty());
    }

    public Person(String name, Person boss) {
        this(name, Optional.of(boss));
    }

    public String getName() {return name;}
    public Optional<Person> getBoss() {return boss;}
}
```

The field and constructor definitions express our intent quite well: name is required, boss is optional. The static factory methods `Objects.requireNonNull` and `Optional.of` prevent objects being constructed with unwanted nulls. And the expression `p.getBoss().getName()` fails to compile: you'll still have to deal with the no-boss case, but you won't be surprised by it. The compiler forces you to deal with it.

The most common usage patterns for optionals are as follows. Let o be an optional, and x its wrapped value, if present. Then:

- o`.isPresent()` returns whether or not o wraps a value.
- o`.ifPresent(`f`)` invokes $f(x)$ if present and does nothing otherwise.

- o.orElse(y) returns x if present and y otherwise.
- o.filter(p) returns o if present and $p(x)$ true; it returns an empty optional otherwise.
- o.map(f) returns $f(x)$ wrapped if present, and an empty optional otherwise.
- o.flatMap(f) is similar; we'll leave this one for the exercises.

Let's get a feel for these methods with a small example:

```
import java.util.Optional;
public class OptionalDemo {
    public static void main(String[] args) {
        Person alice = new Person("Alice");
        Person bob = new Person("Bob", alice);

        bob.getBoss().ifPresent(p -> {assert p == alice;});
        alice.getBoss().ifPresent(p -> {assert false;});
        assert alice.getBoss().orElse(bob) == bob;
        assert bob.getBoss().orElse(bob) == alice;

        Optional<Person> b = bob.getBoss();
        assert b.filter(p -> p.getName().startsWith("A")) == b;
        assert !b.filter(p -> p.getName().startsWith("B")).isPresent();
        assert b.map(Person::getName).orElse("").equals("Alice");
        assert alice.getBoss().map(Person::getName).orElse("").equals("");
    }
}
```

For reference, the public methods of the class Optional<T> (excluding overrides from superclasses) follow. In this relatively small class, we see both upper and lower bounded wildcards, as well as a few functional interfaces:[10]

```
static<T> Optional<T> empty()
static <T> Optional<T> of(T value)
static <T> Optional<T> ofNullable(T value)
T get()
boolean isPresent()
void ifPresent(Consumer<? super T> consumer)
Optional<T> filter(Predicate<? super T> predicate)
<U> Optional<U> map(Function<? super T, ? extends U> mapper)
<U> Optional<U> flatMap(Function<? super T, Optional<U>> mapper)
T orElse(T other)
T orElseGet(Supplier<? extends T> other)
<X extends Throwable> T orElseThrow(Supplier<? extends X> excSupplier)
```

In this section, we've barely scratched the surface of the world of Java generics. We've not had space to cover **unbounded wildcards**, reification, or wildcard capture. To continue exploring this rich area of the Java language, visit the Generics Trail in the Java Tutorial [91] or Angelika Langer's excellent Generics FAQ [74]. And stay tuned for further discussion of generics later in the book when we cover Swift and Rust.

[10]Briefly, the functional interfaces are: Consumer<T>, a function on T returning nothing; Predicate<T>, a function from T to bool; Function<T,U>, a function from T to U; and Supplier, a function of no arguments returning a T.

7.6 THREADS

Concurrency refers to the modeling and coordination of independent, distinct computing activities whose execution spans may overlap in time. [98] When we design a system to be concurrent, we focus on these independent activities and the interactions between them. One classic example is a word processor: in a concurrent design, we define separate agents for input device listening, spell checking, hyphenation, line justification, page breaking, and so on, rather than define a monolithic function that does bits of each activity in different regions of the code. Note that while the lifetime of the activities overlap in time, it is not necessary (for concurrency) that multiple activities execute in **parallel**, i.e., at the same physical instant on separate physical processors or cores.

In previous chapters, we've encountered concurrency in Lua coroutines and Julia remote processes. In this section, we'll cover Java threads. Unlike coroutines, which cooperatively take turns on one line of execution, threads run on their own execution lines, possibly in parallel, and may be **preempted** by a scheduler when not enough physical execution units exist for each of the available threads.[11]

Let's see an example. We'll start 100 threads $t_1 ... t_{100}$, each of which will determine whether i is prime, and print the value if so:

```
public class PrimePrinter {
    public static void main(String[] args) {
        for (int i = 1; i <= 100; i++) {
            final int candidate = i;
            new Thread(() -> {
                for (int divisor = 2; divisor <= candidate; divisor++) {
                    if (divisor == candidate) {
                        synchronized (System.out) {
                            System.out.print(" " + candidate);
                        }
                    } else if (candidate % divisor == 0) {
                        break;
                    }
                }
            }).start();
        }
    }
}
```

Our main thread enters a loop creating 100 instances of the class `java.lang.Thread`. We pass an instance of the functional interface `Runnable` to the thread's constructor. Calling `start` on the thread invokes its runnable on a new thread.[12] Because each thread writes to a **shared resource**—in this case `System.out`—we use a `synchronized` statement to ensure only one thread can call its methods at a time, thus preventing multiple threads from interleaving their outputs.

[11]Whether or not threads may be forcibly preempted at any time is up to an implementation. A scheduler may bump a thread from an execution unit only at specific points, such as when the thread (voluntarily) waits for an event, a resource to become available, or a timer to expire.

[12]In our example, the runnable is a closure. Java has a rule that closed-over variables must be `final`, so we had to introduce a new variable, `candidate` instead of using `i` directly.

Threading naturally introduces **non-determinacy** into an application. One run of the script above produced: `3 7 5 2 11 13 17 19 23 29 31 37 41 43 47 53 59 61 67 71 73 83 79 89 97`; others gave slightly different orderings.

Writing multithreaded code operating on shared, mutable data brings the possibility of starvation, livelock, deadlock, memory consistency errors, and race conditions. This book is not a treatise on programming with threads, so we won't attempt to define these terms; instead, we simply wish to highlight some of the language and library support for concurrency in Java. Let's outline the important concepts, then finish this chapter with a more comprehensive example:

- **Thread objects**. A thread t has a name, a priority, and interrupted status (`true` or `false`). You can wait for a thread to finish, waiting either indefinitely (via *t*`.join()`) or for a maximum period of time (*t*`.join(`*millis*`,`*nanos*`)`). Call *t*`.start()` to **launch** the thread.
- **Monitors**. Every Java object has its own intrinsic lock, known as a **monitor**. A thread executing the statement `synchronized(`*o*`) {...}` acquires the lock for *o* if not already acquired, or blocks indefinitely until a different thread holding the lock releases it.
- **Condition synchronization**. Sometimes you need to execute code of the form "if this condition holds, acquire a lock." However, the condition might change from true to false between the test and the lock acquisition, causing enormous trouble. Java defines the very special `wait`, `notifyAll`, and `notify` methods on every object to allow programming styles to safely, and efficiently, acquire resources conditionally.
- **Executor services and thread pools**. Executor services manage the creation, execution, and termination of one or more **tasks**. Java uses the term "task" for anything designed to be run on a thread, specifically objects implementing one of the functional interfaces `Callable` (value-returning) or `Runnable` (non-value returning). Use the service to either *submit* a single task (asynchronously) or *invoke* a collection of tasks, blocking until all (`invokeAll`) or one (`invokeAny`) of the tasks has completed, or, optionally, until a timer has expired. Submission and invocation both return `Future` objects used to check the pending status of a task, or to wait on the task for completion. The most common type of executor service is a **thread pool**, which allows you to multiplex a huge number of tasks onto a smaller number of threads.
- **Explicit locks and synchronization objects**. The standard library provides a number of different objects for handling various synchronization use cases, including:
  - Locks, reentrant locks, and various locks to coordinate reading and writing;
  - Count down latches, allowing a number of threads to wait until a counter goes down to zero;
  - Cyclic barriers, used when a fixed number of threads need to meet (block) at some point, then all proceed after the last thread arrives.
  - Phasers, barriers with a variable number of threads that can be waiting; Exchangers, defining a point in which two threads can exchange objects; and
  - Semaphores, allowing a maximum number of threads to use a resource at a time.
- **Atomics**. Objects that take advantage of hardware instructions to perform ultra-fast lock-free atomic read and update operations.

- **Concurrent data structures.** Java's `BlockingQueue` and `ConcurrentMap` implementations simplify writing some concurrent applications. In a blocking queue, for example, use `put` to block until the queue has free space before inserting, and `offer` to attempt an insertion either immediately or within a given duration (returning `true` on insertion and `false` if capacity was not available). When removing, `take` blocks until the queue is non-empty, and `poll` expects removal to be possible immediately or within a time frame, returning whether an element was able to be removed.

The classic examples of concurrent programming include Readers and Writers, the Producer-Consumer Bounded Buffer, and Dijkstra's *Dining Philisophers* [59]. We'll break with tradition and build a little simulation, with food as in Dijkstra's problem, but without shared utensils.

Fifteen customers spend their time placing orders with three cooks. The orders are placed in a queue holding only five orders at a time. A customer will wait up to 7 seconds to get an order into the queue, going home for a bit if waiting too long, The cooks prepare the order then deliver the finished meal to the customer, blocking until the customer takes the meal. The simulation runs until each customer has eaten 10 meals.

Our simulation runner provides an abstract base class for our active threads with a couple methods to allow reporting to standard output: `log` to log a message, and `act` which logs a message and then sleeps for a random number of milliseconds (bounded above) to simulate an action (such as cooking or eating). The runner's thread—the application's main thread—creates 18 new threads to simulate the cooks and customers.

```
public class Restaurant {

    static class Agent extends Thread {
        Agent(String name) {
            super(name);
        }
        void log(String action) {
            synchronized (System.out) {
                System.out.println(getName() + " " + action);
            }
        }
        void act(String action, int time) throws InterruptedException {
            log(action);
            Thread.sleep((int)(Math.random() * time/2 + time/2));
        }
    }

    public static void main(String[] args) {
        for (int i = 0; i < 3; i++) {
            new Cook("Cook-" + i).start();
        }
        for (int i = 0; i < 15; i++) {
            new Customer("Customer-" + i).start();
        }
    }
}
```

A Java application finishes when all **non-daemon** threads have finished; **daemon** threads, therefore, are those threads intended to run indefinitely, performing services as needed. Cooks are naturally daemon threads, grabbing orders, cooking, and serving customers. As we'll see in a minute, fetching an order and serving the customer are **synchronous** operations; cooks will block until an order can be fetched, and also block until the customer accepts the handoff of the cooked order.

```
class Cook extends Restaurant.Agent {

    Cook(String name) {
        super(name);
        this.setDaemon(true);
    }

    @Override public void run() {
        try {
            while (true) {
                Order order = Order.begin();
                act("cooking " + order, 12000);
                order.serve();
            }
        } catch (InterruptedException e) {
            log("got fired from the restaurant");
        }
    }
}
```

All blocking operations in Java may throw an `InterruptedException`; this exception must be (1) handled in any method that may throw it, or (2) declared to be throwable. In our simulation, interrupting a cook terminates (pun intended) the employee (thread).

Customers eat ten meals then go home. The attempt to place an order may **time out**. If the order is placed within 7 seconds, the customer will wait for it to be cooked, accept the meal from the chef, then eat. If the order cannot be added to the order queue quickly enough, the customer goes shopping and returns later. Interrupting a customer causes banishment from the restaurant.

```
import java.util.concurrent.BlockingQueue;
import java.util.concurrent.SynchronousQueue;

public class Customer extends Restaurant.Agent {
    private BlockingQueue<Order> meal = new SynchronousQueue<>();

    Customer(String name) {
        super(name);
    }

    void serve(Order order) throws InterruptedException {
        meal.put(order);
    }
```

```
    @Override public void run() {
        try {
            for (int mealsEaten = 0; mealsEaten < 10;) {
                if (Order.place(this, "pancakes", 7000)) {
                    Order order = meal.take();
                    act("eating " + order, 10000);
                    mealsEaten++;
                } else {
                    act("can't place order, may return later", 5000);
                }
            }
            log("going home");
        } catch (InterruptedException e) {
            log("banished from restaurant");
        }
    }
}
```

The handoff of the cooked meal between the cook and the customer employs a **synchronous queue**. A synchronous queue has no capacity: its "head" is the object the cook is inserting; items are thus "removed" only when an attempted insert is in progress. The methods `put` and `take` both block indefinitely, a less-than-ideal situation were one party to disappear while the other was waiting.

Finally we come to the order class. We maintain a single queue to hold all (uncooked) order objects. We've made the constructor private so that customers may create orders only via the static method `place`, which may time out. Cooks grab an order from the queue in a blocking fashion (via `take`), and when finished cooking, take credit for the work by updating the order object and delivering it personally to the customer. We've chosen the size of this queue to be small, relative to the number of customers, so in running the simulation, you'll be likely to see customers getting frustrated (threads will time out) and leaving the restaurant temporarily.

```
import java.util.concurrent.BlockingQueue;
import java.util.concurrent.ArrayBlockingQueue;
import java.util.concurrent.TimeUnit;

public class Order {
    private static BlockingQueue<Order> orders =
            new ArrayBlockingQueue<>(5);

    private String content;
    private Customer customer;
    private Cook cook;

    private Order(Customer customer, String content) {
        this.customer = customer;
        this.content = content;
    }
```

```
    static boolean place(Customer customer, String content, int maxWait)
            throws InterruptedException {
        Order order = new Order(customer, content);
        return orders.offer(order, maxWait, TimeUnit.MILLISECONDS);
    }

    static Order begin() throws InterruptedException {
        return orders.take();
    }

    void serve() throws InterruptedException {
        cook = (Cook)Thread.currentThread();
        customer.serve(this);
    }

    @Override public String toString() {
        return content + (cook == null ? " for " + customer.getName()
                : " cooked by " + cook.getName());
    }
}
```

7.7 METAPROGRAMMING

We'll finish our tour of Java with a very brief stop at its metaprogramming facilities. We've seen Ruby and Julia metaprogramming in the two previous chapters, so the basic ideas by now should be familiar. We would like to treat language constructs *of* our program as objects *in* the program, for both introspection and code generation. On the introspective side, Java provides ample means of inspecting objects and classes. Using our animal classes from earlier in the chapter:

```
public class IntrospectionExample {
    public static void main(String[] args) throws Exception {

        // Check types at runtime
        Animal bessie = new Cow("Bessie");
        assert bessie instanceof Cow;
        assert bessie.getClass() == Cow.class;

        // Get a class object via its name!
        assert Class.forName("Cow") == Cow.class;

        // Inspect a class
        assert Cow.class.getName() == "Cow";
        assert Cow.class.getSuperclass() == Animal.class;
        assert Animal.class.getDeclaredMethods().length == 2;
    }
}
```

Given a class object, we can obtain its superclass, package, and (arrays of) its constructors, fields, methods, implemented interfaces and other structural properties. We can use constructor objects to create new instances, field objects to read and set object properties, and method objects to invoke methods whose names are not available until runtime. The following contrived example illustrates instance creation and method invocation:

```
import java.lang.reflect.Constructor;
import java.lang.reflect.Method;
public class ReflectionExample {
    public static void main(String[] args) throws Exception {
        Class<?> c = Class.forName("Horse");
        Constructor<?> ctor = c.getConstructor(String.class);
        Animal h = Animal.class.cast(ctor.newInstance("CJ"));
        Method m = h.getClass().getMethod("speak");
        assert m.invoke(h).equals("CJ says neigh");
    }
}
```

And here's a very surprising illustration of indirect field manipulation. Objects of class `Integer` have a private field called `value` storing their wrapped primitive int value. The field method `setAccessible` overrides the restricted private access and allows us to change the wrapped value. Enjoy the output of the following script:

```
import java.lang.reflect.Field;
public class ThreeIsFive {
    public static void main(String[] args) throws Exception {
        Field v = Integer.class.getDeclaredField("value");
        v.setAccessible(true);
        v.set(3,5);
        System.out.printf("%d\n", 3);
    }
}
```

It's rare for "regular" applications to use metaprogramming facilities, but tools such as GUI builders, and the popular Spring [99] and Hibernate [104] frameworks, perform introspection and code generation, hiding these powerful operations behind easy-to-use annotations. In addition, libraries such as ASM [93] and javassist [68] allow the Java programmer to manipulate Java bytecode, the binary instructions that run on the JVM. These frameworks and libraries are well beyond the scope of this book, but their existence and popularity points out that Java need not be considered only a "static" language.

7.8 JAVA WRAP UP

In this chapter we explored the Java language. We learned that:

- Programs are made up entirely of classes (and interfaces).
- Java uses static typing nearly everywhere, and makes a strong distinction between type and class.

- Like Lua and JavaScript, Java has both primitive and reference types, though conversion between the two via autoboxing and unboxing is not difficult.
- A Java class can extend only one class but implement multiple interfaces.
- Interfaces can contain only constants or methods. Interface methods can be abstract, static, or default. Language rules prevent confusion or ambiguity surrounding multiple inheritance.
- Interestingly, Java arrays are covariant but collections (from the standard library) are invariant.
- Because of static typing, the language features generic types and methods, allowing parameterization by type. Type parameters sometimes require wildcards to enable certain operations to work on objects that can be parameterized by any type, or any type family.
- Java specifies that `null` belongs to every reference type, so programmers often use optionals from the standard library to get around the "billion dollar mistake."
- Java uses threads for concurrency, and hence provides a large set of both in-language and standard library support for synchronization and thread management.
- Java's rich metaprogramming facilities include support for both introspection and code generation. While indispensable in many frameworks and libraries, typical applications use these techniques only rarely.

To continue your study of Java beyond the introductory material of this chapter, you may wish to find and research the following:

- **Language features not covered in this chapter**. Packages, serialization, processes, volatiles, checked vs. unchecked exceptions, scoping details, type erasure and reification, native members, instance and static initializers, nested classes, enumerations, annotations, and classloaders.
- **Open source projects using Java**. Studying, and contributing to, open source projects is an excellent way to improve your proficiency in any language. You may enjoy the following projects written in Java: Elasticsearch (`https://github.com/elastic/elasticsearch`), Retrofit (`https://github.com/square/retrofit`), Spring Framework (`https://github.com/spring-projects/spring-framework`), and Guava (`https://github.com/google/guava`).
- **Reference manuals, tutorials, and books**. Both Java and the Java Virtual Machine are defined in specifications found at `https://docs.oracle.com/javase/specs/`. Oracle's Java Tutorials at `https://docs.oracle.com/javase/tutorial/` are quite complete. Regarding standard libraries, note that multiple Java Platforms exist, including the Standard Edition, the Enterprise Edition, and a Micro Edition, each with navigable online documentation. Online resources for learning the language and exploring advanced features are plentiful and perhaps too numerous to mention.

EXERCISES

Now it's your turn. Continue exploring Java with the activities and research questions below, and feel free to branch out on your own.

7.1 Install a Java development environment (and if necessary, an associated runtime) and execute the opening three scripts in the chapter.

7.2 Are there other advantages besides the ones mentioned in this chapter for Java's decision to require explicit types on (nearly) everything?

7.3 Write the smallest possible example script that highlights the difference between the stream methods `map` and `flatMap`.

7.4 We glossed over the distinction between a UTF-16 code unit and a Unicode code point. Research the difference.

7.5 Research the primitive type `char`. Can a `char` be automatically converted to an `int`? To a `byte`? How exactly in Java do we represent the character U+1F415 DOG?

7.6 The value `null` belongs to every reference type, but not to any primitive type. What happens if you try to unbox `null`?

7.7 Research Java's instance initializers and static initializers, and write an example class that uses each.

7.8 Find out the difference between nested classes, member classes, local classes, and anonymous classes, and give examples of each.

7.9 In the functional interface example in this chapter, what is the class of the expression `x->x*2`? Use `getClass` to find out.

7.10 What exactly is a method reference? Rewrite the expression `reduce(Math::max)` two ways: (1) using the arrow notation, and (2) by instantiating an instance of a class implementing `java.util.function.BinaryOperator`.

7.11 Write a script to find the first Fibonacci number greater than 10^50 by generating an infinite stream of Fibonacci values and using the `findFirst` method on the `Stream` class. You will have to learn about the `BigInteger` class to do this exercise.

7.12 The following definition of static typing appears at `http://c2.com/cgi/wiki?StaticTyping`: "[Static typing] means that a reference value is manifestly (which is not the same as at CompileTime) constrained with respect to the type of the value it can denote, and that the language implementation, whether it is a compiler or an interpreter, both enforces and uses these constraints as much as possible." Break down and explain the rationale for this definition and show whether it differs from the common understanding of static typing as validating type constraints prior to execution. You will of course need to research the precise definition of reference value.

7.13 Some bloggers have considered the term "unityped" a perjorative and have written in defense of such languages. What are their arguments?

7.14 How is Java's static typing similar to, or different from Julia's allowance of type constraints?

7.15 The class `Class` is parameterized. What is the purpose of this type parameter?

7.16 In the source code for the standard library, in the class `java.util.Objects`, you will find:

```
public static boolean isNull(Object obj) {
    return obj == null;
}
```

When would a programmer likely find this method preferable to the simple expression it invokes?

7.17 We spent a lot of time discussing optionals, almost as if they were encouraged to be used everywhere as a solution to the billion dollar mistake. However, their implementation in Java is somewhat controversial. Read `http://blog.codefx.org/java/dev/design-optional/` and summarize the main points against optionals. What do you think?

7.18 Explain the rationale behind the wildcard parameters in the methods of the `Optional` class listed in this chapter. In particular, why were the lower bounded vs. upper bounded wildcard choices made where they were? Can you think of a general rule-of-thumb for using `extends` vs. `super` based on your analysis?

7.19 Determine if it is possible, prohibited, or required, for Java threads managed by the JVM to be mapped to real operating system threads.

7.20 In addition to threads, the Java standard library supports *processes*. Research how processes are created and managed in a Java application.

7.21 Enhance the restaurant example from this chapter to log a message when customer places order.

7.22 We may wish to finish our restaurant simulation with the message `"The restaurant is closing down"` after all of the customers have gone home. Implement this feature in two ways: (a) using `join`, and (b) using a `CountDownLatch`.

7.23 In this chapter, we saw examples of threads responding to being interrupted, but we never explained how to interrupt a thread! Find out how this is done. Enhance the restaurant simulation so that after launching each of the threads, the main thread interrupts a few customers and one cook after a random amount of time has passed.

7.24 Learn about Java's `ForkJoinPool`. Use one to build a Java version of the π-approximation Julia script we saw in the last chapter.

7.25 We took a little shortcut in our `ReflectionExample` script in this chapter, by declaring that our `main` method could throw any exception. In the Java culture, this is considered bad practice. Rewrite the script so that the throws clause properly lists the individual exceptions that may be thrown.

7.26 Find out about the concept of **type erasure** in Java. Why exactly did Java's designers make this a feature?

7.27 Read about the languages Kotlin, Ceylon, and Fantom. What do they purport to "fix" in Java?

CHAPTER 8

Clojure

Clojure is a "dynamic ... general-purpose language, combining the approachability and interactive development of a scripting language with an efficient and robust infrastructure for multithreaded programming." [56]

First appeared 2007
Creator Rich Hickey
Notable versions 1.0 (2009) • 1.7 (2015)
Recognized for Macros, Lisp-ness, Concurrency support, Java interoperability
Notable uses Data mining, Artificial intelligence
Tags Functional, Dynamic, Concurrent, Homoiconic
Six words or less A Lisp for the JVM

In the 1950s John McCarthy created Lisp. In doing so, he was said to have "[done] for computing something like what Euclid [had done] for geometry," [46] written the "Maxwell's Equations of software," [34] and given the world "the greatest single programming language ever designed." [113] Lisp, remarked Edsger Dijkstra, has "assisted a number of our most gifted fellow humans in thinking previously impossible thoughts." [26]

This chapter introduces Clojure. Clojure is a Lisp—a modern Lisp. It has a richer set of data structures than the original Lisp. It does a better job than most (if not all) other Lisp dialects in isolating side effects and implementing concurrent features. Clojure interoperates with Java, so your Clojure program has access to the entire Java platform. Your Clojure code can call Java libraries. Your Java code can execute Clojure-written code.

After presenting our usual three introductory scripts, we will cover the basic data types of Clojure (there are quite a few of these) and devote some time to **persistent data structures**, and the use of **transients** to isolate mutable operations. We'll then look at Clojure's various approaches to concurrency (one of the language's great strengths) visiting refs, agents, and software transactional memory.

We'll close our tour with a look at the macro system, which takes advantage of Clojure's representation of code as data, and is considered by many to be the essential ingredient of any Lisp dialect.

8.1 HELLO CLOJURE

We begin with our usual three programs. Let's generate right triangle measurements:

```
(doseq [c (range 1 41) b (range 1 c) a (range 1 b)]
  (if (= (+ (* b b) (* a a)) (* c c))
    (printf "%d, %d, %d\n" a b c)))
```

Clojure's ranges are inclusive at the lower bound and exclusive at the upper bound. Like Julia, we can write nested loops with a flattened syntax. The equality operator is (perhaps refreshingly) `=`, not `==` or `===`. Clojure's `printf` works like Java's `printf`.

But oh that syntax. We can't help but notice the use of parentheses (and brackets) for structure. This design choice has some advantages over the control structure syntaxes we've seen up to this point: it is more compact than the terminal `end` of Lua, Ruby, and Julia; avoids the curly-brace style debates plaguing JavaScript, Java, and friends; and does not impose any particular indentation requirements on the programmer, as does CoffeeScript and Python. Clojure (like all Lisps) takes the world of nested parentheticals even further, applying it regularly to all code—control flow and expressions—throughout the entire language, even to the point of placing arithmetic operators in an unfamiliar prefix position. So where we might traditionally see:

```
(-b + Math.sqrt(b * b - 4 * a * c)) / (2 * a)
```

we have, in Clojure:

```
(/ (+ (- b) (Math/sqrt (- (* b b) (* 4 a c)))) (* 2 a))
```

The Clojure text is a direct translation from the abstract syntax tree (see Figure 8.1). This syntax simplifies a couple aspects of language design: operators and named functions are treated exactly the same, and there is no need to define operator precedence or associativity.

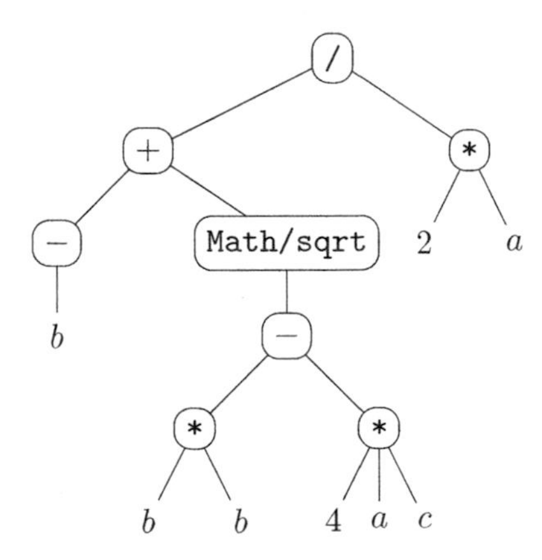

Figure 8.1 An abstract syntax tree

Now let's turn to our second running example: Heap's algorithm. Although this algorithm is defined in terms of successive mutations on a single, global array, we will embrace the Clojure culture and use immutable objects only:

```
(defn swap [a i j]
  (assoc a j (a i) i (a j)))

(defn generatePermutations [v n]
  (if (zero? n)
    (do (println (apply str v)) v)

    (loop [i 0 a v]
      (if (< i n)
        (do
          (let [a (generatePermutations a (dec n))]
            (recur (inc i) (swap a (if (even? n) 0 i) n))))
        (generatePermutations a (dec n))))))

(if (not= (count *command-line-args*) 1)
  (do
    (println "Exactly one argument is required")
    (System/exit 1))
  (let [word (->> *command-line-args* first vec)]
    (generatePermutations word (dec (count word)))))
```

We may take a slight performance hit: the invocation `(swap a i j)` returns a *new* vector like a but with the values at indexes i and j swapped. We'll talk about efficiency later. For now, let's focus instead on the many interesting aspects of the script:

- Clojure uses the syntax `(f x y)` for function calls (such as `inc`, `dec`, and `count`), **special forms** (such as `if`, `let`, and `loop`), and macros (such as `->>`). Clojure evaluates arguments prior to a function call, but does not do so for macros and special forms. You will need to memorize which symbols refer to functions and which do not; there is no marker, such as the `@` in Julia, to distinguish the various kinds of entities.
- The pretty macro `->>` reduces the number of parentheses: `(->> x f g h)` rewrites, for example, to `(h (g (f x)))`.
- Clojure code can reference Java objects: `System/exit` is Java's `System.exit`.
- Several Clojure special forms, including `let`, `loop`, and `doseq`, create bindings local to the form. For example `(let [x 5 y 2] (* x y))` evaluates to 10, with x and y bound only within the form.
- The special form **do** evaluates a sequence of expressions, presumably for their side-effects, and returns the value of the last expression in the sequence.
- The `loop` special form uses `recur` to advance the loop, rebinding its locals to the "arguments" sent by the `recur` form. Our loop starts with i bound to 0 and a bound to our initial vector. Each recurrence of the loop binds i to the next index position $(i + 1)$ and a to the next permutation. The loop ends at $i = n$, since no `recur` form is evaluated at this point.[1]

Next, as usual, we'll count words from standard input:

[1]The `loop/recur` form takes the place of tail-recursive functions in similar languages, and exists because Clojure does not wish to trust that its target virtual machine is smart enough to avoid the pile of memory-consuming frames generated during a chain of recursive calls.

```
(let [counts (apply merge-with +
  (map #(->> % .toLowerCase (re-seq #"[a-z']+") frequencies)
    (line-seq (java.io.BufferedReader. *in*))))]
  (doseq [[word count] (into (sorted-map) counts)]
    (println word count)))
```

This is an extremely dense script, but instructive to study, especially if you are new to functional programming. The global structure of the script says:

> "Let *counts* be the word-count map in a loop that prints the map sorted by key"

Let's look at how the map is produced first, reading from the inside out. The standard input stream `*in*` is decorated with Java's (and hence Clojure's!) `BufferredReader` then turned into a **lazy** sequence of lines with `line-seq`. A lazy sequence is one that produces its elements only when needed, so we need not worry about slurping a large file into memory. Next, we map a word frequency-generating function over each line. Clojure's `map` function produces lazy sequences from lazy sequences, so we are still using memory efficiently. The frequency generating function lowercases the line, splits it into words, then produces a frequency count! Mapping this function over each line gives a sequence of dictionaries, which we merge together, summing the values.

The printing portion of the script is a bit less dense. The special form `do-seq` iterates over the key-value pairs of the sorted map and prints the components with a single space between them.

In cases where the input file is small enough to read entirely into memory, we can build a frequency map in one line:

```
(->> *in* slurp .toLowerCase (re-seq #"[a-z']+") frequencies)
```

8.2 THE BASICS

Understanding Clojure begins with understanding its execution model, defined in terms of the **reader** and the **evaluator**. The reader reads source code from a stream, parsing the text into **forms**. Each form will subsequently be passed to the evaluator for evaluation. Forms known to the reader, and the kinds of objects they represent, include:

- **Integers**: `98`, `98325679988881235624987003`, `0x3F`. All bases between 2 and 36 are supported: we can write 512 as `30rh2` or `7r1331`. The integer types are `Byte`, `Short`, `Integer`, and `Long` from the package `java.lang`; and `BigInt` from `clojure.lang`. Literals have the type `Long` if in the range $-2^{-63}...2^{63}-1$ and `BigInt` otherwise. You can force the type BigInt with the suffix `N` (e.g., `200N`). The functions `byte`, `short`, `int`, `long`, and `bigint` create numbers of a desired type.
- **Doubles**: `2.54`, `3E5`, `5.0`, `-99.73`. The double types are `Float` and `Double` from `java.lang`. Create objects of a desired type with the `float` and `double` functions.
- **Ratios**: `22/7`, `3/10`. These have the type `clojure.lang.Ratio`.
- **BigDecimals**: `3M`, `78.2E900M`. These values have type `java.math.BigDecimal`. BigDecimals can be created from other numbers with the `bigdec` function.
- **Strings**: `"Hello"`. Enclosed in double quotes and may span multiple lines.

- **Characters**: `\z`, `\newline`, `\tab`, `\u263A`. Identical to Java characters in meaning, if not in syntax.
- **Booleans**: `true`, `false`.
- **Nil**: Use the value `nil` as you would `null` in Java. Note, also, as in Lua and Ruby, `nil` and `false` are the only falsy values.
- **Keywords**: `:dog`, `:rat`. A keyword in Clojure evaluates to itself. Always.
- **Symbols**: `factorial`, `+`, `even?`. Symbols refer to global variables, function parameters, and similar entities.
- **Lists**: `(:a 2 (6 8 \q) "ok" 'then)`. Whitespace-separated forms, delimited by parentheses.
- **Vectors**: `[2 5.5 \a [(list 1 2) 3]]`. Whitespace-delimited forms, delimited by square brackets.
- **Sets**: `#{6.3 [8 :k] "hello"}`.
- **Maps**: `{:jo "Amman" :ws "Apia" :ee "Tallinn"}`. Key-value pairs. The keys needn't be keywords, though they tend to be in practice.
- **Functions**: `#(+ (* 2 %1 %2)`. Arguments are `%1`, `%2`, and so on.

The reader also handles comments, metadata maps, syntax quoting and unquoting, anonymous functions, regular expressions, etc. It also deals with syntax sugaring; for example, `@x` desugars to `(deref x)` and `'x` to `(quote x)`. We'll see several of these forms and constructs later in the chapter.

Simple forms other than symbols evaluate to themselves, while vector, set, and map forms evaluate their elements and produce collection objects. Symbols get resolved to a value depending on what the symbol currently refers to. Evaluation of the list $(f_1, f_2, ..., f_n)$ first evaluates the form f_1. If f_1 evaluates to a macro or special form, it operates on the *unevaluated* remaining forms. Otherwise Clojure casts f_1 to a function and applies it to the *evaluations* of $f_2, ..., f_n$. Quoting suppresses the symbol and list evaluations: `'x` evaluates to the symbol itself, and `'(a b c)` evaluates to a three-element list rather than applying a to b and c.

Clojure uses dynamic typing, with a number of symbols predefined to stand for types:

```
(use 'clojure.test)
(is (= (type 3) Long))
(is (= (type 5.0) Double))
(is (= (type 4/7) clojure.lang.Ratio))
(is (= (type 5N) clojure.lang.BigInt))
(is (= (type 5.88M) BigDecimal))
(is (= (type "Hello
  World") String))
(is (= (type \z) Character))
(is (= (type false) Boolean))
(is (= (type nil) nil))
(is (= (type 'dog) clojure.lang.Symbol))
(is (= (type :dog) clojure.lang.Keyword))
(is (= (type Long) java.lang.Class))
(is (= (type java.lang.Class) java.lang.Class))
```

Because symbols need not contain only alphanumeric characters, addition, subtraction, multiplication and friends are not special operators, but rather regular functions. Thanks to Clojure's syntax, they each accept a variable number of arguments:

```
(use 'clojure.test)
(is (= (+ 8 7 3 2 8) 28))           ; sum a lot of numbers
(is (= (+ 8) 8))                    ; yes we can sum just one
(is (= (+) 0))                      ; 0 is the identity for +
(is (< 3 5 7 9 12))                 ; a convenient sort test
(is (not (<= 8 9 10 22 20 8)))
(is (= (max 8 7 14 2 8) 14))
```

The division operator produces longs, doubles, or ratios intelligently:

```
(use 'clojure.test)
(is (= (type (/ 3 9)) clojure.lang.Ratio))
(is (= (type (/ 8 4)) Long))
(is (= (type (/ 3.0 9)) Double))
```

The functions `+`, `-`, `*`, `inc`, and `dec` on longs may overflow, throwing an `ArithmeticException`; use `+'`, `-'`, `*'`, `inc'`, and `dec'` to auto-promote to `BigInt`:

```
(use 'clojure.test)
(let [x (long 1E15) max-long Long/MAX_VALUE]
  (is (thrown? ArithmeticException (* x x)))
  (is (= (*' x x) (bigint 1E30)))
  (is (thrown? ArithmeticException (inc max-long)))
  (is (= (inc' max-long) (+' 1 max-long))))
```

Collections are immutable and implement equality with value—not reference—semantics. Many common operations apply to all collections, such as `count`, `empty?`, `not-empty`, and `conj`, short for **conjoin**. Conjoining a collection c and an element e returns a *new* collection with all the elements c together with e.

```
(use 'clojure.test)
(let [a-list '(1 2 3) a-vec [1 2 3] a-set #{1 2 3} a-map {1 2 3 4}]
  (is (= (empty? a-set)) false)
  (is (not-empty a-vec))
  (is (= (map count [a-list a-vec a-set]) [3 3 3]))
  (is (= (count a-map) 2))
  (is (= (conj a-list 4) '(4 1 2 3)))         ; list: conj to front
  (is (= (conj a-vec 4) [1 2 3 4]))           ; vec: conj adds to end
  (is (= (conj a-set 4) #{3 2 1 4}))          ; order is irrelevant
  (is (= (conj a-map [5 6]) {1 2 5 6 3 4}))) ; adds a new pair
```

These functions, and dozens more like them, operate by internally turning the collection into a **sequence**, a logical list of elements, usually (but not always) lazy. (Recall that a lazy sequence produces its elements on demand, while a non-lazy sequence stores each element in memory). You can produce a sequence directly from a collection c by evaluating `(seq c)`. Let's look at a few more sequence functions:

```
(use 'clojure.test)
(let [a '(10 20 30) b #{5 15 25 35}]
  (is (some #(> % 30) b))
  (is (some (partial > 30) b))
  (is (count (filter (partial >= 20) a)) 2)
  (is (reduce * a) 6000)
  (is (= (map inc a) (seq [11 21 31])))
  (is (take 2 a) (seq [10 20]))
  (is (= (interleave a (sort b)) (seq '(10 5 20 15 30 25)))))
```

The previous script illustrated function values. The expression `#(> % 30)` evaluates to a function of a single argument and returns whether its value is greater than 30. As functions are first-class values, we can associate a symbol with them through a **local binding**.

We can also write function values with the special form `fn`:

```
(use 'clojure.test)
(let
  [average (fn [x y] (/ (+ x y) 2))]
  (is (= (average 9 5) 7)))
(let
  [average #(/ (+ %1 %2) 2)]
  (is (= (average 9 5) 7)))
```

Functions, by the way, support a variable number of arguments. The parameter after the `&` becomes a sequence containing the additional arguments in the call:

```
(use 'clojure.test)
(let [show (fn [x y & z] [x y z])]
  (is (= (show 1 2) [1 2 nil]))
  (is (= (show 1 2 3) [1 2 '(3)]))
  (is (= (show 1 2 3 4) [1 2 '(3 4)])))
```

This brief overview should give a feel for what Clojure code looks like. Now we're ready to move on to some very interesting, and even sophisticated, corners of the language. We'll begin with namespaces.

8.3 NAMESPACES

Namespaces contain mappings of symbols to **vars** and **classes**. Vars refer to mutable storage locations. The special form `def` stores, or **interns**, a var into a namespace. We'll create a second var using `defn`, which gives us a pleasant syntax for interning a var containing a function value (using `def` behind the scenes).

```
(ns library)              ; Going to work in this ns

(def x 5)                 ; Map symbol x to a Var
(defn average [x y]       ; Map symbol average to a Var
  (/ (+ x y) 2))          ; Local x and y are not Vars
(def x 8)                 ; Vars are mutable!
```

```
(ns application)          ; Switch to new namespace

(use 'clojure.test)       ; No need to say clojure.test/is
(is (= library/x 8))
(is (= (library/average 10 20) 15))

(use 'library)            ; No need to say library/x now
(is (= x 8))              ; this is x from library
```

We begin this script by entering a (presumably new) namespace named `library`, and interning two vars. We then modify the contents of (mutate) the var. The symbol `x` *remains mapped to the same var*; only the contents of the var changes. See Figure 8.2. We then switch to a new namespace named `application`; to access the objects in the other namespace we use a qualified symbol. After invoking `use`, however, the symbol mappings within the library namespace become available without qualification.

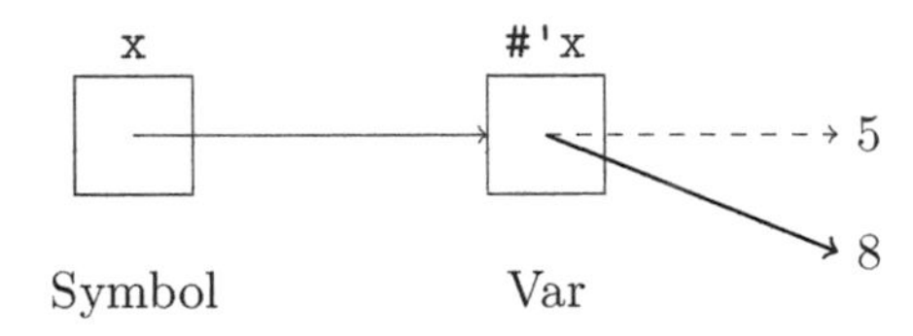

Figure 8.2 Symbols, vars, and values

Vars are Clojure objects. We denote the var object mapped to the symbol x by `#'x`, or equivalently, `(var x)`. Evaluating the symbol produces the value held in the mapped var. To get this value from the var object itself, we use `deref`, or the shorthand notation with the `@` prefix. Study the following script carefully:

```
(use 'clojure.test)
(def x 100)                          ; create a Var, map x to it
(def y #'x)                          ; the actual Var that x maps to
(is (= (type y) clojure.lang.Var))   ; the Var is an object with a type
(is (= (deref y) 100))               ; deref is the value in the Var
(is (= @y 100))                      ; @y is shorthand for (deref y)
(def x 200)                          ; Update the var through the symbol
(is (= (deref y) 200))               ; See the change directly
(is (identical? #'x (var x)))        ; #'x is shorthand for (var x)
```

Prior to encountering vars, all of the Clojure objects we've seen, including collections, were immutable. Immutability is certainly the norm in Clojure; the *only* mutable objects are vars, refs, agents, and Java object fields (Clojure "interops" with Java quite nicely). Immutability makes threaded programming easier by removing race conditions and other concurrent problems arising from shared state. However, threads do need to cooperate via the changing state of a system. Fortunately, vars, refs, and agents have been carefully designed to manage this cooperation quite well. We'll get to refs and agents shortly, but first, we have an important question to answer. If collections are immutable, with add, remove, and update operations generating new collection objects, how can such operations possibly be efficient?

8.4 PERSISTENT DATA STRUCTURES

Consider a 1000-element list, referred to by the symbol a. The expression `(conj a x)` creates a brand new 1001-element list, whose first element is the value of x. Should this new list contain copies of each of the elements of a, or can the two lists share structure, as in Figure 8.3? While conjoining by sharing is as efficient as possible, the downside is that changes to one list would (surprisingly) affect the other. However, as long as the lists are immutable, sharing works! No surprises are even possible.

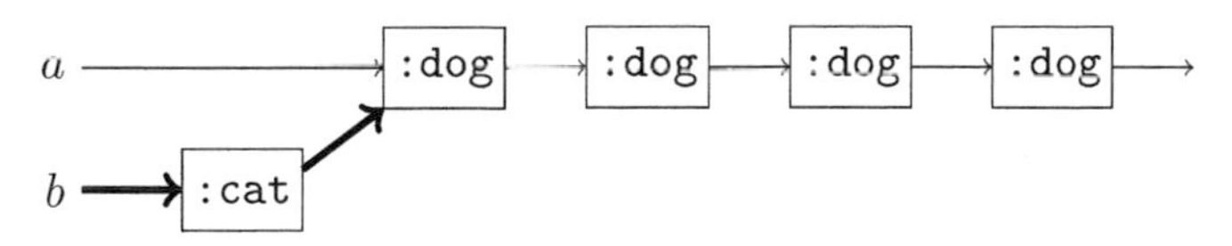

Figure 8.3 Conjoining to a Persistent List

We can verify that Clojure does indeed share:

```
(def a (list (repeat 1000 :dog)))     ; 1000-dog list
(def b (conj a :cat))                 ; new list, cat at front

(use 'clojure.test)
(is (identical? a (rest b)))
(is (= (type a) clojure.lang.PersistentList))
```

The expression `(rest b)` refers to the sequence of all elements except the first element of b, which our script tells us is indeed `identical` to (the same internal object as) a. Since the list a remains, and was not destroyed by modification, and still remains accessible, we call it a **persistent** list.

Adding to the front of a list poses no efficiency concerns, but how can we add to the rear, as vector conjoining requires? Copying each of the vector elements appears to be necessary, but we can get by with a logarithmic number of copies if we index the vector elements using a tree structure. In Figure 8.4 our vector v has the value `[1 2 3 4 5 6]`. Elements reside in the leaf nodes from left to right. To compute `(conj v 7)`, we walk down the path to the node at the end of the vector, copying internal nodes as we go, sharing structure where we can. This technique of **path copying** applies also to functions like `assoc`, which we encountered in our anagrams example at the beginning of the chapter.

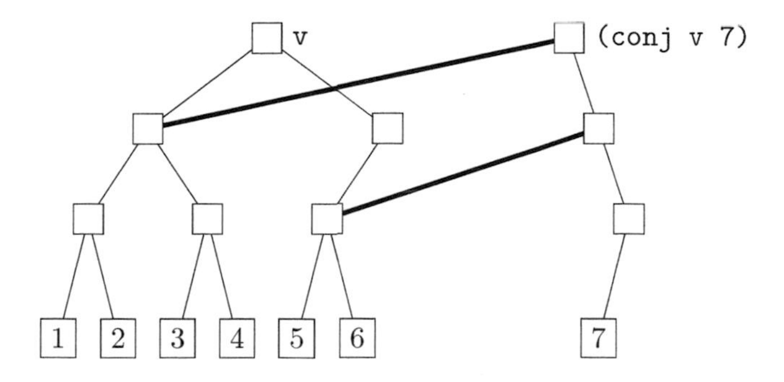

Figure 8.4 Conjoining to a Persistent Vector

We've drastically simplified things here. The actual persistent vector implementation used by Clojure is far more sophisticated: internal tree nodes have up to 32 children, using bit vectors to manage the tree structure. We're not concerned in this book with specific implementation details; our goal is simply to point out the existence of sublinear algorithms, so that language designers may embrace immutability and developers may gain a sense of the cost of using these objects. See Okasaki [90] for a comprehensive treatment of persistent data structures.

8.5 TRANSIENTS

Persistent structures pay for the advantages of immutability (thread safety, sharing) by sacrificing performance. However, while *constructing* an object—before it ever needs to be seen by the rest of the program—you can take advantage of internal, mutable operations that can speed things up significantly. Clojure's **transient data structures** operate this way; they are built up in an efficient, mutable fashion, then turned into a persistent structure when ready to use. The following example constructs a (non-lazy) vector of Fibonacci numbers:

```
(defn fib [n]
  (loop [a 0 b 1 result (transient [0])]
    (if (> b n)
      (persistent! result)
      (recur b (+' a b) (conj! result b)))))

(use 'clojure.test)
(is (= (fib 0) [0]))
(is (= (fib 1) [0 1 1]))
(is (= (fib 5) [0 1 1 2 3 5]))
(is (= (fib 144) [0 1 1 2 3 5 8 13 21 34 55 89 144]))
(is (= (fib 200) [0 1 1 2 3 5 8 13 21 34 55 89 144]))
(println (fib (bigint 1E200)))
```

We begin our loop binding *a* to 0, *b* to 1, and *result* to `(transient [0])`, a transient vector containing the single element 0. We add new values via `conj!`, not `conj`; the bang (`!`) alerts the reader that something unusual is taking place, in this case mutation. After the vector has been filled, we create, and return, a new persistent vector.

Clojure will not prevent you from misusing transients by accessing them from multiple threads at the same time. Mutability demands responsibility in a multithreaded environment. However, Clojure offers several facilities for making concurrency safe. We'll turn to these now.

8.6 SOFTWARE TRANSACTIONAL MEMORY

When running on the JVM platform, Clojure threads are Java threads. The translation of Java's `new Thread(r)` for runnable *r* to Clojure is `(Thread. r)` (the trailing dot is part of the name of the symbol denoting the constructor). And the Java expression `t.start()` becomes `(.start t)` in Clojure—method symbols have a leading dot. The following script creates 10 threads, each of which prints a list of asterisks, in a completely unsynchronized, free-for-all fashion:

```
(dotimes [i 10]
  (.start (Thread. (fn [] (println (repeat i "*"))))))
```

This script's output may vary each time you run it; one output was:

```
((((((**  * ** *  *(* ) **)*

((* **  ** *  * **)
(*)*
 * ** )*
 **))
* * **  ***  * * )

* * * ** )*
 * *)
```

We can apply a simple fix, using the `locking` macro to print only while the current thread holds the intrinsic lock for the standard output object:

```
(dotimes [i 10]
  (.start (Thread. (fn []
    (locking *out* (println (repeat i "*")))))))
```

This produces, on one run:

```
()
(* * * * * * * * *)
(* * * * * * * *)
(* * * * * * *)
(* * * * * *)
(* * * * *)
(* * * *)
(* * *)
(* *)
(*)
```

But Clojure offers sophisticated concurrency support that makes explicit use of threads and locks very rare. In this section we'll explore **software transactional memory**, or STM, and cover agents in the next section. Transactional memory is implemented in Clojure with **refs**, mutable storage locations that can only be updated inside a **transaction**, specified in Clojure with `dosync`. Transactions prevent concurrent updates of references from other threads.

STM solves one of the classic problems in concurrency—transferring money between accounts—quite nicely. First, let's note that transferring, say, 100 units of currency from a savings account to a checking account generally requires at least 6 steps:

1. Read the balance of the savings account (suppose it is 10000), into a machine register
2. Decrement the register by 100
3. Write the decremented value (99900) back to memory
4. Read the balance of the checking account (suppose it is 500), into a machine register
5. Increment the register by 100
6. Write the incremented value (600) back to memory

If another thread were able to access and mutate either of these account balances during the sequence of transfer instructions, new money could be gained or lost. Suppose thread t_1 is currently executing this six-statement sequence. Further suppose thread t_2, deposits lottery winnings into the savings account *between* t_1's steps 1 and 3. Step 3 (on t_1) overwrites the huge deposit. Placing the entire transfer in a transaction prevents such situations from happening:

```
(use 'clojure.test)

(def savings (ref 10000))
(def checking (ref 500))

(is (and (= @savings 10000) (= @checking 500)))

(dosync
  (let [amount 100]
    (alter savings - amount)       ; alter savings by subtracting amount
    (alter checking + amount)))   ; alter checking by adding amount

(is (and (= @savings 9900) (= @checking 600)))
```

Refs work well when you need to coordinate access to multiple objects in a synchronized fashion; the language mandates that any attempt to alter refs outside of a transaction throw an exception. For a different scenario—safe access to a *single* object in an *asynchronous* fashion—Clojure offers **agents**.

8.7 AGENTS

Like vars and refs, agents store values which you retrieve via `deref` (or `@`). To change the value managed by the agent, you dispatch an **action**. An action is simply a function call that, provided the optional validation test passes, produces the new value for the agent. The function executes asynchronously on a thread from thread pools managed by Clojure; the programmer never starts any threads or performs any locking. The threads are non-daemon threads, however, so the programmer must explicitly shutdown the pool. The following trivial example sends 50,000 increment actions to a counter agent with a validation function ensuring the counter never becomes negative. It prints the value while, presumably, the queued actions are still being executed, then after a two-second pause giving the queued actions more time to execute, prints the value again. Finally, the script shuts down the agent system, terminating any as yet unprocessed actions.

```
(def counter (agent 0 :validator #(>= % 0)))

(dotimes [i 50000] (send counter + 1))

(println @counter)    ; Will likely be < 50000
(Thread/sleep 2000)
(println @counter)    ; Will likely be 50000
(shutdown-agents)
```

Note how Clojure's agent-based programming exhibits strong philosophical differences from "object-oriented" programming. In OOP, we focus on objects with internal state and behavior defined for that object in particular or objects of a certain type. Agents, on the other hand, have state but no behavior! No restrictions are placed on the functions you can send to the agent. Indeed, the actions are the center of attention.

Emphasizing functions over objects does not mean we can't model iconic object-oriented tasks. In the next section we take a look at what Clojure offers.

8.8 THE EXPRESSION PROBLEM

In previous chapters we've seen various implementations of cows, horses, and sheep, each making their own type-specific sound. Generally the animals were defined with classes, with a common superclass `Animal` specifying that each animal should have a name and speak by returning a message saying that the animal (noted by its name) makes its sound. In the following script we create these animal classes in Clojure, and, rather than inheriting from a superclass, we define a **protocol** for sound-making that each class will implement. Unlike Java, however, but like Python, each "method" requires an explicit parameter for the receiver. As protocols specify only behavior and no state, methods that would ordinarily belong a superclass are, in Clojure, just functions:

```
(defprotocol Sounder
  (sound [sound-maker]))

(defrecord Horse [name] Sounder
  (sound [this] "neigh"))
(defrecord Cow [name] Sounder
  (sound [this] "moooo"))
(defrecord Sheep [name] Sounder
  (sound [this] "baaaa"))

(defn speak [animal]
  (str (.name animal) " says " (sound animal)))
```

The `defrecord` macro accepts a class name, fields, protocols that it should implement, and the bodies of any methods it needs. Assuming a Clojure installation running on a JVM, Clojure records are real Java classes (instances of `java.lang.Class`), obeying the usual Java interop syntax for construction, e.g. `(Horse. "CJ")`, and for field access, e.g. `(.name h)`. Let's verify this:

```
(use 'clojure.test)
(def h (Horse. "CJ"))
(is (= (speak h) "CJ says neigh"))
(def c (Cow. "Bessie"))
(is (= (speak c) "Bessie says moooo"))
(is (= (speak (Sheep. "Little Lamb")) "Little Lamb says baaaa"))

(is (= (type h) user.Horse))
(is (= (type Horse) java.lang.Class))
```

Our animal system now has three classes and one class-specific method. But we can imagine extending our project with both new kinds of animals—dogs, perhaps—or new kinds of functionality, such as producing an emoji character. See Figure 8.5. The **expression problem** refers to the situation in some languages that extending a system in one direction is easy but extending in the other requires touching existing code. Languages with classes clearly have no trouble extending a system with new classes; Clojure is no exception:

```
; Add a new class
(defrecord Dog [name] Sounder
  (sound [this] "woof"))
```

Adding new functionality to existing classes (without modifying the source code for existing classes) requires that the language has a mechanism for **type extension**. The `extend-type` macro solves the expression problem in Clojure:

```
; Add new functionality
(defprotocol EmojiDescribable (emoji [this]))
(extend-type Horse EmojiDescribable (emoji [this] "\uD83D\uDC0E"))
(extend-type Cow EmojiDescribable (emoji [this] "\uD83D\uDC04"))
(extend-type Sheep EmojiDescribable (emoji [this] "\uD83D\uDC11"))
(extend-type Dog EmojiDescribable (emoji [this] "\uD83D\uDC15"))

(println (emoji h))
(println (emoji c))
(println (emoji (Sheep. "Woolie")))
(println (emoji (Dog. "Spike")))
```

Protocols enable the creation of objects with shared behaviors in a Ruby-esque or Java-esque fashion. But Clojure *also* supports a style similar to, but not quite the same as, Julia's generic functions. A Clojure **multimethod** specifies a dispatching function and one or more actual methods that dispatch based on the result of applying the dispatching function to the method arguments. Here is our animals example, implemented with a multimethod `sound`, dispatching based on the result of applying `:Animal` to the object we'd like to hear a sound from. Note that in this example, we have *not* introduced any new types. The animal objects are just plain maps!

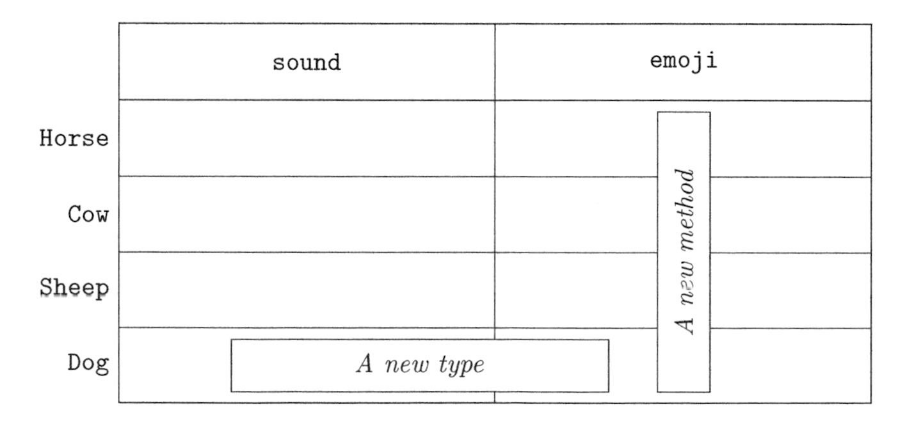

Figure 8.5 The Expression Problem

```
(defmulti sound :Animal)
(defn speak [animal]
  (str (:name animal) " says " (sound animal)))

(defn horse [name] {:Animal :horse :name name})
(defmethod sound :horse [h] "neigh")

(defn cow [name] {:Animal :cow :name name})
(defmethod sound :cow [c] "moooo")

(defn sheep [name] {:Animal :sheep :name name})
(defmethod sound :sheep [s] "baaaa")

(use 'clojure.test)
(def h (horse "CJ"))
(is (= (speak h) "CJ says neigh"))
(def c (cow "Bessie"))
(is (= (speak c) "Bessie says moooo"))
(is (= (speak (sheep "Little Lamb")) "Little Lamb says baaaa"))
```

Notice we have no expression problem with multimethods either:

```
; Add a new kind of animal
(defn dog [name] {:Animal :dog :name name})

; Add a new method
(defmulti emoji :Animal)
(defmethod emoji :horse [h] "\uD83D\uDC0E")
(defmethod emoji :cow [c] "\uD83D\uDC04")
(defmethod emoji :sheep [s] "\uD83D\uDC11")
(defmethod emoji :dog [d] "\uD83D\uDC15")
(doseq [a [h c (sheep "Wooly") (dog "Spike")]] (println (emoji a)))
```

Though our animals example did not take advantage of this capability, multimethods allow dispatch on (any or all of its) arguments, while protocol-based dispatch considers (only) the *receiver*. We saw this difference before, when contrasting Ruby and Julia. Clojure and Julia implement multiple dispatch a little differently, however; Julia considers only the types of the method arguments, while Clojure executes a user-defined function on the argument values to determine the method to dispatch. You'll explore this highly flexible feature in the end-of-chapter exercises.

8.9 MACROS

We've saved the best of Clojure for the last section.

You've undoubtedly noticed that programs (code) in Clojure are just lists (data). Programs can easily output lists and then evaluate them. This makes metaprogramming feel just right: In most other languages we generate code as strings (which need compilation) or byte code (which is unreadable for average humans), Clojure simply generates lists. Julia is a bit like

Clojure in that it generates expression objects, but Clojure's generated code *looks* exactly like the code you'd write yourself. Clojure is a **homoiconic** language.

As in Julia, a macro accepts unevaluated expressions, performs some computations, and evaluates the result of these computations. Three classic examples come to mind here. The first "evaluates" an expression written in infix form:

```
(defmacro eval-infix [form]
  (list (second form) (first form) (second (rest form))))

(use 'clojure.test)
(is (= (eval-infix (20 * 7)) 140))
(is (= (macroexpand '(eval-infix (20 * 7)))
      '(* 20 7)))
```

Because macro invocations do *not* evaluate arguments before the call, invoking `(eval-infix (20 * 7))` passes a 3-element list to the macro. The macro returns a new three-element list, specifically `(* 20 7)`, which is immediately evaluated to produce 140. The `macroexpand` function will show you the forms produced by the macro without evaluation.

The second popular example, highlights the use of macros to "extend the syntax" of the language. We define a new form, called `unless`:

```
(defmacro unless [condition this that]
  (list 'if (list 'not condition) this that))

(use 'clojure.test)
(is (= (unless (< 1 2) 10 7) 7))
(is (= (unless (< 2 1) 10 7) 10))
(is (= (macroexpand '(unless (< 1 2) 10 7))
      '(if (not (< 1 2)) 10 7)))
```

We can write this macro another way, using the **syntax quote** (`` ` ``):

```
(defmacro unless [condition this that]
  `(if (not ~condition) ~this ~that))
```

The syntax quote is similar to the regular quote except that (1) symbols inside syntax quotes are fully-qualified, e.g. `(unless 1 2 3)` expands into `(if (clojure.core/not 1) 2 3)`, and (2) we can *unquote* within syntax quoted forms. Unquoting simply tells Clojure to evaluate certain forms inside of a quoted form. The use of syntax quoting tends to produce more readable macros, constructing the resulting forms from patterns rather than from complex expressions piecing together lists.

The third popular example comes from the Clojure core library itself—the `or` macro. Clojure's `or` evaluates arguments from left-to-right, returning the first truthy value encountered, or where no truthy operands exist, returning the last argument—or `nil` if no arguments. This means the expression `(or false nil "hello" (/ 1 0))` produces `"hello"`, without ever evaluating the fourth argument.[2] Here's the code:

[2] Rght here we have one of several answers to "Why do we have macros?" We can't easily write short-circuit-or as a function since functions evaluate their arguments before a call. Granted we could *explicitly* pass in quoted forms and invoke `eval`, but such a mechanism does not integrate nicely with the rest of the language.

```
(defmacro or
  ([] nil)
  ([x] x)
  ([x & next]
    `(let [or# ~x]
      (if or# or# (or ~@next)))))
```

We're seeing a definition by cases for the first time; we specify what the macro should produce when given zero arguments, one argument, or more than one argument. Let's look at the third case closely. The macroexpansion

```
(macroexpand '(or alpha beta gamma delta))
```

yields

```
(let* [or__4469__auto__ alpha]
  (if or__4469__auto__
    or__4469__auto__
    (clojure.core/or beta gamma delta)))
```

Suffixing a identifier with a hash character causes the macroexpansion to generate a new symbol guaranteed not to conflict with any other; this prevents the macro from accidentally shadowing an existing binding. Also note the use of **unquote splicing** (`~@`). Had we simply written `~next` rather than `~@next`, our expansion would have ended with `(clojure.core/or (beta gamma delta))`. Splicing helps a great deal in keeping macros easy to read.

Macros are not arcane features, but very common in Clojure. A large number of the definitions in the standard library are macros, including `if-not`, `when`, `delay`, `and`, `or`, `locking`, `->`, `->>`, `defmulti`, `defmethod`, `dotimes`, and `doseq` and literally dozens more. You can find the source code for these macros at [57].

Because macros provide a means to transform code prior to execution, they give us the ability to create powerful language extensions in a way that is qualitatively different from adding functions to a standard library. There are no limits to the number of macros you can create. However, as in any high-level language, you will see some forms that are so low-level, so wired in to the inner workings of Clojure itself, that they could not be (possibly or conveniently) written even as macros. There are a few, and are known in Clojure as special forms. We've listed for reference, a few of these, in Table 8.1.

Clojure, like all Lisps, use special forms where other languages would use syntactic structures such as `if` and `while` statements, function declarations, and all kinds of punctuation. In Clojure, there really is no surface syntax; all is abstract syntax, where function calls, macro calls, and special form invocations all look the same. This does make macro writing a bit less cumbersome than say, in Julia, where the programmer must mindfully construct the abstract syntax tree of the code being generated. It will always, perhaps, remain a design choice that is both very popular and very unpopular. But there's something compelling, perhaps profound, about this integration of code and data, the homoiconicity, and the ease of metaprogramming in the family of Lisp languages. Something worthwhile, too, as Eric S. Raymond writes:

> Lisp is worth learning for the profound enlightenment experience you will have when you finally get it; that experience will make you a better programmer for the rest of your days, even if you never actually use Lisp itself a lot.

`def`	Sets the root binding of a variable, creating it if necessary.
`if` e_1 e_2 e_3	Evaluates e_1. If truthy, evaluates and produces e_2. Otherwise evaluates and produces e_3.
`do` $e_1 ... e_n$	Evaluates the expressions in order, producing e_n.
`let`	Evaluates e in the context of local bindings.
`quote` *form*	Produces *form* without evaluating it.
`var` *sym*	Produces the var bound to *sym*, not the value in the *var*.
`fn`	Defines a function.
`loop`	Like `let`, but sets up a **recursion point**.
`recur`	Transfers control to the recursion point with bindings rebound to the arguments of the `recur` form.
`throw` e	Similar to Java's `throw`.
`try`	Similar to Java's `try-catch-finally`.

Table 8.1 Some Clojure Special Forms

8.10 CLOJURE WRAP UP

In this chapter we briefly looked at Clojure. We learned that:

- Clojure is a Lisp, with powerful metaprogramming facilities gained from code-as-data, such as macros.
- Clojure programs are executed with a reader and evaluator. The reader reads source text and creates forms, which can be symbols, keywords, literals, or a variety of interesting data structure representations. The evaluator evaluates forms. Symbols and lists are evaluated specially. Quoting suppresses evaluation, allowing symbols and lists to be referred to directly.
- The `let` special form creates local bindings, and for convenience, so do a few other forms.
- Instead of tail recursive functions, the language provides the special forms `loop` and `recur`.
- Clojure favors dynamic typing.
- Code appears as (nested) lists, with the head treated as the operator (function, macro, or special form), and the tail (the rest of the list elements) its operands.
- The collections—lists, vectors, sets, and maps—are all immutable, persistent, and have value semantics for equality. They can each be treated as lazy sequences, upon which dozens of predefined operations can be performed.
- Namespaces are used to organize large programs. Namespaces hold vars and classes.
- Vars are one of the few mutable objects in Clojure.
- Transients speed up the creation of certain persistent structures, taking advantage of mutability that programmers are expected to isolate.
- Refs are mutable but can only be updated within a transaction, providing synchronous coordinated access to multiple values. Agents allow multiple threads to access a single value asynchronously. Both agents and refs, like vars, wrap a value which the programmer accesses via `deref`.

- Protocols and multimethods solve the expression problem, allowing both new types and new operations to be added to a system without modifying existing code.
- A powerful macro system together with homoiconicity characterizes Lisp dialects, including Clojure, like no other feature. The interplay between code and data is considered by some to be an enlightening experience once fully understood. Taking a more pragmatic view, Clojure's macro system allows for relatively clean language extensions, especially when employing syntax quoting, unquoting, and splicing.

To continue your study of Clojure beyond the introductory material of this chapter, you may wish to find and research the following:

- **Language features not covered in this chapter**. Tagged literals, Reader conditionals, Documentation strings, Metadata, Destructuring, Transducers, Reducers, Condition maps, and quite a few others.
- **Open source projects using Clojure**. Studying, and contributing to, open source projects is an excellent way to improve your proficiency in any language. You may enjoy the following projects written in Clojure: LightTable (`https://github.com/LightTable/LightTable`), Overtone (`https://github.com/overtone/overtone`), Cascalog (`https://github.com/nathanmarz/cascalog`), and Incanter (`https://github.com/incanter/incanter`).
- **Reference manuals, tutorials, and books**. Clojure's home page is `https://clojure.org/`, which contains convenient links to the documentation on the language rationale, usage guide, a complete reference, and documentation for various APIs. An excellent community-curated list of books, tutorials, and learning materials, including video presentations, is found at `http://clojure.org/community/resources`. You may find the Clojure Koans (`http://clojurekoans.com/`) a good source of both learning and practice.

EXERCISES

Now it's your turn. Continue exploring Clojure with the activities and research questions below, and feel free to branch out on your own.

8.1 Locate the Clojure home page and the documentation section. Read the language Rationale page. Clojure's creator, Rich Hickey, has written an excellent piece on state versus identity. Read this essay. Give a concrete example of Java, for example, conflating identity and state.

8.2 Experiment with the Clojure REPL.

8.3 Find the class of each of the following expressions. Invoke `(class` *e*`)` in the REPL to find out.

```
=> (* 3 8756287748365654783476538465M)
=> (/ 9 3)
=> (do (def x 0) #'x)
=> #(* %2 %1 %3)
=> #"abc"
=> *in*
=> (quote quote)
```

8.4 In this chapter we encountered the `->>` macro. A similar macro, `->` also exists. How, exactly do they differ? Give an example in which `->` must be used over `->>`.

8.5 The word count program in the first section of this chapter was described without too much detail. In particular, neither `merge-with` nor `into` were explained. Research these forms and use them in an example script of your own.

8.6 Extend the word count script from this chapter to consider the full range of Unicode letters as words.

8.7 We've seen the special syntax for vector ($[v_1 \ldots v_n]$), map ($\{k_1\, v_1 \ldots k_n\, v_n\}$), and set ($\#\{v_1 \ldots v_n\}$) literals. We can't use literals for collections with thousands of items, however. Give Clojure expressions for: (a) the vector of the first thousand primes, (b) the set of the first thousand multiples of 3, starting at 0, and (c) a mapping from the characters `\a`...`\m` to `\n`...`\z` and `\n`...`\z` to `\a`...`\m`.

8.8 We've seen that `(+)` evaluates to 0 because the identity of addition is 0. Similarly `(*)` produces 1, the identity for multiplication. What are the identities of `max` and `min`, if any? If none, how does Clojure evaluate `(max)` and `(min)`?

8.9 Experiment with the application of the less-than function `<` to operands of differing types. Do numeric types freely cast? How do values like `nil` compare with numbers and strings? Can strings be compared with vectors or lists? Are booleans comparable?

8.10 Compare Clojure's equality function (`=`) with that of Java. Does it operate more like Java's `==` operator or its `equals` method?

8.11 How does `interleave` work with sequences of different sizes?

8.12 What is the difference, if any, between `(partial < 7)` and `#(< % 7)`? In which situations would `partial` be preferred to a function expression?

8.13 What function is this?

```
(defn mystery [n]
  (loop [a 0 b 1 i n]
    (if (zero? i) a (recur b (+' a b) (dec i)))))
```

8.14 This is not a factorial function:

```
(defn fact-maybe [n]
  (let [f (fn [x a] (if (<= x 1) a (f (dec x) (*' a x))))]
    (f n 1)))
```

What exactly does the function do? Make a small adjustment to make the function compute the function of its argument.

8.15 Use a transient to build up a vector of the first 1000 primes.

8.16 The following script:

```
(def x 0)
(let [
  t1 (Thread. #(dotimes [_ 10000] (def x (inc x))))
  t2 (Thread. #(dotimes [_ 10000] (def x (dec x))))]
  (.start t1)
  (.start t2)
```

```
  (.join t1)
  (.join t2)
  (println x))
```

need not print 0. Rewrite the script so that increments and decrements are done in a transaction, and verify that multiple runs of your program print 0.

8.17 Redo the previous problem using agents.

8.18 What problems could occur if a mutable value was stored in a ref?

8.19 Extend the expression problem example from Section 8.8 using protocols to both (a) add a type for pigs, and (b) add a function to produce the binomial name (the Latin, or scientific, name consisting of the genus and species) of the animal.

8.20 Repeat the previous problem to extend the animal example using multimethods.

8.21 Learn about the library `core.async`. Develop a simple script that defines an unbuffered channel in which one process writes a message and a second process reads and prints the message from the channel.

8.22 Find out how your Clojure implementation autogenerates symbols in macros, or if perhaps the language definition itself specifies a particular format. Knowing this, can you or can you not guarantee that you would never get unlucky and have a name clash with a generated symbol?

8.23 Are Clojure macros hygienic? If so, show how they fit a commonly accepted definition of the term. If not, are they (1) not hygienic at all, or (2) perhaps partially hygienic? In which situations do they fall short of being fully hygienic?

8.24 Is it possible to determine at runtime, whether a symbol refers to a function, a macro, or a special form?

8.25 Run the following script on your Clojure installation (we've taken it straight from the Clojure documentation, with slight modifications, so you might have seen it before):

```
(defn len [x] (.length x))

(defn fast-len [^String x] (.length x))

(doseq [f [len fast-len]]
  (time (reduce + (map f (repeat 1000000 "dogs")))))
```

The parameter of `fast-len` is annotated with a **type hint**. How exactly do type hints explain the large performance difference between the two functions?

8.26 Read about Clojure transducers, and give an example of their use.

CHAPTER 9

Elm

Elm is a functional language ideal for interactive applications.

First appeared 2011
Creator Evan Czaplicki
Notable versions 0.10 (2013) • 0.14 (2014) • 0.17 (2016)
Recognized for Interactive functional programming
Notable uses User interfaces, Games
Tags Functional, Statically-typed, Subscription-oriented
Six words or less Functional programming for the web

Evan Czaplicki created Elm in 2011 as a language for building browser-based applications in the style of functional reactive programming, or FRP. Like CoffeeScript, Elm applications transpile to JavaScript. But while CoffeeScript looks like Python or Ruby—with dynamic typing and assignable variables—Elm is part of the statically-typed "ML family" of languages, and borrows a little from Haskell. Elm will be the only ML language we will feature in this book, so take the time to enjoy the many features it offers.

In this chapter we'll visit many of the ideas originating in the design and development of the ML languages, including **type variables** and extensive type inference. We'll also encounter pattern matching and destructuring. We'll even see a bit of Haskell-inspired syntax. Haskell programmers may find much of Elm quite familiar.

But Elm is not an ML or Haskell knock-off. It adds many new features of its own and fuses old and new ideas quite nicely. While Elm has moved on from its FRP roots, it now aims to supersede JavaScript for HTML applications on the client side by replacing JavaScript's callback-heavy architecture with a **subscription**-based approach to interactivity. Elm also differs from JavaScript by living very far to the "pure" side of the functional language spectrum. Data structures are persistent, like those of Clojure. You bind variables rather than assigning to them.

We'll begin our exploration of Elm with our three standard introductory programs, running in the browser, since Elm doesn't usually script standard input and standard output. We'll then look at base language elements including modules, functions, and types. We'll cover Elm's approach to type inference (borrowed, of course, from its ancestors in the ML family).

We'll spend time with advanced types such as tagged unions and a particularly interesting record type with some neat syntax for its associated operations. We'll close with an overview of subscriptions and their role in interactive programs.

9.1 HELLO ELM

Let's get right to the traditional first program:

```
import Html exposing (text, ul, li)
import List exposing (map, concatMap, filterMap, repeat)

main =
  [1..40] |> concatMap (\c ->
    [1..c-1] |> concatMap (\b ->
      [1..b-1] |> filterMap (\a ->
        (if a*a+b*b==c*c then Just (a,b,c) else Nothing)))
    |> map (toString >> text >> repeat 1 >> li [])
    |> ul []
```

There are no for-loops in Elm (nor statements for that matter), so we generate our right-triangle triples via mapping and filtering. Rather than writing to standard output, we build an HTML unordered list to display in a web browser.

Despite its size, this little program highlights quite a few areas that benefit from some explanation:

- Function application does not require parentheses, so `f x` means $f(x)$. Unlike CoffeeScript, Elm's application is left-associative, so `f x y` means $f(x)(y)$.
- `x |> f` means $f(x)$. Similarly, `x |> f |> g |> h` means $h(g(f(x)))$.
- Anonymous functions employ a backslash and a thin arrow, so `\x -> e` denotes the function with parameter x and body e.
- `g >> f` is the function that applies g then f, i.e., `(g >> f)x` means $f(g(x))$.
- `concatMap` maps a list-producing function over a list, and flattens (concats) the result. This ensures that we pass a complete list of triples, rather than several lists of triples, to the test for inclusion in our output.
- `filterMap` maps an optional-producing function over a list and keeps only the present ("`Just`") values. Elm's optionals are called **maybes**. `If`-expressions always have both `then` and `else` parts, so we had to combine maybes with filtering in order to select the appropriate triples.
- The `ul` and `li` functions produce HTML `<ul>` and `<li>` elements. Functions in the `Html` module take a list of attributes and a list of child elements. For example, the Elm expression `div [id "notes"] [(text "hello"), (em [] [text "world"])]` yields the HTML `<div id="notes">hello<em>world</em></div>`.
- The function call `repeat n x` produces a list of n x's. So the expression `repeat 1` evaluates to a function that creates a one-element list containing its argument.
- Elm automatically renders the value bound to `main`.

Now let's generate permutations. We'll take our input from an HTML text box, so we'll write the application in two parts to separate the UI from the core computation. First we have a **module** that exposes a function returning the anagrams of a string in a list:

```
module Anagrams exposing (anagrams)

import String exposing (toList, fromList)
import List exposing (concatMap, map, foldr)

insertEverywhere : a -> List a -> List (List a)
insertEverywhere x xs =
  case xs of
    [] -> [[x]]
    (y::ys) -> (x::y::ys) :: map ((::)y) (insertEverywhere x ys)

permutations : List a -> List (List a)
permutations =
  foldr (concatMap << insertEverywhere) [[]]

anagrams : String -> List String
anagrams s =
  s |> toList |> permutations |> map fromList
```

This little module defines three functions but exports only the one named on the module declaration line. We've supplied **type annotations** for each of them. Though unnecessary, it's good practice to provide the annotations, as they add useful documentation and serve as a check against getting something wrong. Our `anagrams` function, of type `String -> List String` produces a list of anagrams from a given string. The `permutations` function produces a list of all permutations of a given list. Note the **type variables** in the function type. Elm's `List a` serves the same purpose as Java's `List<T>`, namely defining the type of lists whose elements all have the same type. And `insertEverywhere` produces a list of lists, each with a given element in a different place in a given list, such that all places are covered. Examples should help to clarify things a bit:

`anagrams "dog"` ⇒ `["dog","odg","ogd","dgo","gdo","god"]`

`permutations [1,2,3]` ⇒ `[[1,2,3],[2,1,3],[2,3,1],[1,3,2],[3,1,2],[3,2,1]]`

`insertEverywhere 1 [2,3]` ⇒ `[[1,2,3],[2,1,3],[2,3,1]]`

Our module illustrates a few characteristics of **functional programming**: there are no assignments, no mutable variables (we have only immutable function parameters), and no control flow other than function combination. It also introduces new items from the core library: function composition (`<<`), list prefixing (`::`, pronounced "cons" for historical reasons), and folding from the right (`foldr`). These functions behave as follows:

$(f \texttt{<<} g)\ x \Rightarrow f(g(x))$

$x \texttt{ :: } [a_0, ..., a_n] \Rightarrow [x, a_0, ..., a_n]$

$\texttt{foldr}\ f\ x\ [a_1, a_2, a_3, a_4] \Rightarrow f(a_1, f(a_2, f(a_3, f(a_4, x))))$

We now turn to the main application, which runs in a web browser. It displays a textfield and responds to each change in the box's content by rendering all of the permutations directly on the page.

```
import Anagrams exposing (anagrams)
import List exposing (map)
import Html exposing (div, text, input)
import Html.Attributes exposing (placeholder, value, maxlength)
import Html.Events exposing (onInput, targetValue)
import Html.App as App

type alias Model = String
type Message = ChangeTo String

main =
  App.beginnerProgram { model = "", view = view, update = update }

update (ChangeTo newModel) model =
  newModel

view model =
  div []
    ((input
      [ placeholder "Text to anagram"
      , value model
      , maxlength 6
      , onInput ChangeTo
      ]
      []) :: map (\s -> div [] [text s]) (anagrams model))
```

The entire program may have been longer than you expected. Some of the length stems from having a large number of imports, but this is an indication of emphasizing modularity in the core libraries. Many languages have string and list functions built right in, but we need to explicitly import them in Elm. The HTML `<div>` and `<input>` elements, along with their associated attributes and events, have representations in Elm which we must also import. And the application itself follows the *Elm architecture*, a style of writing code made from:

- A model, representing our application's "state" (here just the string we wish to anagram),
- A view function, which takes in the current model and produces the HTML to render. The HTML may include elements that generate messages. In our case, we have an HTML `<input>` element that generates `oninput` events. We've arranged that these events send the message `ChangeTo` *s*, where *s* is the string in the textfield, to the `update` function.
- An update function, which receives a message and the current model, and produces the new model. In our app, the message tells us exactly what to do to: update the model with the string in the message.

Our traditional third program has been to count words from standard input. To keep things simple for now, we'll hardcode the source text. We'll leave it to you as an exercise to generalize the script with HTML textareas and file uploads.

```
import List exposing (map, foldl, repeat)
import Dict exposing (Dict, insert, get, toList)
import String exposing (toLower)
import Regex exposing (regex, find)
import Maybe exposing (withDefault)
import Html exposing (table, tr, td, text)
import Html.Attributes exposing (style)

message =
  "The fox. It jumped over/under...it did it did the Fox."

countWord : String -> Dict String Int -> Dict String Int
countWord word counts =
  insert word (withDefault 0 (get word counts) + 1) counts

tableCell data =
  td [] [text (toString data)]

main =
  find Regex.All (regex "[A-Za-z']+") message
  |> map (.match >> toLower)
  |> foldl countWord Dict.empty
  |> toList
  |> map (\(word, count) -> tr [] [tableCell word, tableCell count])
  |> table [style [("border", "1px solid black")]]
```

Once again, we've used quite a few of Elm's core modules. The script itself introduces few new linguistic features, but the following demand a little explanation:

- Extracting words uses Elm's regular expression matching, which returns matches in a list of Elm **records**. We'll discuss records later in the chapter.
- We're storing words and their counts in a **dictionary**, which Elm provides in the core `Dict` module. Dictionaries are persistent data structures, which we introduced in the last chapter. Invoking `get key dict` produces a `Maybe` value, and the function `withDefault` (from the `Maybe` module) operates on maybes. This is in contrast to Java, Python, Ruby and many other languages in which the dictionary itself must have special knowledge of missing keys. Java, in particular, has methods on the dictionary class named `getOrDefault` and `putIfAbsent`. Elm dictionaries need no such knowledge: they simply deliver maybes on lookup.
- Elm dictionaries require comparable keys, and are always processed in key order when operated on by functions such as `keys`, `values`, `toList`, and `foldl`.

9.2 THE BASICS

As a nearly pure functional language, Elm is quite small. So small, in fact, we've encountered most of the syntax in our opening example programs. It features five basic types: `Bool`, `Int`, `Float`, `Char`, and `String` and four mechanisms for defining new types:

- **Tuples**. The value `(7,"abc")` has type `(Float, String)`. Tuple types are Elm's product type. They can have any number of constituent types, including zero.[1]
- **Records**. The value `{ x = 3.5, y = 5.88, color = "red" }` has type `{ color : String, x : Float, y : Float }`. Note that the record type *includes* the field names; records are essentially tuples with labeled components.
- **Tagged Unions**. Values of a tagged union type (sometimes called a **discriminated sum type**) contain a tag together with a value from another type—or, just a tag. Elm's unions solve the same problem sum types solve, and can beautifully model enumerations, state machines, and type hierarchies from the popular "object-oriented languages" as we will see momentarily. The requirement for including tags with values allows for code more readable and expressive than your basic sum type.
- **Functions**. The value `\x -> (x,x)` has type `Int -> (Int, Int)`.

Did you expect to see lists? If you're wondering where the lists and dictionaries went, don't worry—they've been built with tagged unions. Let's move straight to examples, beginning with tuples.[2] Our first example also illustrates the use of a **type alias**. An alias does not create a new type; rather, it simply provides a convenient name for a more complex type expression.

```
import ElmTest exposing (suite, test, runSuite, assertEqual)

f : (Int, Int, Int) -> Int
f (a, _, c) =
  3 * c + a

type alias Vector = (Float, Float)

magnitude : Vector -> Float
magnitude v =
  let (x, y) = v in
    sqrt(x * x + y * y)

main =
  runSuite <| suite "Tuples"
    [ test "Value Semantics for =" ((6,8) `assertEqual` (7-1,4*2))
    , test "Pattern match" (f(1,2,3) `assertEqual` 10)
    , test "Destructuring" (magnitude(3,4) `assertEqual` 5.0)
    , test "Tuple return" ((\x -> (x,x))1 `assertEqual` (1,1))
    ]
```

Tuples are accessed via pattern matching, which works both in passing the function argument and in `let` bindings. Note we said *the* function argument. Like its relatives in the ML family, function types have the form $T_1 \to T_2$ for types T_1 and T_2. Instead of thinking about "multiple parameters" and "multiple return types," think about accepting and returning tuple values.

[1] The empty tuple type, `()`, is commonly called the **unit type**.

[2] Elm does not have a built-in assertion library, but you can easily install a third-party package. We've used the popular `ElmTest` package. You will need to install with a command such as `elm-package install elm-community/elm-test`.

But wait! None of the examples in the opening section of this chapter used tuples, though they appeared to take "multiple arguments." They looked quite different, like the example on the right below:

```
plus: (Int, Int) -> Int
plus (x, y) =
  x + y
```

```
plus: Int -> Int -> Int
plus x y =
  x + y
```

The example on the left returns the sum of its two tuple components, but what of the example on the right? What is the type `Int -> Int -> Int`? The function type operator `->` is right-associative, so we actually have `Int -> (Int -> Int)`. The function accepts an integer and returns a function from integers to integers. And although it appears to take two parameters, the definition on the right simply sugars the direct "one parameter" definition that returns a function:

```
-- Sugared definition
plus: Int -> Int -> Int
plus x y =
  x + y
```

```
-- Desugared definition
plus: Int -> (Int -> Int)
plus x =
  \y -> x + y
```

The act of rewriting a function of type `(a,b)->c` into an equivalent of type `a->(b->c)` is called **currying**; the reverse process is called **uncurrying**. Curried functions allow something called **partial application**. For example, although we might normally write

```
plus 3 4
```

we can just as easily write

```
plus 3
```

to obtain the function that adds three to its argument. We are free to apply this function at some later time to obtain essentially a sum of two integers.

Elm provides two forms of sugaring for curried functions. A function of type `a->(b->c)` can be written between its two "arguments" when enclosed in backticks. We saw this in our previous example with the function `assertEqual`: The expression ``x `assertEqual` y`` sugars `assertEqual x y`. Similarly a function defined to appear in the infix position can be used in prefix for by enclosing in parentheses, for example `(+) 5 8` evaluates to 13. This feature was not written to make Clojure programmers happy; rather it gives us a way to use the function as a value in its own right. Study the two functions in the example below:

```
import ElmTest exposing (runSuite, suite, test, assertEqual)

main =
  let (input, expected) = ([10, 20, 30], [15, 25, 35]) in
    runSuite <| suite "Use + in prefix position"
      [ test "Arrow" (List.map (\x->x+5) input `assertEqual` expected)
      , test "Prefix" (List.map ((+)5) input `assertEqual` expected)
      ]
```

To define an infix function, supply a function definition with the name in parentheses (as if it were a prefix function). But note that using functions as infixes introduce questions of precedence and associativity. In Elm, precedences are integers from 0 (lowest) to 9 (highest), and associativities are left, right, or none The common infix functions from the packages `Basics` and `List` (from the Elm core library) follow.

Operator(s)	Prec	Assoc	Description
`<<`	9	R	composition: $(f \ll g)x = f(g(x))$
`>>`	9	L	composition: $(f \gg g)x = g(f(x))$
`^`	8	R	exponentiation
`* / //` `% rem`	7	L	multiply, divide, floor-divide, modulo, remainder
`+ -`	6	L	add, subtract
`:: ++`	5	L	cons, append
`< <= > >=` `= /=`	4	None	less, less or equal, greater, greater or equal, equal, not equal
`&&`	3	R	(short-circuit) logical and
`||`	2	R	(short-circuit) logical or
`<|`	0	R	application: $f \lhd x = f(x)$
`|>`	0	L	application: $x \rhd f = f(x)$

By default, operators you define yourself have precedence 9 and are left-associative, but you can set these values yourself. Here we create two operators that compute $2x + y$, one with a high precedence (8) and one with low precedence (3). We use this operator in an expression with multiplication (precedence level 7) to show there's a difference.

```
import ElmTest exposing (runSuite, suite, test, assertEqual)

(<-*->): Int -> Int -> Int
(<-*->) x y =
  2 * x + y

(>*<): Int -> Int -> Int
(>*<) x y =
  2 * x + y

infix 8 <-*->
infix 2 >*<

main =
  runSuite <| suite "Test own operators with precedence"
    [ test "high" <| (8 * 3 <-*-> 5) `assertEqual` 88   -- 8*(2*3+5)
    , test "low" <| (8 * 3 >*< 5) `assertEqual` 53      -- 2*(8*2)+5
    ]
```

We've chosen to make our new operators non-associative; to set left or right associativity we would use `infixl` or `infixr`, respectively.

Note that the standard operators (addition, multiplication, less-than, etc.) are not wired in to the language but are simply defined as plain Elm functions in the standard library. Indeed, examining the source code for the Elm compiler we see:

```
(/) : Float -> Float -> Float
(/) =
  Native.Basics.floatDiv

infixl 7 /
```

The implementation of floating point division will execute fast since the Elm compiler can access lower-level code through its `Native.Basics` module,[3] but the point is that you won't see basic operators appearing in the syntax definition for the language; they are simply functions.

A final note on precedence: prefix function application *always* binds tighter than infix application. While very simple, the rule does lead to a few gotchas that the programmer coming to Elm from other languages might want to keep in mind:

```
negate 3 ^ 4   parses as   (negate 3) ^ 4
show x * y     parses as   (show x) * y + z
```

You can, of course, use parentheses as expected, or make use of the function application operator `<|` from precedence level 0:

```
negate <| 3 ^ 4   parses as   negate(3 ^ 4)
show <| x * y     parses as   show(x * y + z)
```

Surprisingly, that covers most of the basics. In the remainder of the chapter we will cover two fascinating items that are characteristic of the ML family, a powerful type inference facility and tagged unions; one characteristic of modern ML family languages, extensible records; and Elm's approach to interactivity (subscriptions and commands).

9.3 TYPE INFERENCE

Elm performs static typing: it checks the types of expressions at compile time and refuses to execute programs that may have type errors. It is the second such language in this text; Java was the first.[4] In order to typecheck prior to execution, the compiler must be given information about the intended types of expressions or be smart enough to *infer* types. Java is capable of relatively little **type inference**, requiring type annotations on nearly all variables, method parameters and return values, fields in classes and interfaces, etc. Elm can infer the type of nearly every expression. The three other statically-typed languages[5] in this book, Go, Swift, and Rust, fall somewhere in the middle of this spectrum.

Elm's type inferencing ability, like that of all ML-family languages, has its theoretical roots in the **Hindley-Milner type sytem** [135, 86, 22]. Elm's compiler computes (or deduces), the least general type of any expression not already explicitly annotated. Let's start with a very simple example. Consider the Elm function:

```
import String exposing (repeat)

mystery x s =
  repeat (round(sqrt x)) s
```

[3] Needless to say, most Elm programs will never touch `Native.Basics`, or its cousins such as `Native.Utils`, `Native.String`, and `Native.Time`.

[4] Though technically, you may recall, Java does leave a tiny amount of type checking to runtime in some cases.

[5] Colloquially, we say a *language* is statically-typed if it performs the majority of type checking prior to execution.

None of the identifiers have type annotations, but Elm can infer them. How? Roughly, we proceed as follows:

1. Working from left-to-right, we start by assigning **type variables**: `x` gets type α, `s` gets β, and therefore `mystery` gets $\alpha \to \beta \to \gamma$.
2. We know the type of `repeat` is `Int -> String -> String`, so the type of `round(sqrt x)` must be `Int`.
3. We know the type of `round` is `Float -> Int`, so the type of `sqrt x` must be `Float`.
4. We know the type of `sqrt` is `Float -> Float`, so the type of `x` must be `Float`. Since we had previously assigned the type variable α to `x` we can now **unify** α and `Float`.[6]
5. The second argument of `repeat` (`s`) must be a string, so we unify β with `String`.
6. Because `repeat` produces a string, we unify γ with `String`.
7. Therefore, the type of `mystery` is `Float -> String -> String`.

Let's look at a more interesting case:

```
mystery (x, y) z =
  (y / 2.0, z, [x, x, x])
```

Here we begin by mapping the type variable α to `x`, β to `y`, γ to `z`, and thus $(\alpha, \beta) \to \gamma \to \delta$ to `mystery`. Because we know the type of float division (`/`) we can infer that y has type `Float` and unify β with `Float`. We can't get any new information for `z` or `x`, but we can determine the return type of the function: $(\texttt{Float}, \gamma, \texttt{List}\ \alpha)$. While it is customary to use Greek letters for type variables when discussing type inference, Elm sticks to lowercase letters for its type variables, and reports the type of `mystery` as `(a, Float) -> b -> (Float, b, List a)`.

Generally, type variables are allowed to be unified with any type, but Elm has three[7] special type variables that unify *only* with particular types:

- `number`, which unifies only with `Int` and `Float`, to support functions such as `(+)`, `(-)`, and `(*)`;
- `comparable`, which unifies only with `Int`, `Float`, `Char`, `String`, tuple types containing comparables, and list types containing comparables, to support `(<)`, `(<=)`, `(>)`, `(>=)`, `min`, `max`, and `compare`; and
- `appendable`, which unifies only with `String`, `Text`, and list types, to support `++`.

Examples of these special type variables from Elm's core package include:

```
-- From Basics
(>) : comparable -> comparable -> Bool
abs : number -> number
(++) : appendable -> appendable -> appendable
```

[6]In this case unification appears to be simply instantiation of a type variable with a concrete type; however, in more complex cases, we might discover that two distinct type variables must refer to the same type, and thus need to be "unified" into one variable.

[7]Perhaps strangely, a fourth such variable, `compappend`, arises internally when type checking expressions that must both be comparable and appendable. You may see it "leaked" into your own world when the compiler tells you that it has inferred the type of `f` in `f x y = if (x < y) then (x ++ y) else (y ++ x)` to be `compappend -> compappend -> compappend`; however, the compiler will not respect your use of this variable. You will carry out further explorations in the exercises.

9.5 RECORDS

Just like tagged unions generalize sum types by labeling variants, Elm's **records** generalize product types (tuples) by labeling components. Let's take a look:

```
import ElmTest exposing (runSuite, suite, defaultTest, assertEqual)
import List exposing (map)

boss = {name = "Alice", salary = 200000 }
worker = {name = "Bob", salary = 50000, supervisor = boss}
newWorker = {worker | name = "Carol" }

payrollTax : {a | salary : Float} -> Float
payrollTax {salary} =
  salary * 0.15

main =
  runSuite <| suite "Exploring records" <| map defaultTest
    [ "Alice"  `assertEqual` .name boss
    , boss `assertEqual` worker.supervisor
    , 7500.0 `assertEqual` payrollTax worker
    , boss `assertEqual` newWorker.supervisor
    ]
```

We've created three records: the first of type `{ name : String, salary : number }`, the second with type `{ name : String, salary : number, supervisor : { name : String, salary : number } }`, and the third with the same type as the second. We defined our third record, `newWorker`, to be just like `worker` except with the name `"Carol"` (instead of `"Bob"`). Next we defined a function to compute a 15% payroll tax, illustrating pattern matching for record arguments. Read the type of the function, `{a | salary : Float} -> Float` as "a function that accepts any record that has a field named `salary` of type `Float`, and produces a `Float`." That's pretty flexible. We never introduced any type names at all, never subclassed employees into workers and managers, and yet Elm ensures the entire program is fully type-safe.

You may find Elm's **type aliases** useful in describing records. An alias is *not* a new type, but rather an abbreviation for a type:

```
type alias Person = { name : String, id : Int}
type alias Widget = { name : String, id : Int}

type PrimaryColor = Red | Blue | Green
type alias PixelColor = PrimaryColor

p : Person
p = { name = "Alice", id = 239 }
w : Widget
w = p                -- Disappointing perhaps, but legal
g : PixelColor
g = Green
```

While tagged union definitions (e.g., `PrimaryColor` in the example script), *do* introduce new, distinct, named types, we can only name record types via aliasing. In the lingo of programming languages, record types live in a world of **structural typing** rather than **nominative typing**. Only the record's structure—the names and types of its fields—is used to determine its type. We can, however, use tagged unions with a single variant to make the distinction:[9]

```
type Person
  = Person { name : String, id : Int }
type Widget
  = Widget { name : String, id : Int }

p : Person
p = Person { name = "Alice", id = 239 }
-- w : Widget; w = p -- would be a syntax error now.
```

We can use type aliasing to describe a record type that need have only certain fields:

```
import ElmTest exposing (runSuite, suite, defaultTest, assertEqual)

type alias Positioned a =
  { a | x : Float, y : Float, z : Float }

move : Positioned a -> Float -> Float -> Float -> Positioned a
move object dx dy dz =
  { object | x = object.x + dx, y = object.y + dy, z = object.z + dz}

main =
  let
    robotAtStart = { name = "Mari", x = 3, y = 8, z = -2}
    robotAtEnd = move robotAtStart 7 -2 9
    expectedEnd = { name = "Mari", x = 10, y = 6, z = 7}
  in
    runSuite <|
      defaultTest (expectedEnd `assertEqual` robotAtEnd)
```

We just moved a robot. But we can, in fact, move *anything* that has `x`, `y`, and `z` fields, without the need to include some kind of position field within the object, or derive our object from a position object or class.

9.6 EFFECTS

One of the more useful dimensions on which to characterize and compare programming languages is the extent to which the language's design, and its culture, encourage, discourage, contain, or try to eliminate **side-effects**. Roughly, a side effect occurs when a function changes a program's state in a way other than producing an explicit return value. Side-effects, and side-causes—entities manipulated by a function other than its explicit pa-

[9]This technique also applies when you need to avoid recursive records, which you'll have an opportunity to explore in the end-of-chapter exercises.

rameters [69]—are associated with mutable variables, files, and time sensitive features such as user interaction and animation. Languages we think of as *functional* go to great lengths to discourage and contain side-effects. Clojure allows controlled mutation via vars, refs, and agents, and even lets us drop into Java. Relatives of Elm, such as Standard ML and OCaml, provide references that wrap mutable values.

Some languages, like Haskell, work to contain even I/O effects. Think back to the way operations on persistent data structures never mutate a structure but rather produce a new structure. If we wrap (snapshots of) our files and variable values into a "state" object, then a function may accept a state as an argument and produce a new state as a result. In general, an input state can also contain *event histories* such as key presses, clicks or gestures on an input device, canvas resizings, or the passing of time. Older versions of Elm directly supported these history sequences, calling them **signals**. The signal `Mouse.position`, of type `Signal (Int, Int)`, for example, would have a value such as:

```
(3,4) (7,5) (11,4) (13,3) (19,6)
```

Signals, like all sequences, could be mapped (to perform an operation on each event), filtered (to ignore events of no concern), merged (to handle events from multiple sources), and folded (to accumulate state). To sketch on a canvas in the early days of Elm, we folded a line segment generation function over the mouse position signal![10] Dozens of other signals like `Keyboard.Arrows` and `Window.Dimensions` lived in the standard library. Manipulating asynchronous streams with higher-order functions is known as **functional reactive programming**, or FRP.

Elm has left its FRP heritage behind and has removed signals from its core libraries, replacing them with two kinds of **managed effects**: **commands** and **subscriptions**. A command object requests an action: "fetch geolocation data" or "send this message over a websocket." A subscription expresses interest in an event: "tell me when the mouse moves" or "tell me when an amout of time has passed."

Effects are managed within programs. A program is a record with four components:

1. `init: (model, Cmd msg)` is the initial model, or state, of the application, together with a command to launch on startup.
2. `view: model -> Html msg` produces HTML for a given model. The HTML may generate messages of the type *msg*.
3. `subscriptions: model -> Sub msg` produces, for the given model, a subscription to sources that can produce messages (of type *msg*) that will be routed to the `update` function.
4. `update: msg -> model -> (model, Cmd msg)` produces, given a model and message, the new model (next state) and a command to invoke.

Elm provides functions for creating subscriptions on windows, keyboards, mice, web sockets, geolocation changes, timers, and the browser's rendering refresh cycle. You'll also find functions for generating commands to fire random numbers and perform various asynchronous tasks.

Let's start with a simple example. We'll render an image at the current mouse position: the JavaScript logo when the mouse button is up and the Elm logo when the mouse button is held down. The model is the mouse position together with the name of the image. The view

[10]The folding operation was called neither `foldl`, folding from the left, nor `foldr`, folding from the right, but `foldp`, folding from the past.

function renders the image at the given position. We subscribe to mouse movements and the up/down state of the mouse button. We won't use any commands in this example.

```
import Html exposing (Html, img)
import Html.Attributes exposing (style, src)
import Html.App as App
import Mouse

main =
  App.program
    { init = init
    , view = view
    , update = update
    , subscriptions = subscriptions
    }

type alias Model =
  { x : Int
  , y : Int
  , image : String
  }

type Msg
  = Down
  | Up
  | MoveTo Int Int

init : (Model, Cmd Msg)
init =
  ({x = 0, y = 0, image = "js"}, Cmd.none)

subscriptions : Model -> Sub Msg
subscriptions model =
  Sub.batch
    [ Mouse.moves (\{x, y} -> MoveTo x y)
    , Mouse.downs (\_ -> Down)
    , Mouse.ups (\_ -> Up)
    ]

update: Msg -> Model -> (Model, Cmd Msg)
update msg model =
  case msg of
    Down ->
      ({model | image = "elm"}, Cmd.none)
    Up ->
      ({model | image = "js"}, Cmd.none)
    MoveTo x y ->
      ({model | x = x, y = y}, Cmd.none)
```

```
view: Model -> Html Msg
view model =
  img
    [ style
      [ ("position", "absolute")
      , ("left", toString (model.x+2) ++ "px")
      , ("top", toString (model.y+2) ++ "px")
      ]
    , src (model.image ++ "-logo.png")
    ]
    []
```

Following the convention outlined in the Elm Architecture, we've defined all possible messages in a tagged union type. In our subscriptions function, we turn the interesting system-generated events (mouse down, mouse up, mouse move) into message objects that Elm will route to our update function, where we generate the new model. Our view function renders the model (the correct image at the correct position) when necessary.

We'll close with an illustration of time effects, and throw in a bit of Elm's support for Scalable Vector Graphics, or SVG. Our next script counts down from 10 to 0. It subscribes to the current time via `Time.every second`, which gives us notifications every second and sends a `Tick` message, together with the current epoch time[11] to our update function. We're not interested in the real time, but we do want to decrement a counter. When the counter reaches zero, we'll "stop subscribing" by setting our subscriptions to none.

```
import Html exposing (Html)
import Html.App as Html
import Svg exposing (svg, circle, text', text)
import Svg.Attributes exposing (..)
import Time exposing (Time, second)

main =
  Html.program
    { init = init
    , view = view
    , update = update
    , subscriptions = subscriptions
    }

type alias Model = Int

type Msg
  = Tick Time

init : (Model, Cmd Msg)
init =
  (10, Cmd.none)
```

[11]The number of seconds elapsed since midnight, January 1, 1970 UTC, or 1970-01-01T00:00Z.

```
subscriptions : Model -> Sub Msg
subscriptions model =
  if model <= 0 then
    Sub.none
  else
    Time.every second Tick

update : Msg -> Model -> (Model, Cmd Msg)
update action model =
  case action of
    Tick _ ->
      (model - 1, Cmd.none)

view : Model -> Html Msg
view model =
    svg [ viewBox "0 0 100 100", width "400px" ]
      [ circle [ cx "50", cy "50", r "45", fill "#A88428" ] []
      , text' [x "50", y "50", fontSize "64", textAnchor "middle",
        dominantBaseline "central"] [text <| toString model]
      ]
```

Our examples covered only the basic outline of the Elm architecture with keyboard, mouse, and time subscriptions. You will want to continue your study by looking into animation frames, web sockets, asynchronous HTTP requests, HTML local storage, browser histories, geolocation tracking, and other platform services.

9.7 ELM WRAP UP

In this chapter we briefly looked at Elm. We learned that:

- Elm is a nearly-pure functional programming language from the ML family of languages, with a Haskell-inspired syntax.
- Elm applications generally run in a web browser. Production-quality Elm apps will be organized around the textslElm Architecture, a convention for writing programs centered around models, updates, views, and subscriptions.
- Like Clojure, Elm greatly facilitates functional programming: functions are first class values, data structures are immutable and persistent, and there is no assignment statement.
- The basic types are `Int`, `Bool`, `Float`, `Char`, and `String`. New types are created via tuples, records, and tagged unions.
- Lists and options (known as maybes) are defined as tagged unions.
- Curried functions are the norm.
- You can create your own infix operators in Elm, defining precedence and associativity.
- Unary operators always bind tighter than binary operators.
- Elm type inference is based on the Hindley-Milner type system. Type annotations are practically never required.

- Elm (currently) has no type classes, but has a few special variables that approximate a type class facility.
- Tagged unions and record serve as labeled sum and product types, respectively.
- Elm was once a language that facilitated functional reactive programming. It featured signals, sequences of values generated over time in response to events, such as keyboard or mouse events, or the passage of time. Signals would be mapped, filtered, merged, and especially folded. Folding over signals was the manner in which application state could be accumulated or stepped forward.
- Modern applications subscribe to event sources. The Elm Architecture ensures that subscribed events are turned into customized messages routed to a centralized update function to step the state of the application forward.

To continue your study of Elm beyond the introductory material of this chapter, you may wish to find and research the following:

- **Language features not covered in this chapter**. Tasks, commands, JavaScript interoperation, caching of effect managers.
- **Open source projects using Elm**. Studying, and contributing to, open source projects is an excellent way to improve your proficiency in any language. You may enjoy the following projects written in Elm: Sketch-n-sketch (`https://github.com/ravichugh/sketch-n-sketch`), Gipher (`https://github.com/matthieu-beteille/gipher`), Hop (`https://github.com/sporto/hop`), and Elm WebGL (`https://github.com/elm-community/elm-webgl`).
- **Reference manuals, tutorials, and books**. Elm's home page is `http://elm-lang.org/`, from which prominent links will take you to documentation, a playground, community resources, and the official blog. The portal *Awesome Elm*, at `https://github.com/isRuslan/awesome-elm`, contains an up-to-date curated set of links to articles, examples, videos, tutorials, and many other resources.

EXERCISES

Now it's your turn. Continue exploring Elm with the activities and research questions below, and feel free to branch out on your own.

9.1 Try out several of the examples at `http://elm-lang.org/try`.

9.2 Install Elm on your own machine. Find the REPL. Experiment with various kinds of expressions, including records. Generate a few type errors.

9.3 We've not provided type annotations for `main` in our examples. Find out what types are allowed.

9.4 What is the type of the Elm expression `\f g x y. f(g x)y`?

9.5 What are the types of the operators `<|`, `|>`, `<<` and `>>`?

9.6 Trace out the evaluation of the expression `insertEverywhere 5 [1,2,3]` on pencil and paper. The function `insertEverywhere` is defined in the anagrams module in the opening section of this chapter.

9.7 Suppose you wished to fold the `++` operator over a list of lists. Does it matter if you fold from the left or the right? Why or why not?

9.8 If you are acquainted with HTML and CSS, stylize the anagrams web application from earlier in the chapter with a modern look and feel. This exercise will give you some practice with Elm's `Html` library.

9.9 Research some of the scenarios in which the unit type `()` is used in practice.

9.10 What is the curried version of the function `\(x,(y,z)) -> z - y * x`?

9.11 See if you can redefine the precedence of the `+` operator to have higher precedence than multiplication. Were you able to do so?

9.12 Locate a description of the function known as the Y Combinator. Can this function be defined in Elm? Why or why not? Can it be defined in Clojure? Why or why not?

9.13 Read an article on the Hindley-Milner type system and give an explanation of unification.

9.14 There is a chance that by the time you read this chapter, Elm may have gained type classes. If so, create your own comparable type and demonstrate that values of this type can be sorted. If type classes have not been introduced into Elm, research the online discussion in the community and summarize the arguments for and against their inclusion in the language.

9.15 (Mini-project) Build your own set type backed by a binary search tree defined by a tagged union. Include unit tests.

9.16 Define a tagged union for a user data type, where users can be anonymous "guests" or logged-in users with names and credentials.

9.17 Do a side by side comparison of Java optionals and Elm maybes. Be as extensive as possible.

9.18 In this chapter, we saw a `Shape` type defined as a tagged union of circles and rectangles, with separate area and perimeter functions performing pattern matching to perform the correct computation. Explain how this design illustrates the expression problem. Research approaches to solving the expression problem in Elm.

9.19 Explain why the function `andThen` is needed to take the tail of the tail of a list instead of just using composition (i.e., `tail << tail`).

9.20 Given

```
type Person = Alice | Bob | Chi | Dinh | Emmy | Faye | Guo

supervisor: Person -> Maybe Person
supervisor p =
  case p of
    Alice -> Nothing
    Bob -> Just Alice
    Chi -> Just Alice
    Dinh -> Just Bob
    Emmy -> Just Dinh
    Faye -> Just Emmy
    Guo -> Just Chi
```

Write the function `bossOfBossOfBoss: Person -> Maybe Person` to return the supervisor of the supervisor of the supervisor of its person argument, or `Nothing` if appropriate. Your function should return `Just Alice` for argument `Emmy`, and `Nothing` for `Chi`.

9.21 Define type aliases called `Percentage` (aliasing the `Float` type), and `Employee` aliasing records containing, among other fields, a name field and a salary field. Write a function called `giveRaise` of type `Person -> Percentage -> Person`.

9.22 Determine if Elm's records are value types or reference types? Does this question even make sense in Elm? Why or why not?

9.23 Implement a "recursive record" in Elm. Specifically, create a representation of a type for an employee, containing a name, a date of hire, and a supervisor (who is also an employee).

9.24 Learn about the Haskell language. How does it perform I/O? Does it admit that file manipulation inherently involves side-effects, or is this language able to completely contain these effects within pure functions? To help you in your research, read Kris Jenkins's two-part series on functional programming [69] and the section on Basic Input/Output in the Haskell Language Report [96].

9.25 Our logo dragging example does not begin well. A logo is rendered in the upper-left corner of the display area and jumps to the current mouse position when the mouse is first moved. Rewrite the example so that this does not happen.

9.26 Implement a sketching program. Experiment with Elm's graphics capabilities such as changing the thickness and colors of the lines.

9.27 Create a simple game in which a player navigates an avatar through various obstacles.

9.28 How is Elm's use of subscriptions both like and unlike traditional callback-style interactive programming? How is it both like and unlike the use of promises?

CHAPTER 10

Erlang

Erlang was designed for building fault-tolerant, long-lived applications that handle a large number of concurrent activities, meet real-time constraints, scale across large distributed networks, and are able to be modified while running.

First appeared 1986
Developed at Ericsson
Notable versions R15B (2011) • 17.0 (2014) • 19.0 (2016)
Recognized for Concurrency support
Notable uses Telecom
Tags Concurrent, Scalable
Six words or less For incredibly high-availability applications

Organizations and personnel responsible for service availability like to talk about nines. "Five nines" means up and running 99.999% of the time, down only five minutes per year. Seven nines means down only three seconds per year (or two minutes every forty years). Can a *language* help its users (developers) achieve high availability? Erlang can, and does. It's almost entirely functional. Processes are lightweight and share nothing, communicating only via messages. Processes distribute over machines transparently. And if you need to replace or upgrade software, you do it while the code is running.

Like Java, Erlang arrived with its own virtual machine. And just as hundreds of languages have been targeted to Java's JVM, a number of up-and-coming languages (including Elixir, Joxa, and LFE) are designed to run on Erlang's BEAM.

In this chapter, we'll explore Erlang. You may have heard of its "unusual" syntax but this just adds to the reasons why we'd wish to study it. We'll encounter some new vocabulary as we learn about atoms, references, binaries, and pids, and we'll see pattern matching used in place of `if` statements. Messaging, though, is what makes Erlang Erlang, and we'll see a number of examples. We'll learn, too, about the "let it crash" philosophy.

Erlang is the oldest of the 13 languages on our tour, but new versions arrive regularly. We see it used in modern applications like CouchDB, RabbitMQ, WhatsApp, and Whisper. If you've haven't yet seen a language like it before, chances are you will find a great deal here you can't wait to try out.

10.1 HELLO ERLANG

Here's our traditional first program. We've set it up to run as a command line script, admittedly not the most common use case for Erlang—telecom and messaging networks don't usually write to standard output. Still, the standard distribution comes with a little utility called `escript`, which runs the second through last lines of a file:

```
% First line is ignored when running with escript
main(_) ->
  lists:foreach(fun (C) ->
    lists:foreach(fun (B) ->
      lists:foreach(fun (A) ->
        if
          A*A + B*B == C*C -> io:format("~p, ~p, ~p~n", [A,B,C]);
          true -> ok
        end
      end, lists:seq(1,B))
    end, lists:seq(1,C))
  end, lists:seq(1,40)).
```

This script shouldn't appear too unusual. Requiring a `main` function is not odd (we saw that in Java), nor is the `foreach`-with-callback model of nested loops (we saw that with Ruby). We qualify items from external modules with a colon: `seq` and `foreach` come from the `lists` module and `format` comes from the module `io`. The `if`-expression must produce a value for all possible clauses, even though we ignore the result of the second clause. Formatting differs from the traditional `printf`: `~p` formats values in their standard form, and `~n` produces a new line. Commas separate small things (e.g., parameters in a function call), semicolons separate medium things (clauses in a case expression) and periods end big things (such as top-level declarations). Variables, interestingly, *must* begin with a capital letter or underscore; we'll see why later.

We can get away with a shorter script using **list comprehensions**, though this approach needlessly constructs a list of return values from `io:format`:

```
% First line is ignored when running with escript
main(_) ->
  [io:format("~p, ~p, ~p~n", [A,B,C]) ||
    C <- lists:seq(1,40),
    B <- lists:seq(1,C),
    A <- lists:seq(1,B),
    A*A + B*B == C*C].
```

Note that Erlang's list comprehensions differ only slightly from Python's:

```
% Erlang
[exp || i <- sequence1, j <- sequence2, condition]
```

```
# Python
[exp for i in sequence1 for j in sequence2 if condition]
```

Now let's move to anagrams. Erlang's lists and strings aren't mutable, so we'll bypass Heap's Algorithm, and go with a simple, but inefficient, recursive solution:

```
% First line is ignored when running with escript
main([S]) ->
  lists:foreach(fun(X) -> io:format("~s~n", [X]) end, permutations(S));
main(_) ->
  io:format("Exactly one argument is required\n").

permutations([]) -> [""];
permutations(S) -> [[H|T] || H<-S, T<-permutations(S--[H])].
```

Function definitions use pattern matching for arguments. Running the script with a single argument sends a one-element list of strings to `main` and generates and prints the desired permutations. Anything other than a one-element list prints an error message. Semicolons separate the clauses, while a period ends the entire function definition.

If the `permutations` function looks a little cryptic, note that it simply expresses the basic well-known recursive definition, which comes from the following observation. The permutations of the string `"rat"` consists of those strings

- starting with the letter `r` followed by all permutations of `"at"`,
- starting with the letter `a` followed by all permutations of `"rt"`, and
- starting with the letter `t` followed by all permutations of `"ra"`.

Erlang strings are simply lists of characters, and lists with first element `H` and remaining (list of) elements `T` are written `[H|T]`. The `--` operator does list subtraction. You can thus read the second clause of our permutations definition as:

```
The list of a strings with first letter H and rest of letters T
  for each H from: the original string
    for each T from: the permutations of the original string minus H
```

Now for the word counting script, which we'll present in the functional style, with no mutation:

```
% First line is ignored when running with escript
main(_) ->
  print_counts(process_lines(orddict:new())).

process_lines(Counts) ->
  case io:get_line("") of
    eof -> Counts;
    Line ->
      Options = [global,{capture,first,list}],
      Words = case re:run(string:to_lower(Line), "[a-z']+", Options) of
        {match, Captures} -> lists:map(fun erlang:hd/1, Captures);
        nomatch -> []
      end,
      process_lines(lists:foldl(
        fun (W,D) -> orddict:update_counter(W,1,D) end, Counts, Words))
  end.

print_counts(Counts) ->
  orddict:map(fun (K,V) -> io:format("~s ~p~n", [K,V]) end, Counts).
```

This script introduces quite a bit of Erlang! We'll assume the first and third functions (`main` and `print_counts`) aren't too opaque—you should be comfortable with `map` and anonymous functions from ample previous chapters. The module `orddict` supports **ordered dictionaries**—dictionaries that produce their pairs in key-order during iteration.

The line processing function shows us quite a few aspects of Erlang, among them:

- **No loops**. Erlang does not have `for` or `while` loops. So `process_lines` first reads a single line, if available. If no lines remain to be read, the function returns; otherwise the function processes the line and then calls itself to read the rest of the file.
- **Immutable data**. Instead of continually growing a single word count dictionary via destructive updates, we employ the `update_counter` function to produce a new dictionary like the old one but with the value of a single key incremented. The reducer `foldl`, which we saw in Elm, provides the functional means to count multiple words.
- **Matching**. Our script calls `io:get_line` to read a file and `re:run` to find words in a string. What happens if no lines are found, or if there are no matches? Rather than throwing exceptions, Erlang varies the return values to indicate success or failure. If a file has been fully read, `get_line` returns the **atom** `eof`; if it reads a line, it simply returns that line. If `re.match` (when, as above, using the options asking for all matches to be returned in a list) finds no matches, it returns the atom `nomatch`. It returns a positive result in the **tuple** `{match,`*matches*`}`.[1] In each case, we handle the variance of return values not with an `if` statement, but rather by pattern matching on the resulting terms.

Our brief introduction didn't venture into concurrency, so stay tuned.

10.2 THE BASICS

Erlang data values, or **terms**, are the following:

- **Atoms**, which, like Ruby symbols and Clojure keywords, simply stand for themselves. They should appear in single quotes unless beginning with a lowercase letter and containing only letters, numbers, `_`, and `@`. Examples: `success`, `last_name`, `'last name'`.
- **Integers**, such as `18`, `-722`, `16#FFC8` (hex), `2#11011` (binary), and `34#D9NBW1` (base-34). The language allows any base between 2 and 36, inclusive. Erlang does not have a character type, so character literals evaluate to their code points: `$c` is just 99, and `$\t` (the horizontal tab character) is just 9.
- **Floats**, such as `8.83e-20`, and `-3.14`. The base selection syntax (*base*`#`*value*) does not apply to floats, only integers.
- **Binaries**, representing chunks of raw memory.
- **References**, (almost) unique values generated at runtime by calling `make_ref`. References support only a single operation: an equality test.
- **Pids**, process identifiers.
- **Ports**, values used by Erlang to communicate to the outside world.

[1]The form of the regular expression matches vary depending on the options; our script produces a list of lists of words, which we must extract via mapping the `hd` (or "head") function over the list of matches.

- **Funs**, Erlang's functions.
- **Lists**, which are sequences of terms, such as `[sparky,spot,spike]`. We can also write the list with head H and tail T as `[H|T]`. Interestingly, Erlang has no string type: the expression `"café"` is exactly the list (of code points) `[99,97,102,233]`.
- **Tuples**, which collect other terms. Example: `{dog,"Nika",'SHEP',13}`.
- **Maps**, which associate keys with values. Example: `#{name => "Nika", breed => 'SHEP', age => 13}`.

Types are not themselves terms, but several built-in functions test whether an expression is of a type. Let's explore. Our trivial "assertion mechanism" in this chapter will attempt to match an expression with its expected value (a hack, perhaps, but very unobtrusive):

```
% First line is ignored when running with escript
main(_) ->
  true = is_atom(ten),
  true = is_integer($a),
  true = is_float(-3.55e-8),
  true = is_function(fun (X) -> X*X end),
  true = is_reference(make_ref()),
  true = is_tuple({dog, "Nika", 5, 'G-SHEP'}),
  true = is_list("a string"),
  true = is_map(#{name => "Nika", age => 5, breed => 'G-SHEP'}).
```

We'll explore binaries, references, pids and ports in subsequent sections. And in case you are wondering, there's no boolean type: `true` and `false` are just atoms; however, the built-in function `is_boolean` recognizes them. Speaking of built-in functions, or BIFs, Erlang provides a few dozen, including a few shown in the following script:

```
% First line is ignored when running with escript
main(_) ->
  17 = abs(-17),
  dog = element(2, {cat, dog, rat, bat}),
  78.0 = float(78),
  "22.1300" = float_to_list(22.13, [{decimals, 4}]),
  Dogs = [rex, sparky, spike],
  rex = hd(Dogs),
  [sparky, spike] = tl(Dogs),
  3 = length(Dogs),
  98 = list_to_integer("98"),
  22 = max(8, 22),
  {10,20} = setelement(2, {10,30}, 20).
```

We see that function names are atoms. Not any string of characters can form an atom, however; the following words are reserved and cannot be used as atoms: `after`, `and`, `andalso`, `band`, `begin`, `bnot`, `bor`, `bsl`, `bsr`, `bxor`, `case`, `catch`, `cond`, `div`, `end`, `fun`, `if`, `let`, `not`, `of`, `or`, `orelse`, `receive`, `rem`, `try`, `when`, and `xor`.

Erlang includes thirty-some operators, arranged into approximately a dozen precedence levels. From highest to lowest precedence we have:

Operator(s)	Assoc.	Description
`:`		Selection from module
`#`		*used in record expressions—not covered in this text*
`+ -` `bnot not`	L	unary plus, unary negation, bit complement, logical not
`* / div rem` `band and`	L	multiplication, division, integer division, integer remainder, bitwise and, logical and
`+ - bor bxor` `bsl bsr or xor`	L	addition, subtraction, bitwise or, bitwise xor, bit shift left, bit shift right, logical or, logical xor
`++ --`	R	list addition, list subtraction
`== /= =< <` `> >= =:= =/=`	L	equal, not equal, less, less or equal, greater, greater or equal, exactly equal, exactly not equal
`andalso`	L	short-circuit and
`orelse`	L	short-circuit or
`= !`	R	pattern match, message send
`catch`		catch

Like Clojure and Elm in the two previous chapters, Erlang emphasizes functional programming over object-oriented programming. Lists, tuples, and maps are immutable. Variables represent values not by having values assigned to them, but by being bound once, and only once. There are no loops; instead, programmers use tail recursion frequently:

```
factorial(N) -> factorial(N, 1).
factorial(0, Acc) -> Acc;
factorial(N, Acc) -> factorial(N-1, Acc*N).
```

Erlang isn't a **pure functional language**;[2] it can generate side effects through, among other things, throwing **exceptions**. Exceptions cause a process to **crash**, (exit with an error code), though we can **catch** them if our expression is part of a catch-expression:

```
% First line is ignored when running with escript
main(_) ->
  10 = (catch 7 + 3),
  {'EXIT', {badarith, _}} = (catch 10 * dog),
  {'EXIT', {badarg, _}} = (catch dog ! 100),
  {'EXIT', {if_clause, _}} = (catch if false -> ok end),
  {'EXIT', {customerror, _}} = (catch error(customerror)),
  {'EXIT', 205} = (catch exit(205)),
  205 = (catch throw(205)).
```

If no errors occur, the expression `catch` *e* produces the value of *e*. Otherwise it produces some error-indicating expression. Runtime errors occur for many reasons, including badly-typed operands, calling a function with the wrong number of arguments, match failures, timeouts, and calling the functions `error`, `exit` or `throw`.

Like nearly every other language, Erlang features modules for program decomposition. Modules define entities, and selectively export the ones needed by its clients. Here's an Erlang version of the vectors module we saw in a few previous chapters:

[2]A pure functional language is one that is free of side-effects.

```
-module(vectors).
-export([create/2,magnitude/1,add/2,mul/2,as_string/1]).

create(I, J) ->
  {vec, I, J}.

magnitude({vec, I, J}) ->
  math:sqrt(I*I + J*J).

add({vec, I1, J1}, {vec, I2, J2}) ->
  {vec, I1+I2, J1+J2}.

mul({vec, I1, J1}, {vec, I2, J2}) ->
  I1*I2 + J1*J2;
mul(K, {vec, I, J}) when is_number(K) ->
  {vec, K*I, K*J}.

as_string({vec, I, J}) ->
  lists:flatten(io_lib:format("<~w,~w>", [I, J])).
```

Modules consist of **attributes** and **function declarations**. The first attribute, `-module`, is required. It provides the module a name, which must agree with the file name containing the module. The second attribute, `-export`, lists the functions available to clients of the module; any functions not exported are thus private to the module. Functions in the export list are given with their arity (number of parameters), as Erlang differentiates functions with the same name but different arities.

After compiling, we use our module just like any of the pre-existing modules:

```
% First line is ignored when running with escript
main(_) ->
  U = vectors:create(3, 4),
  V = vectors:create(-5, 10),
  {vec, 3, 4} = U,
  {vec, -5, 10} = V,
  {vec, 15, 20} = vectors:mul(5, U),
  25 = vectors:mul(U,V),
  5.0 = vectors:magnitude(U),
  "<-2,14>" = vectors:as_string(vectors:add(U,V)).
```

We've represented a vector as a tuple whose first element was an atom indicating its "type." While alternative representations exist, this common and idiomatic technique leads us nicely into the world of pattern matching, a topic to which we now turn.

10.3 MATCHING

Unlike assignment statements common in many languages, Erlang variables become bound through **pattern matching**. Once bound, they cannot be rebound. We can see this in the REPL:

```
$ erl
1> X = 10.
10
2> X = 20.
** exception error: no match of right hand side value 20
3> X = 10.
10
```

Our first entry matches the unbound variable X to 10, binding X to 10. The second attempts to match 10 (the value to which X is already bound) with 20, which fails. Our third match succeeds, as it matches 10 with 10.

Pattern matching in Erlang can bind multiple variables at a time:

```
% First line is ignored when running with escript
main(_) ->
    S = {student, {name, "Alice"}, {scores, [93,100,81,89]}},

    % Pattern match to bind two variables
    {_, {name, Name}, {scores, [_|[Midterm|_]]}} = S,

    % Asserts
    100 = Midterm,
    "Alice" = Name.
```

Here we've used the special variable `_`, which doesn't bind to anything, and therefore matches everything. It's the perfect placeholder for positions within a pattern that are of no concern at the moment.

Matching occurs not only with the match operator (=) but in function definitions and case expressions. These structures have multiple clauses, separated by semicolons. Erlang performs the matching in the order of the clauses, taking action as soon as it finds a match:

```
% Pattern matching in function clauses
area({circle, R}) -> math:pi() * R * R;
area({rectangle, W, L}) -> W * L.

% Pattern matching in case expressions
roll() ->
    case {rand:uniform(6), rand:uniform(6)} of
        {D, D} -> {"You rolled doubles", D+D};
        {X1, Y1} -> {"Uninteresting roll", X1+Y1}
    end.

main(_) ->
    40 = area({rectangle, 8, 5}),
    {_, _} = roll().
```

Sometimes we need to write code for which matching would be inconvenient. Determining whether a value is between 2000 and 32000 would appear to need 30,000 clauses. And we can't directly express a match that succeeds if, say, our term was of a particular type, so Erlang allows **guards**. We'll illustrate with an example:

```
%% Returns whether X is (1) Y or (2) is in the list Y
at (X, Y) when is_list(Y) -> lists:member(X, Y);
at (X, Y) -> X =:= Y.

grade(Score) when Score >= 90 -> 'A';
grade(Score) when Score >= 80 -> 'B';
grade(Score) when Score >= 70 -> 'C';
grade(Score) when Score >= 60 -> 'D';
grade(_) -> 'F'.

main(_) ->
    ['C','A','F','B','D'] = [grade(S) || S <- [71,99,7,80,60]],
    true = at(dog, dog),
    true = at(rat, [dog, rat, cat]),
    false = at(bat, [dog, rat, cat]).
```

Not every boolean-producing expression is allowed in a guard; guards must be free of side-effects. The *Erlang User's Guide* [31] has the complete list of allowable expressions.

Erlang has an `if`-expression for those times that you don't need pattern matching but you do need guards:

```
% First line is ignored when running with escript
main(_) ->
    X = 38,

    Temperature1 = case X of
        _ when X < 0 -> cold;
        _ when X < 20 -> cool;
        _ when X < 45 -> fine;
        _ -> hot
    end,

    Temperature2 = if
        X < 0 -> cold;
        X < 20 -> cool;
        X < 45 -> fine;
        true -> hot
    end,

    % Test
    [fine, fine] = [Temperature1, Temperature2].
```

10.4 MESSAGING

Erlang was originally created to solve problems in telephony, so it was designed from the start for concurrency. Telephony applications demand massive scale and high reliability. Threads, which share memory, and coroutines, which cooperate to share a single processor's time, may not meet these demands as well as Erlang's **processes**. Processes have their own private memory and cooperate by **message passing**. Processes send messages asynchronously to (the **mailboxes** of) other processes; there are no locks. Because processes share no memory, they can crash without leaving the system in an inconsistent state; because there are no locks, no process can hang indefinitely waiting for a lock to be released by a crashed process. Doing without locks enabled the designers of Erlang, and its virtual machine, BEAM, to implement extremely lightweight processes, easy to startup and shutdown, allowing tens of thousands of processes—or more—to run concurrently.

Our first look at concurrent Erlang features just two processes. The main process spawns a new process then immediately waits for a message. The spawned process eventually sends the string `"Hello"` to the main process. The built-in `self()` returns the **process identifier**, or **pid** of the currently executing process. The main process captures its own pid into a variable, which the spawned process uses to send the message:

```
% First line is ignored when running with escript
main(_) ->
  Main = self(),
  spawn(fun() -> Main ! "Hello" end),
  receive
    S -> io:format("~s\n", [S])
  end.
```

This little script suffices to illustrate four basic elements of Erlang's process design:

- As soon as p spawns q, q runs concurrently with the rest of p.
- When p sends message x to q ($q\,!\,x$), p simply drops the message in q's mailbox *without* waiting for q to receive it.
- Evaluating a `receive` expression blocks until a matching message is in the mailbox.
- An Erlang application terminates as soon as the main process terminates, even if hundreds of other processes are active; therefore, it is quite common to see the main process end with a `receive` expression waiting for some kind of a "done" signal.

Our next script illustrates this last item. It starts 1,000 processes—999 to each check whether a specific number in the range 2..1000 is prime and one responsible for printing to standard output:[3]

```
% First line is ignored when running with escript
main(_) ->
  Max = 1000,
  Printer = spawn(printer, print_server, [self()]),
  lists:foreach(
    fun (N) ->
      spawn(prime_checker, is_prime, [N, Printer])
    end,
    lists:seq(2, Max)),
  wait(Max-1).

wait(0) -> io:format("~n");
wait(N) -> receive _ -> wait(N-1) end.
```

[3]Prime finding is a fun exploration of concurrency, but in practice, more appropriate sequential solutions exist for this task. This particular solution is especially naive and inefficient.

The built-in function `spawn`(*module*, *function*, *args*) creates a process to run *module*:*function*(*args*) asynchronously, immediately returning the pid of this new process. We pass to each of the prime checking processes, at construction, the pid of the printer process. Each prime checker sends the printer the number to print if prime, and `false` if composite. We construct the printer process with the pid of the main process itself, so the printer can send a message to main after handling each value. `main` waits for 999 messages from the printer before finishing.[4]

The prime checkers are interesting in their own right, as we have an opportunity to illustrate recursive closures:[5]

```
-module(prime_checker).
-export([is_prime/2]).

is_prime(N, Observer) ->
  (fun Check(D) ->
    if
      D * D > N ->              % No more divisors
        Observer ! N;
      N rem D == 0 ->           % Composite
        Observer ! false;
      true ->                   % Keep looking
        Check(D+1)
    end
  end)(2).
```

The prime checking function illustrates the typical implementation of "loops" with tail recursion. We check possible divisors from $2...\sqrt{N}$, notifying an observing process of our value (if prime) or an indication of non-primeness (`false`). Our printing process repeatedly writes integers to standard output, notifying its observer whether or not it printed a value.

```
-module(printer).
-export([print_server/1]).

print_server(Observer) ->
  receive
    N when is_integer(N) ->
      io:format("~p ", [N]),
      Observer ! true;
    _ ->
      Observer ! false
  end,
  print_server(Observer).
```

Don't rely on the order in which processes will finish; one run of this script began: 23 179 181 191 193 29 197 199 31 71 73 211 37 223 227 229 233 239 241 43 251 61 257 263 269 67 271 277 281 283 79 101 103 83 293 89 307 311 313 317 97 107 109 331 3 5 337 7 11 13 17 19 113 2 ... and other runs were just as scrambled.

[4]We're not very picky about *which* messages are being sent; cleaning this code is a good exercise for the reader, as is research additional, more idiomatic ways to wait for a group of processes to finish!

[5]Because the closed-over inner function `Check` is recursive, we had to give it a name.

Erlang evaluates the `receive` expression as follows. It examines the first message in the mailbox (messages are queued in the order they arrive), and attempts to match the message against each receive pattern, in order. Just as with case expressions and functions, receive patterns may also have guards. If a matching pattern with a true guard is found,[6] the corresponding body is evaluated to produce the result of the receive expression. If no match is found, the second message in the mailbox is considered (and so on through the entire mailbox). If no message applies, the receive expression blocks until a good message arrives—or until a timeout expires, if one is provided (as we shall soon see).

Let's revisit our cooks and customers simulation from the Java chapter, with a couple modifications for variety. Three cooks serve twelve customers. Each customer eats 10 times and then leaves. Customers place an order in an order pool; cooks fetch orders from the pool. Cooks cook the orders and deliver the prepared meals directly to the customer's table. If customers do not get timely service, they go shopping and return later, where they will find, and eat, cold food.

In our simulation, the order pool is not a shared, global data structure but is itself a process. A customer places an order by sending to the pool the message `{place, {`*pid*`, `*food*`}}`, where *pid* is the process id of the customer. Processes frequently send their own pid so they can be called back. A cook who is ready to work sends to the pool the message `{ready, `*pid*`}`; if any orders have been placed, the pool sends the cook the message `{prepare, {`*customerpid*`, `*food*`}}`, instructing the cook to prepare a meal for the given customer. After preparing the meal, the cook messages the customer with `{serve, `*meal*`}`. Figure 10.1 describes the scenario in picture form.

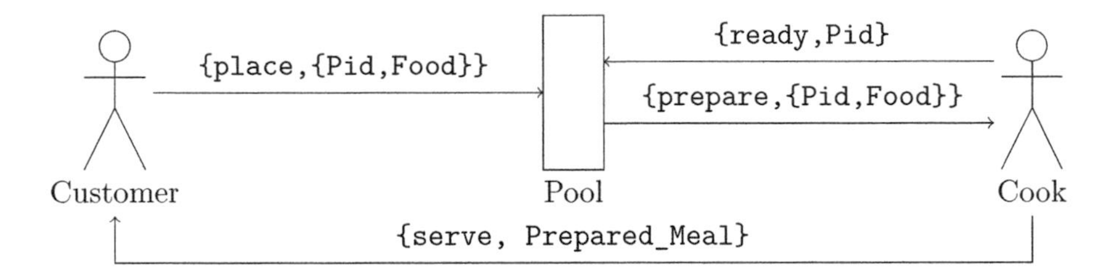

Figure 10.1 Message passing in the Erlang restaurant simulation

Our script creates the order pool and passes its pid to each of the customer and cook processes upon their creation. After each of the processes are spawned, the script sleeps for three minutes, simulating the amount of time the restaurant stays in business. Instead of waiting for completion messages from our actors, our simulation uses the fact that when the main process stops in an `escript` application, the entire process stops. No business lasts forever.

```
% First line is ignored when running with escript
main(_) ->
  Orders = spawn(order_pool, run, [[]]),
  [spawn(cook, run, [Name, Orders]) || Name <- ["Aly","Rob","Chi"]],
  [spawn(customer, run, [Name, Orders]) || Name <- ["Xia","Ann","Bo",
    "Rai","Ali","Kat","Eshe","Sen","Ami","Lina","Ara","Zuri"]],
  timer:sleep(180000).
```

[6]Clauses without guards work as if they have a true guard.

Now let's look at the order pool. The pool can accept an order from a customer at any time, and "adds" the order to the pool. Erlang is a functional language, so "adding" means recursing with a *new* pool just like the original with the new order appended. The pool process will also accept requests from cooks to take orders, but only when orders exist, so we use a guard. If an order is ready, the pool messages the cook with the food to prepare, "dequeueing" the oldest order.

```
-module(order_pool).
-export([run/1]).

run(Orders) ->
  receive
    {place, Order} ->
      run(lists:append(Orders, [Order]));
    {ready, Cook} when length(Orders) > 0 ->
      Cook ! {prepare, hd(Orders)},
      run(tl(Orders))
  end.
```

Cooks repeatedly message the order pool when ready to cook then immediately wait to be messaged with the order. The cook passes its own pid via the built-in function `self`. The cook receives the order as a tuple containing both the food to cook and the pid of the customer who placed the order. After cooking, the cook sends the prepared meal directly to the customer and loops back to being ready to prepare the next order.

```
-module(cook).
-export([run/2]).

run(Name, Orders) ->
  Orders ! {ready, self()},
  receive
    {prepare, {Customer, Order}} ->
      Meal = cook(Name, Order),
      Customer ! {serve, Meal},
      run(Name, Orders)
  end.

cook(Name, Food) ->
  io:format("~s is cooking ~s~n", [Name, Food]),
  timer:sleep(rand:uniform(5000)),
  Food ++ " cooked by " ++ Name.
```

Customers eat ten times and then go home. For brevity, our customers only order soup. They place their order, then wait up to 7 seconds to be messaged by a cook with the prepared meal. The appearance of the `after` portion of the receive expression makes this a **timed wait**: if no message is received within the timeout duration, the `receive` expression executes the body of the `after` clause and then terminates. Whether the customer gets the meal or gives up, the process loops back to waiting for the next meal, ordering again, or cease patronizing the restaurant after ten meals have been eaten.

```
-module(customer).
-export([run/2]).

run(Name, Orders) ->
  Orders ! {place, {self(), "soup"}},
  eat(Name, Orders, 10).

eat(Name, _, 0) ->
  io:format("~s going home~n", [Name]);
eat(Name, Orders, Times) ->
  receive
    {serve, Meal} ->
      io:format("~s eating ~p~n", [Name, Meal]),
      timer:sleep(rand:uniform(5000)),
      Orders ! {place, {self(), "soup"}},
      eat(Name, Orders, Times - 1)
  after 7000 ->
    io:format("~s waiting too long, going shopping~n", [Name]),
    timer:sleep(rand:uniform(5000)),
    eat(Name, Orders, Times)
  end.
```

Our examples so far have highlighted three interesting aspects of the concurrent Erlang philosophy:

- Processes manage state between calls by passing persistent data structures to recursive invocations of the function containing `receive` expressions.[7]
- Process communication is data-driven rather than procedure driven. Our cooks do not, in our example, call a *serve* method on a customer. Instead, the cook places a message in the customer's mailbox with the "action" as one of the fields in the message. The only *verbs* used in communication are *send* and *receive*. Any operations required by the processes (such as *serve*, *place order*, and *cook*) are determined by examining the message internals and pattern matching. However, in production-quality applications, you may wish to create domain-specific functions in order to keep the processes from having to know the details of messages. Doing so allows changes to message formats without breaking the high-level interaction between the actors.
- We've not made use of any process pools; we just create processes as needed and let Erlang worry about how many processes are too many processes and how they are to be scheduled.

There's one more piece of Erlang's approach to concurrency we'll cover in our brief tour of the language: error handling. A process may (seemingly at any time) **crash**—that is, terminate by generating a run time error. We don't want an error to bring down the rest of the system, so errors ideally should be trapped. But how? Must the programmer wrap all process bodies in catch expressions, checking every last expression for possible failure? That adds a lot of complexity to the code, and complexity is at odds with high reliability. Erlang, in fact, suggests that our processes focus only on what the process is supposed to

[7]In the interest of full disclosure, (1) processes have their own **process dictionary** for local storage supporting `get`, `put`, and `erase` operations, and (2) Erlang itself comes with a storage system called ETS. We're not covering either in this chapter, however.

do, and *not* be littered with error-checking `if` or `case` or `catch` expressions. We are told to "Let it crash."

The "Let it crash" philosophy says that error detection and recovery take place *outside* the process, after a crash. In Erlang, we can **link** processes together; when one process in a group of linked processes crashes, it sends an **exit signal** to the other processes. By default this will cause the linked processes to exit as well, but a process may declare its intent to **trap** exit signals. In this case the monitoring process will get the **exit reason** from the crashed process, and may then take corrective action, including possibly restarting the process! Here are the possible exit reasons:

- If the process runs all its code, the exit reason is the atom `normal`. This is not a crash.
- If an error occurs, the exit reason is a tuple {*r*,*s*}. *r* is either (1) an atom such as `badarg`, `badarith`, `function_clause`, `if_clause`, `undef`, `noproc`, or (2) a tuple such as `{badmatch, V}` or `{case_clause, V}` (where *V* is the value that did not match). *s* is the **stack trace**[8] from the beginning of the process to the exit location.
- The process can terminate itself by calling `exit(`*r*`)`, where *r* is any term. The exit reason is the term *r*.
- The process can terminate itself by calling `erlang:error(`*r*`)`, with optional additional arguments. The exit reason is {*r*,*s*} where *s* is the stack trace.

Let's go over a classic example of processing monitoring and restarting after a crash. We have a number of minesweepers (Erlang processes) that work hard trying to clear minefields but occasionally fail—and blow up. Each minesweeper is launched by a process that traps exit signals and establishes a link to the minesweeper process. When the sweeper dies, the launcher receives an exit message in its mailbox, together with the exit reason. The launcher will announce the sweeper's demise and then launch a new sweeper. Here's the code:

```
% First line ignored when running with escript
sweep(42) ->
  io:format("BAM~n"),
  exit(destroyed);
sweep(_) ->
  io:format("."),
  timer:sleep(10),
  sweep(rand:uniform(100)).

launch() ->
  process_flag(trap_exit, true),
  receive
    start ->
      register(sweeper, spawn_link(fun () -> sweep(0) end));
    {'EXIT', From, Reason} ->
      io:format("Sweeper ~p ~p, starting new one~n", [From, Reason]),
      self() ! start
  end,
  launch().
```

[8] A stack trace, or traceback, is a list of representations of the active frames in the current process, useful in debugging.

```
main(_) ->
  register(launcher, spawn(fun () -> launch() end)),
  launcher ! start,
  timer:sleep(5000),
  io:format("~nSimulation over~n").
```

And here's a sample run:

```
$ escript sweeper.erl
..........................BAM
Sweeper <0.33.0> destroyed, starting new one
..............................BAM
Sweeper <0.34.0> destroyed, starting new one
...........................................................................
...........................................................................
...................................................BAM
Sweeper <0.35.0> destroyed, starting new one
.....................................................BAM
Sweeper <0.36.0> destroyed, starting new one
..................................................BAM
Sweeper <0.37.0> destroyed, starting new one
.......
Simulation over
```

We've chosen to use the convenience function `spawn_link` rather than invoking `spawn` and `link` separately. In addition, we've introduced process registration. Rather than capturing and passing around pids, we may register a process with an atom, and use the atom to refer to the process. It's not at all pure functional programming, but Erlang is not beholden to any rules prohibiting storing state in this fashion.

We've only scratched the surface of Erlang in this chapter. As you explore further, make sure to look into both (1) the mechanism for distribution across computing nodes, and (2) the popular OTP framework. In this section, we've focused entirely upon the language mechanisms underlying message passing, but you will necessarily turn to the OTP framework for higher-level functionality used in professional Erlang development.

10.5 ERLANG WRAP UP

In this chapter we explored Erlang. We learned that:

- Erlang is functional programming language designed for highly-reliable, scalable, concurrent applications. Everything is immutable, there are no looping statements, and there are no classes or prototypes.
- Erlang applications are comprised of modules to which the programmer assigns attributes. It is also possible to write Erlang scripts.
- The language is dynamically typed. Types are not first-class values, though built-in functions exist to check that an expression has a particular type.
- Erlang data values are called terms. Terms can be atoms, integers, floats, binaries, references, pids, ports, funs, lists, tuples, and maps. Strings, perhaps surprisingly, are lists of integers.

- Variables must begin with a capital letter, and can be bound to a value only once. There is no assignment; variables are bound during matching.
- In addition to expressions involving the match operator (`=`), matching occurs in function definitions and case expressions.
- Pattern matches can have guards. The `if`-expression uses guards alone.
- Concurrent programming in Erlang is built upon lightweight processes that share nothing and communicate only via message passing. There are no explicit locks. Each process has a unique identifier, known as its `pid`. A message can be any Erlang term.
- A processes sends messages to the mailbox of a receiving process. The receiving process consumes messages from its mailbox according to a specified algorithm which tries to match messages in the order that they were placed in the mailbox. Match attempts occur in the order they appear in the code.
- A process reads messages via a `receive` expression. If no messages can be read, either because the mailbox is empty or no messages match the patterns in the receive clause, or all guards on valid matches are false, the process normally blocks. However, the receive may include a timeout in milliseconds, including zero, to avoid waiting indefinitely.
- Processes can evaluate the built-in function `self` to obtain their own pid, which they frequently send inside a message so the recipient can call the original process back.
- Communication is always asynchronous, but a process can immediately follow a send with a `receive` to simulate a synchronous exchange.
- The Erlang virtual machine, the BEAM, takes care of scheduling.
- Processes can be set up to monitor other processes, so that if a process crashes (exits with an error), the monitor can clean up and restart it if necessary. Monitoring processes keep the regular processes clean and free of extensive error checking and recovery code, isolating these concerns into the monitors.

To continue your study of Erlang beyond the introductory material of this chapter, you may wish to find and research the following:

- **Language features not covered in this chapter**. We skipped coverage of most of the module attributes, including macros, and covered very little of the standard library. We did not cover bit packing, try expressions, type and function specifications, source file encodings, ports, and port drivers. We also omitted records, as they have been replaced by maps.[9] When covering concurrency, we looked only at multiprocessing on a single machine, leaving you to explore how to create, and communicate with, processes on other machines.
- **Open source projects using Erlang**. Studying, and contributing to, open source projects is an excellent way to improve your proficiency in any language. You may enjoy the following projects written in Erlang: OTP (`https://github.com/erlang/otp`), RabbitMQ (`https://github.com/rabbitmq/rabbitmq-server`), Cowboy (`https://github.com/ninenines/cowboy`), and Disco (`https://github.com/discoproject/disco`).
- **Reference manuals, tutorials, books—and movies**. Erlang's home page is

[9]Erlang's creator Joe Armstrong wrote: "Records are dead—long live maps!" [8]

`https://www.erlang.org/`. The docs pages, `https://www.erlang.org/docs` and `http://erlang.org/erldoc`, contain links to a Getting Started Guide, the extremely useful User's Guide, and an Efficiency Guide. Online resources for learning the language and exploring advanced features are plentiful. The books *Introducing Erlang* [112], *Programming Erlang* [7], and *Learn You Some Erlang for Great Good* [54] are notable. Ericsson produced the short film *Erlang: The Movie* [21], and Garrett Smith produced the followup *Erlang The Movie II: The Sequel* [111].

EXERCISES

Now it's your turn. Continue exploring Erlang with the activities and research questions below, and feel free to branch out on your own.

10.1 Install Erlang and experiment with its REPL.

10.2 Try out the early examples in this chapter via the `escript` utility.

10.3 The word count example in the opening section of this chapter used the function `erlang:hd`—a built-in function. Locate and peruse the complete list of Erlang's built-in functions.

10.4 See if you can improve the word count example to match Unicode letters rather than just the Basic Latin letters A–Z.

10.5 Research the `foldl` and `update_counter` functions from the `orddict` module.

10.6 Unlike Elm, JavaScript, and Python, Erlang uses entities from external modules without explicitly importing (or requiring) them. How is it able to do so? What do you think of this design decision?

10.7 You might hear some people call Erlang a *single-assignment* language. What does this term mean? Does this label even make sense, given that what appear to be assignment statements in Erlang are actually matching expressions? What do you think of this terminology?

10.8 How do `==` and `=:=` differ, exactly? (No pun intended.)

10.9 Using the pattern-matching fibonacci function from this chapter as a guide, write a tail-recursive formulation of a function to find the nth Fibonacci number.

10.10 We've seen only two module attributes in this chapter (`-module` and `-export`). Learn about a few more.

10.11 Experiment with the vectors module from this chapter. What happens when you call `vectors:add(1,2)`? Execute this call in the shell within a catch expression. What are the components of the exception you caught?

10.12 Evaluate the following two expressions in the shell.

```
catch if id:format("",[]) -> true; true -> false end.
catch 8 + dog.
```

How do you account for the differences? (You may need to consult the section on Errors and Error Handling in the Erlang User's Guide.)

10.13 The vectors module from this chapter used tuples to represent vectors. Rewrite the module using maps.

10.14 Can an Erlang pattern match bind unbound variables on the right-hand side? What happens when you try to do so?

10.15 Find out and explain the difference between the guard expression `A orelse B` and `A ; B`.

10.16 The opening script in this chapter used an `if`-expression. Would a `case`-expression have worked better? Could the innermost function have been written with multiple clauses? Experiment with different approaches and make the case for the approach you like best.

10.17 Our prime number printer in this chapter checked divisors inefficiently in order to determine primality. Create new versions of this application using enhancements such as (1) trying odd divisors only (but don't forget 2), and (2) trying only prime divisors.

10.18 In our cooks and customers example, the main process simply sleeps for three minutes before terminating the program. If all customers finish early, the application idles before shutting down. If not, the application terminates with customers still eating. Enhance the application to shutdown cleanly. After the last customer finishes eating, the restaurant should be notified that it can now exit cleanly. Experiment with different ways to carry this out.

10.19 Rewrite the cooks and customers application so that processes use the built-in function `register`. What are the advantages and disadvantages of registration?

10.20 Erlang allows a special type of linking via verb!erlang:monitor!. Find out how monitors differ from the usual kind of process linking, and give an example of their use.

10.21 Execute the following three expressions in the REPL:

```
> self().
> exit(exiting_the_shell).
> self().
```

What happened? Hint: Think about the minesweeper example from the chapter.

10.22 The Erlang documentation states that "Application processes should normally not trap exits." This is exactly what we've done in our minesweeping example. What, then are the alternatives?

10.23 Research and experiment with OTP, the Open Telecom Platform. You may wish to rewrite the concurrent examples from the chapter using OTP.

CHAPTER 11

Go

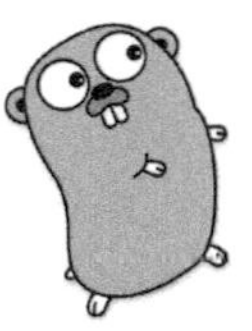

Go is "an open source programming language that makes it easy to build simple, reliable, and efficient software." [37]

First appeared 2009
Designers Robert Griesemer, Rob Pike, Ken Thompson
Notable versions 1 (2012) • 1.5 (2015)
Recognized for Being simpler than C++, Goroutines
Notable uses Servers
Tags Statically Typed, Concurrent, Channel-Oriented
Six words or less Google's language for Google-sized problems

Go was conceived, designed, and first implemented at Google to help address Google-sized problems. It powers large, scalable, distributed systems running on thousands of machines. Though a garbage-collected language, it targets applications traditionally written in systems languages such as C++. Implementations are known for lightning fast compilation and lightweight concurrent processes.

Go is statically typed, but feels lighter than Java. You'll find a good deal of type inference—not as much as in Elm—but plenty nevertheless. You create concurrent programs with **goroutines**, **channels**, and a powerful `select` statement; not with threads, semaphores, polling loops, or other low-level constructs. Go often feels like a dynamic language. Your types conform to **interfaces** not via explicit `implements` clauses, but simply by defining the stated methods. You have easy access to run-time type information when needed.

Go's designers strove to keep the language simple, omitting not only threads but implementation inheritance, compile-time generics, conditional expressions, and operator overloading. You won't encounter exception objects in Go: operations that can fail generally return two values, one of which signals an error. You can **panic**, though, when you deem an error condition too unwieldy to handle immediately, and recover from it elsewhere. You get safety, too: simply include **deferred operations** to ensure resources are cleaned up even during panics.

Our tour of Go begins with a look at our three traditional scripts and coverage of the basics. Go is intentionally a small language, so we won't have a great deal to discuss here.

We will instead spend time on several of Go's interesting features: pointers, arrays and slices, its approach to interfaces, and panics. We'll conclude with Go's signature mechanisms for concurrent programming and a brief look at its reflective capabilities.

11.1 HELLO GO

Here we..., no.

```
package main

import "fmt"

func main() {
	for c := 1; c <= 40; c++ {
		for b := 1; b < c; b++ {
			for a := 1; a < b; a++ {
				if a * a + b * b == c * c {
					fmt.Printf("%d, %d, %d\n", a, b, c)
				}
			}
		}
	}
}
```

The entry point for a Go program is the function `main` within package `main`. The `Printf` function from the package `fmt` works like that of C; its name is capitalized because capitalized entities are exported from a package while lowercased ones are private to the package—convention over configuration! However, the `package` declaration is required, as Go will not assume a default package. And like Python but unlike Java, you must explicitly `import` a package in order to use entities from it.

Next we visit string permutations via the now familiar Heap's Algorithm:

```
package main

import (
	"fmt"
	"os"
	"strings"
)

func main() {
	if len(os.Args) != 2 {
		fmt.Println("Exactly one argument is required")
		os.Exit(1)
	}
	word := os.Args[1]
	generatePermutations(strings.Split(word, ""), len(word)-1)
}
```

```
func generatePermutations(a []string, n int) {
    if n == 0 {
        fmt.Println(strings.Join(a, ""))
    } else {
        for i := 0; i < n; i++ {
            generatePermutations(a, n-1)
            if n % 2 == 0 {
                a[0], a[n] = a[n], a[0]
            } else {
                a[i], a[n] = a[n], a[i]
            }
        }
        generatePermutations(a, n-1)
    }
}
```

Split and Join are functions from the strings package, rather than methods on string objects. The os package provides access to command line arguments and an exit function, as in Python. We must declare types for parameters, but the types of local variables (like i and word) can be inferred. Type names come *after* variables, and complex type expressions have prefix operators: []string reads as "slice of strings." The := symbol introduces Go's **short variable declaration**. We see a verbose if-statement because Go lacks a conditional expression. Thankfully, parallel assignment keeps the script from getting too long!

Now let's count words from standard input:

```
package main

import (
    "fmt"
    "bufio"
    "os"
    "regexp"
    "sort"
    "strings"
)

func main() {
    counts := make(map[string]int)
    scanner := bufio.NewScanner(os.Stdin)
    r := regexp.MustCompile(`[a-z\']+`)
    for scanner.Scan() {
        line := strings.ToLower(scanner.Text())
        for _, word := range r.FindAllString(line, -1) {
            counts[word] += 1
        }
    }
    report(counts)
}
```

```
func report(counts map[string]int) {
    var words []string
    for word := range counts {
        words = append(words, word)
    }
    sort.Strings(words)
    for _, word := range words {
        fmt.Println(word, counts[word])
    }
}
```

Our script begins by defining a **map** (Go's term for a dictionary) to hold the counts for each word, a scanner to read standard input line by line, and a regular expression for matching words. Scanners by default break streams into *lines* for reading, though you can also configure them to split by words or even define a custom splitter. As we've done in previous chapters, we break each line into words and count them in a map. We need not check whether a word is already in the map: the value for a nonexistent key in a Go map is the "zero" element of the type! For counting purposes, 0 could not be more convenient.

A second function outputs our word counts. We want the output sorted by word, but Go's maps do not guarantee any traversal order at all. As in Lua, we need to collect the keys in a separate structure, sort them there, then iterate the new sorted structure to write the output.

11.2 THE BASICS

Go predeclares a few boolean, numeric and string types:

- The type `bool` with values `true` and `false`.
- Numeric types with explicit bit sizes: the signed integer types `int8` (and its alias `byte`), `int16`, `int32`, and `int64`; the unsigned integer types `uint8`, `uint16`, `uint32`, and `uint64`; the floating point types `float32` and `float64`; and the complex types `complex64` (made from two `float32` values) and `complex128` (from two `float64`s). Go does not have a character type: you process characters by manipulating their code points, normally referred to as **runes**. The type `rune` aliases `int32`.
- Numeric types with implementation-specific sizes: `int`, `uint`, and `uintptr`, designed to correspond to the underlying machine's native data and address bit sizes. These *do not* alias any other type.[1]
- The type `string`, for sequences of bytes—not runes. The operation `len(s)` produces the byte length, not the character length. The byte sequence of a string *s* representing text will generally be a UTF-8 byte sequence, from which you can obtain the sequence of code points (as a rune slice) via the expression `[]rune(s)`.

None of the basic types are nullable. Every number, boolean, and string has an actual value, which can never be `nil`. When declaring a variable without an initial value, Go assigns the **zero value** for the type: `0` or `0.0` for numbers, `false` for booleans, and `""` for strings.

[1]Specifically, `int64` and `int` are distinct types, even on a 64-bit machine.

The following script illustrates several of the basic types in action.[2]

```
package main

import "fmt"

func ExampleSimpleTypes() {
    var a uint8 = 0x22
    var b int64 = -9377888
    var c rune = '\u03c0'
    var d string
    var e bool
    var f complex128 = 2.5-3i

    fmt.Println(a, b, c, d, e, f)
    // Output: 34 -9377888 960  false (2.5-3i)
}
```

Go has eight kinds of composite types: arrays, functions, structures, maps, pointers, slices, interfaces, and channels. These complex types depend on constituent types. Array types, interestingly, also depend on their size: the types `[10]string` and `[20]string` are distinct. Let's look first at arrays:

```
package main

import "fmt"

func ExampleArrays() {
    var flags [3]bool
    powers := [5]int{1, 2, 4, 8, 16}
    identity := [2][2]float64{{1,0},{0,1}}

    fmt.Println(flags, powers, identity)
    // Output: [false false false] [1 2 4 8 16] [[1 0] [0 1]]
}
```

Go's functions are first-class. When specifying a function value, you must mention the types of the parameters and return type:

```
package main

import (
    "math"
    "fmt"
)
```

[2]Many of the examples in this chapter will use Go's *Runnable Examples* feature. Store the code in a file whose name ends with `_test.go` and run with `go test`, which will find and execute functions whose names begin with `Example`. The output of these functions will be captured and compared with text in comments beginning with `Output:`. Unlike the other languages featured in this book, Go does not have a built-in assertion facility. The runnable examples do work quite well, and have the advantage that the standard *Godoc* utility place runnable test code directly into autogenerated documentation.

```
func twice(f func(float64)float64, x float64) float64 {
    return f(f(x))
}

func square(x float64) float64 {
    return x * x
}

func ExampleFunctions() {
    addTwo := func(x float64)float64 {return x + 2}
    fmt.Println(twice(square, 4))
    fmt.Println(twice(addTwo, 100))
    fmt.Println(twice(math.Log2, 256))
    // Output: 256
    // 104
    // 3
}
```

Functions can have more than one return value. This isn't the same as returning a value of a tuple type; Go functions actually do produce multiple result values. Additionally, you may give names to return values; these act as local variables to which you may assign values and have them returned through a simple return statement:

```
package main

import "fmt"

func multipleReturnValues() (bool, int) {
    return true, 5
}

func namedReturnValues(x int) (a int, b string, c float64) {
    a = 8 * x
    b = "ok"
    return          // note no expressions after `return`
}

func ExampleWithMoreFunctions() {
    p, q := multipleReturnValues()
    r, s, t := namedReturnValues(10)
    fmt.Println(p, q, r, s, t)
    // Output: true 5 80 ok 0
}
```

Structs work pretty much as expected, with little syntactic sugar. You may specify struct literals using (1) only the values in order of appearance in the type declaration, or (2) field names and values, in any order. If you omit a field, Go gives you the zero value for the type. Similarly, if you fail to initialize a variable of a structure type itself, Go gives you a struct filled with zero values.

```
package main

import "fmt"

type Point struct {
    X, Y float64
    Color string
}

func ExampleStructs() {
    p := Point{8, 5, "green"}
    q := Point{Color: "blue", Y: -2.0}
    var r Point
    s := struct{Name string; Ok bool}{"ABC", true}
    fmt.Println(p, q, r, s, p.Y)
    // Output: {8 5 green} {0 -2 blue} {0 0 } {ABC true} 5
}
```

As with other identifiers, capitalized field names will be automatically exported from the current package, while other field names will stay private unless explicitly exported.

Go's structs have a few additional capabilities, including anonymous fields and tags; we'll run into both of these later in the chapter.

Map types have the form `map[`K`]`V for key type K and value type V. Reading a map returns not an option value, but two values: the value read (or the zero value if the key is not present) followed by a boolean indicating whether the key was present.

```
package main

import "fmt"

type Counts map[string]int

func ExampleMaps() {
    m := Counts{"A": 2, "B": 3, "C": 10, "D": 5}

    m["E"] = 8                // add key-value pair
    delete(m, "D")            // remove key-value pair

    b, hasB := m["B"]         // returns value, wasPresentFlag
    z, hasZ := m["Z"]

    fmt.Println(m["A"], len(m), b, hasB, z, hasZ)
    // Output: 2 4 3 true 0 false
}
```

When invoking operations such as reading from maps that return both a value and a present indicator, as well as similar operations returning a value and an error indicator, you will often take advantage of Go's ability to create local variables just prior to the condition in an `if`-statement. The following shows the idiomatic way to handle both the present and non-present cases for keys in a map:

```
package main

import "fmt"

func Example() {

    capitals := map[string]string{
        "NSW": "Sydney",
        "VIC":
        "Melbourne",
        "TAS": "Hobart",
        "WA": "Perth",
        "QLD": "Brisbane",
        "SA": "Adelaide",
    }

    for _, state := range [2]string{"NSW", "AL"} {
        if capital, known := capitals[state]; known {
            fmt.Printf("The capital of %s is %s\n", state, capital)
        } else {
            fmt.Printf("I don't know the capital of %s\n", state)
        }
    }
    // Output:
    // The capital of NSW is Sydney
    // I don't know the capital of AL
}
```

We'll postpone discussion of pointers, slices, interfaces, and channels to future sections, and close our tour of the basics of Go with a look at its statements. Go features a curly-brace syntax without too many noisy parentheses or semicolons. While semicolons terminate most statements, the compiler will insert them for you when the last token on a line "looks like" the end of a statement.[3] You do need semicolons to separate multiple statements on a line (Ruby and Python require this, too), though you might rarely encounter multi-statement lines in practice.

Go's statements, which we'll only list for now, are as follows:

- Declarations, of types, variables, constants, functions, and methods.
- Empty statements, which do nothing.
- Expression statements, which execute expressions, only for their side effects.
- Assignments, such as `x = 3` or `x++`. Assignments are not expressions in Go.
- The usual compound statements—`for`, `if`, `switch`—and control transfer statements `break`, `continue`, `fallthrough`, `return`, and `goto`.
- Defer statements, which schedule a function call to be executed at the time the surrounding function ends (normally or when panicking).

[3] In particular, Go inserts a semicolon when a line ends with an identifier (including type names); number, string or other literals; a right parenthesis or right brace; a `++` or `-` operator; or one of the keywords `return`, `break`, `continue`, or `fallthrough`.

- Concurrency-related statements: `go` to launch a goroutine, a statement to send a value to a channel, and `select`, for coordinating sends and receives. We'll cover concurrency in Section 11.7.

Did the designers miss a `while` statement? Not at all. Look at what the `for` statement can do:

- `for` {*body*} — executes body "forever."
- `for` *condition* {*body*} — executes *body* while *condition* is true.
- `for` *init* ; *condition* ; *after* {*body*} — functions as the classic for-statement in Java, JavaScript, and the C family of languages.
- `for` *x* `:= range` *e* {*body*} — iterates through the indices of array, slice, or string *e*; or through the keys of map *e*, or the elements of receive channel *e*.
- `for` *i*, *x* `:= range` *e* {*body*} — iterates with the index *and* value (or key and value, for maps) of array, slice, string,[4] or map *e*. Use _ for the index variable to iterate *only* with the values.

Any statement may be labelled, allowing it to be referenced by a `break`, `continue`, or `goto`. We've seen `break` and `continue` before, but what is a `goto`? *The Go Programming Language Specification* states: "A `goto` statement transfers control to the statement with the corresponding label within the same function" and furthermore prohibits the transfer of control from causing "any variables to come into scope that were not already in scope at the point of the `goto`." [39] In other words, a `goto` mustn't leave its function jump over the declaration of any new variables.

So we can't use a `goto` to transfer control to *anywhere*. Long ago, in other languages, we could, and did. This prompted Dijkstra to write, in his famous 1968 letter, *Go To Statement Considered Harmful*:

> For a number of years I have been familiar with the observation that the quality of programmers is a decreasing function of the density of **go to** statements in the programs they produce. More recently I discovered why the use of the **go to** statement has such disastrous effects, and I became convinced that the **go to** statement should be abolished from all "higher level" programming languages (i.e. everything except, perhaps, plain machine code).
>
> ...The go to statement as it stands is just too primitive; it is too much an invitation to make a mess of one's program. One can regard and appreciate the clauses considered as bridling its use. [25]

"Bridling" unrestricted control flow means that program flow should be expressed wholly in terms of high level structured statements such as `if` and `for` (the genesis of the term **structured programming**), and perhaps the limited control transfers through `break` and `continue`. Dijkstra's letter was quite influential: `goto` statements are now quite rare, and languages that permit them tend to place significant restrictions on their use.

This concludes our brief coverage of the basics. We now turn to some of the characteristic features of the language, beginning with something new: pointers.

---

[4]Iterating through the values of a string produces the string's runes, not its bytes.

11.3 POINTERS

Go, unlike the other languages in the book we've seen so far, *copies* arrays and structs when assigning or passing as parameters. Any changes made through a parameter will apply to the copy only, and not affect the passed argument:

```
package main

import "fmt"

func useless(a [5]int, b struct{x, y int}) {
    a[0] = 10
    b.x = 5
}

func ExampleCopies() {
    z := [5]int{0, 0, 0, 0, 0}
    p := struct{x, y int}{1, 2}
    useless(z, p)
    fmt.Println(z, p)
    // Output: [0 0 0 0 0] {1 2}
}
```

Copying the entire array or struct can carry a significant performance overhead, compared to the assignment and passing of references. But you *can* pass references in Go, just as you can create copies in other languages. While our previous languages copied and passed references implicitly and required explicit function invocation to make copies, Go implicitly makes copies but gives you a way to explicitly make references. Go's explicit references are called **pointers**.

Pointer values have types. The type of pointers that reference objects of type t is denoted $*t$. To make a pointer to an object x, write $\&x$; to **dereference** a pointer p, write $*p$. Figure 11.1 should help to clarify.

Figure 11.1 Pointers in Go

Passing pointers as arguments gives us efficiency, but for better or worse allows mutations through parameters to affect the corresponding arguments:

```
package main

import "fmt"
```

```
func mutateThroughPointer(a *[5]int, b *struct{x, y int}) {
    (*a)[0] = 10
    (*b).x = 5
}

func ExampleMutation() {
    z := [5]int{0, 0, 0, 0, 0}
    p := struct{x, y int}{1, 2}
    mutateThroughPointer(&z, &p)
    fmt.Println(z, p)
    // Output: [10 0 0 0 0] {5 2}
}
```

You won't be using pointers *only* to avoid making copies. You'll use pointers when creating modifying arrays and structs at runtime—in other words, for building dynamic data structures. Let's go through a common exercise familiar to most students of programming: writing a type for binary trees. We'll restrict our trees to integers for now, and see how to make them more generic in an upcoming section:

```
package binarytree

import "fmt"

type Tree struct {
    Value int
    Left, Right *Tree
}

func (tree *Tree) Size() (count int) {
    if tree == nil {
        return 0
    }
    return 1 + tree.Left.Size() + tree.Right.Size()
}

func (tree *Tree) String() string {
    if tree == nil {
        return "()"
    }
    return fmt.Sprintf("(%s%d%s)", tree.Left, tree.Value, tree.Right)
}
```

We snuck in a few things in this brief example. First, we wrote `tree.Left` rather than `(*tree).Left`. That's right: field access in Go automatically dereferences pointers. The same is true for arrays; we can, and do, write `a[0]` for `(*a)[0]`. Second, we placed, in the function definitions of `Size` and `String`, a parameter *before* the function name! This is how Go defines **methods**. The early parameter is the receiver. As we've seen many times already, method call syntax uses prefix notation, e.g. `tree.Size()`. Third, we've introduced the value `nil`. Every pointer type has a value `nil`. A `nil` pointer doesn't point to anything, and so can't be dereferenced.

Now let's take a look at how we use our trees, via the following test:

```
package binarytree

import "fmt"

func ExampleTree() {
	s := &Tree{7, &Tree{Value:2}, &Tree{8, &Tree{Value:1}, nil}}
	fmt.Println(s)
	fmt.Println(s.Size())
	// Output: ((()2())7((()1())8()))
	// 4
}
```

So we see that an empty tree is simply `nil`, and a one-element tree can be written as either `&Tree{Value:1}` or `&Tree{1,nil,nil}` And calling `fmt.Println` executed the `String` method! How? Keep that in mind when we get to the section on interfaces.

The mechanisms of assigning and passing pointers to arrays and structs do not differ from the languages of the previous ten chapters. The pointers are copied while the referents live somewhere in a shared memory space visible to both the caller and the callee. And at some point a **garbage collector** will come by and free the memory storing the referent when it is no longer used. What Go does offer that the other languages don't, however, is the ability to reference any object whatsoever—even simple numbers, booleans, and strings, and objects stored within local variables. This flexible state of affairs might get a little tricky, but responsible software engineers should know the basics of entity lifetimes and memory management, so let's get technical for a minute.

Recall from Chapter 1 that every call of a function creates an activation record, or frame, to hold parameters, local variables, temporary scratch space for computation, and other assorted information. Go allocates frames on function entry and deallocates them when the function returns. The active frames thus behave like a (last-in, first-out) stack; in fact, we call the memory holding the frames the **stack**. However, when we create pointers to local variables, or local expressions, and pass those pointers *out of the function* via assignment to non-local variables or via a return statement, the referents now have a **lifetime** exceeding that of the function creating them. We say that these long-lived referenced entities have **escaped** the function. Because the runtime system deallocates the frame when a function returns, these escaped entities should *not* be allocated inside the frame. The official language FAQ [38] states:

> When possible, the Go compilers will allocate variables that are local to a function in that function's stack frame. However, if the compiler cannot prove that the variable is not referenced after the function returns, then the compiler must allocate the variable on the garbage-collected heap to avoid dangling pointer errors.[5]

An example will help. In the following, the function assigned to f defines three local variables and an anonymous struct. The function makes a pointer to the local variable b and assigns it to a non-local variable. It also makes a pointer to the struct and returns the pointer from the function. The Go compiler detects both of these cases during **escape analysis**. It allocates b and the struct on the heap. When f finishes execution, its frame (including

[5]A dangling pointer points to memory that was once valid but has since been deallocated.

locals *a* and *c*) go away, but *b* and the struct survive on the heap. These will be deallocated at some later time by the garbage collector.

```
package main

import "fmt"

type Point struct{x, y int}

func ExampleEscape() {

    var g *int                           // g means "global(ish)"

    f := func(x int) *Point {
        a := 5                           // stays local
        b := 8
        c := &b
        g = c                            // b escapes!
        return &Point{x, a * b}          // this escapes too!
    }

    fmt.Print(*f(2))
    fmt.Println(*g)
    // Output: {2 25}8
}
```

Figure 11.2 provides a useful visual for this scenario.

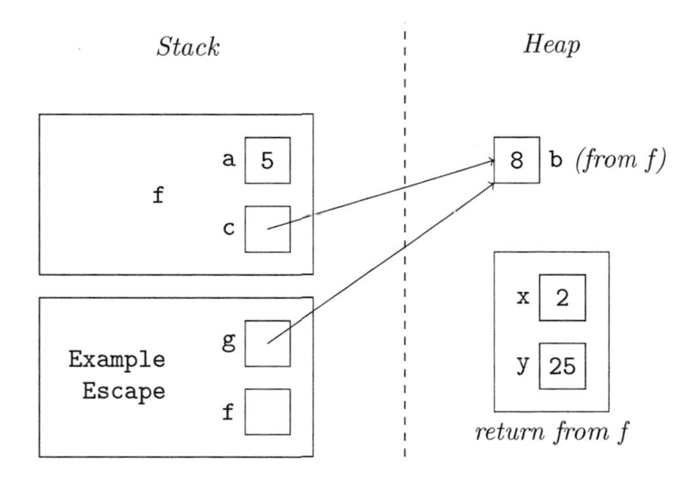

Figure 11.2 Escaped objects allocated on the heap

You may have noticed we made a point of mentioning that arrays and structs were copied on assignment and argument passing. But what about maps? It turns out that when maps are copied or passed, Go actually *copies or passes a reference*, so Go maps are like maps in nearly every language. In fact `nil` is a legal value for a map. Nil maps differ from an explicitly allocated zero-element map in that nil maps are immutable.

11.4 SLICES

So we can build a dynamic collection of structs wired together with pointers. But what about dynamic arrays? If an array's length is part of its type (i.e., `[10]int`), how can we write a function that accepts an integer array of any size? Go's notion of the array-of-any-size is a **slice**. A slice is a range of elements from an underlying physical array.

Figure 11.3 shows an array a and slices s = `a[2:8]` and t = `a[6:9]`. We form slices with the inclusive start index of the array and the exclusive end index. The slices themselves contain (1) a pointer to the starting element, (2) the number of elements in the slice, *len*, and (3) the capacity of the slice, *cap*, derived from the length of the underlying array.[6]

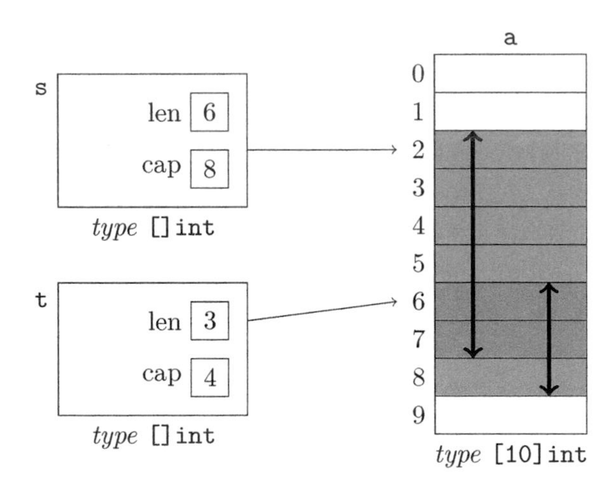

Figure 11.3 Go Slices

Because slices contain pointers, they share storage with the underlying array and any slices that happen to overlap. For example, in Figure 11.3, `s[4]`, `t[0]`, and `a[6]` are aliases of each other.[7] And like a pointer, a slice can have the value `nil`. A nil slice has zero length and zero capacity...and is the zero value of slice variables.

In practice, we generally construct slices via **slice literals** or the built-in function `make`. Evaluating the slice literal `[]int{1, 1, 2, 3, 5, 8, 13}` builds a 7-element array and a slice covering the entire array (a slice of length 7 and capacity 7). Evaluating `make([]int, 5, 8)` creates an 8-element array, and a slice of length 5 (and capacity 8) overlaying indexes 0 through 4 of the array. You can also write `make([]T, n)` to make a new slice whose length and capacity are both n. And as you'd probably expect, `make` will initialize the slice values to the proper zero value.

You can add elements to the end of a slice with the built-in function `append`, which will continue to use the underlying array if the capacity exists, otherwise a new array will be allocated and the elements copied over prior to appending. If you'd like to copy elements between slices yourself, use the built-in `copy` function.

This is a lot to take in, so we'll provide an annotated example to close out this section.

[6]The capacity is the length of the underlying array minus the start index of the slice plus one.

[7]Go shares slice syntax with Python, including the shorthands `a[:n]` for `a[0:n]` and `a[k:]` for `a[k:len(a)]`.

```
package main

import "fmt"

func Example() {
    a := [5]int{0, 10, 20, 30, 40}
    s := a[1:3]                          // overlaid on a
    fmt.Println(s, len(s), cap(s))
    s = append(s, 700, 900)              // append (still in a)
    fmt.Println(s, len(s), cap(s))
    s = append(s, 5, 5, 5)               // append, overflows
    fmt.Println(s, len(s), cap(s) > 5)   // note capacity grown
    t := make([]int, 10)                 // new 10-element slice
    copy(t, s)                           // copy elements to new location
    fmt.Println(t)                       // note the copy
    // Output: [10 20] 2 4
    // [10 20 700 900] 4 4
    // [10 20 700 900 5 5 5] 7 true
    // [10 20 700 900 5 5 5 0 0 0]
}
```

11.5 INTERFACES

Let's now turn to the Go way of specifying types of "objects that know how to do x and y," namely, the **interface**. We'll start with examples. A runner is something that runs:

```
type Runner interface {
    run()
}
```

A finder is anything that looks for a pattern in a source, return the position at which it was found, and a boolean stating whether it was found:

```
type Finder interface {
    find(pattern string, source string) (int, bool)
}
```

A locker is anything that locks, yielding a key, and that unlocks with a key:

```
type Locker interface {
    Lock() string
    Unlock(string)
}
```

Interfaces are types, so you can declare variables and parameters with interface types. Go will check at compile time that whatever you assign to such a variable or pass to such a parameter, actually supports those methods. Unlike Java, you need not explicitly state that your implementing type "implements the interface." If your type has all the methods, congratulations, it implements the interface. Let's see this in action with shapes. A shape is anything that can tell you its area and its perimeter:

```
package main

import (
    "fmt"
    "math"
)

type Shape interface {
    perimeter() float64
    area() float64
}

type Circle struct {
    radius float64
}

func (c Circle) perimeter() float64 {
    return 2.0 * math.Pi * c.radius
}

func (c Circle) area() float64 {
    return math.Pi * c.radius * c.radius
}

type Rectangle struct {
    length, width float64
}

func (r Rectangle) perimeter() float64 {
    return 2.0 * (r.width + r.length)
}

func (r Rectangle) area() float64 {
    return r.width * r.length
}

func showDetails(s Shape) {
    fmt.Printf("%T perimeter=%g area=%g\n", s, s.perimeter(), s.area())
}

func Example() {
    showDetails(Rectangle{5.5, 20.0})
    showDetails(Circle{4})
    // Output: main.Rectangle perimeter=51 area=110
    // main.Circle perimeter=25.132741228718345 area=50.26548245743669
}
```

Next up we'll look at our familiar animals example. Recall that this example requires several types of animals each having a name and making their own sound, and a generic `speak` method operating over all animal types. We'll have an interface `Sounder` for things making a sound; horses, cows, and sheep will all make sounds and thus satisfy the interface. An `Animal` will be a sounder with a name. We get this effect not with inheritance, but by **embedding** the interface in the struct:

```
package main

import "fmt"

type Animal struct {
    Sounder
    name string
}

type Sounder interface {
    sound() string
}

func speak(a Animal) string {
    return fmt.Sprintf("%s says %s", a.name, a.sound())
}

type Horse struct {}

func (h Horse) sound() string {
    return "neigh"
}

type Cow struct {}

func (c Cow) sound() string {
    return "moooo"
}

type Sheep struct {}

func (s Sheep) sound() string {
    return "baaaa"
}

func ExampleAnimals() {
    h := Animal{Horse{}, "CJ"}
    fmt.Println(speak(h))
    c := Animal{Cow{}, "Bessie"}
    fmt.Println(speak(c))
    fmt.Println(speak(Animal{Sheep{}, "Little Lamb"}))
    // Output: CJ says neigh
    // Bessie says moooo
    // Little Lamb says baaaa
}
```

Interfaces are ubiquitous in the Go libraries. Here's one from the package `fmt` you should know:

```
type Stringer interface {
    String() string
}
```

This is used within the body of `fmt.Printf`, `fmt.Println`, and similar functions to produce a string representation of a value. We've implemented a `String()` method earlier in the chapter—for binary trees—and you'll likely find yourself implementing this method on your own data types from time to time, much like implementing `toString` in JavaScript or Java, defining `__toString` in Lua, defining `__str__` in Python, or the equivalent in many (most?) other languages we've seen.

The interface

```
type error interface {
    Error() string
}
```

also finds its way into many Go programs. This type is not defined in any package; rather it's a **predeclared identifier**, like `int`, `true`, `nil`, `len`, and `delete`. A common Go idiom, for functions that may either succeed or fail, is to return two values from the function: the first the function's actual result (if successful), and the second an object implementing this interface. The package `encoding/base64`, for example, contains the function

```
func (enc *Encoding) DecodeString(s string) ([]byte, error)
```

Given a string that supposedly represents a Base64 encoding [117] of a byte array, this function will produce either (1) the original string and the value `nil`, indicating, per convention, no error, or (2) some object implementing the `error` interface as the second value, in which case we don't care about the first value:

```
package main

import (
    "encoding/base64"
    "fmt"
)

func Example() {
    strings := []string{"SGVsbG8sIHdvcmxk", "Bad&^)#****Chars"}
    for _, s := range strings {
        if bytes, err := base64.StdEncoding.DecodeString(s); err != nil {
            fmt.Println("Oops:", err.Error())
        } else {
            fmt.Println("Decoded:", string(bytes))
        }
    }
    // Output:
    // Decoded: Hello, world
    // Oops: illegal base64 data at input byte 3
}
```

Notice that we used the value `nil` in a context where an object of type `error` was expected. This is perfectly fine. `nil` is not just for pointers and maps, as we've already seen, but also can be used for—and plays the role of the zero value for—slice, function, interface, and channel values.

Here's a final thought for this chapter. Go's design choice of a flavor of (static) duck typing,[8] namely allowing types to implement interfaces not by mentioning the interface name in the type declaration, but simply by virtue of implementing the interface's method, means that *every* type in Go implements `interface{}` (the empty interface). There is no need to implement a **cosmic supertype**, a type (or class) from which all other types derive, such as Java's `Object` or Ruby's `BasicObject`. `interface{}` is the perfect type for a parameter that truly accepts any value at all. You can find many uses of this type in the source code of the Go Standard Packages. We've seen at least one already:

```
func Println(a ...interface{}) (n int, err error)
```

And now we can officially add Go to the list of languages that can pack a variable number of arguments into a single parameter.

11.6 PANICS

We've seen that idiomatic Go makes heavy use of multiple return values for handling errors that *might* occur—opening or reading from files, formatting data, or decoding streams. After all, it's perfectly reasonable that someone may ask for a non-existent file to be read, or supply a malformed byte stream to be decoded. These are not really "exceptional conditions," so handling them by throwing and catching exceptions is arguably a misuse of the exception facility. But this kind of programming is practically encouraged by JavaScript, Java, and similar languages, that even allow try-catch-finally blocks right in the middle of a function body.

Go doesn't have exceptions, but it does have a way to handle the truly exceptional cases—the ones that really shouldn't happen or the ones you don't expect to recover from. When something truly exceptional happens, the program—technically, the currently executing goroutine—panics. A panic stops the current function, runs any deferred operations, then returns. In the caller, the panic process is repeated, until some deferred operation is able to *recover*, or the entire chain of calls has been "unwound" and the program **crashes** (exits with a non-zero return code).

To see the mechanism in action, we've built a little contrived demo in where the function `Example` calls `f` which calls `g`, which panics. `g` will run its deferred operation but will not recover from the panic, so `f` panics too. `f` runs its deferred operation but recovers, and returns normally to `Example`.

[8]The moniker "duck typing" gained popularity in programming languages due to the oft-heard phrase "if it looks like a duck, walks like a duck and quacks like a duck, then it's a duck." Rather than declaring that something *is* a duck (as in the Java-esque `class D implements Duck`), a duck-typed language simply ensures the objects in question can do the things ducks can do, i.e., that they implement the corresponding methods. The term "duck typing" is not precisely defined, however; you may sometimes hear the argument that duck typing applies only to dynamically typed languages. Following this line, we would describe Go as having "structural typing" or "implicitly implemented interfaces." The intent, though, is the same: the interface name does not appear.

```
package main

import "fmt"

func g() {
    defer fmt.Println("Cleaning up g")
    panic("Oh no")
    fmt.Println("This message will not be displayed")
}

func f () {
    defer func() {
        fmt.Println("Cleaning up f")
        if r := recover(); r != nil {
            fmt.Println("Recovered the panic value", r)
        }
    }()
    g()
    fmt.Println("This message will not be displayed")
}

func Example() {
    defer fmt.Println("Cleaning up main")
    f()
    fmt.Println("f completed normally")
    // Output: Cleaning up g
    // Cleaning up f
    // Recovered the panic value Oh no
    // f completed normally
    // Cleaning up main
}
```

If called during a panic, `recover` returns the value passed to `panic`. Otherwise, if no panic is in progress, `recover` returns `nil`.

The accepted convention in Go libraries is to limit panicking and recovery to the internals of a library, while returning explicit error values to clients of the library. [35] Such a design leads to a simpler interface between callers and services.

11.7 GOROUTINES

The preceding sections have not covered the best part of Go.

Go is a concurrent language. This is a good thing, because the world is concurrent, i.e., made up of independent, interacting actors. **Concurrent programming** is the art of structuring software systems by composing and coordinating these independent computations. Concurrency isn't the same as parallelism. Parallelism is the state of things happening at the same time; concurrency is about managing activities that overlap in time, whether or not these activities interleave execution on a single processor or run in parallel on distinct processors. [98]

Go and Erlang share a bit of concurrent programming philosophy: they don't generally rely on shared memory and locks. The concurrent execution units called processes in Erlang and **goroutines** in Go, are just functions, with their own local variables, who communicate with other functions via message passing.[9] Implementations of both languages support tens of thousands of concurrent processes or goroutines. But while Erlang's processes send messages by naming the receiving process, Go's goroutines communicate via **channels**.

Let's start with a (literal) hello world example. We will create a channel that can send and receive strings. We'll launch a goroutine that writes `"Hello world"` to the channel. The main goroutine will read the channel and print the value received:

```
package main

import "fmt"

func Example() {
    ch := make(chan string)
    go func() {ch <- "Hello, world"}()
    fmt.Println(<-ch)
    // Output: Hello, world
}
```

Here we see that goroutines are launched by prefixing the `go` keyword to a function call; channels are typed (e.g. the type `chan string` can send and receive strings only); channels are created with `make`; *ch* `<-` *x* writes *x* to channel *ch*; the expression `<-`*ch* reads from channel *ch*. Two more facts about channel communication are essential to know:

- Channels created with `make(chan` *T*`)` are synchronous. A sender will block until a receiver is ready to consume a message; and conversely, a receiver will block until a sender sends a message.
- As in Erlang, when the main goroutine finishes, the application finishes, even if other goroutines are still running.

Let's dial up the complexity ever so slightly. Here's a famous example in which two goroutines sum each half of an array—note the use of a slice parameter here to avoid copying the entire array—and write their half-sums to a channel. The main goroutine reads the half-sums from the channel, adds them, and prints the result.

```
package main

import "fmt"

func sum(a []int, ch chan<- int) {
    result := 0
    for _, value := range a {
        result += value
    }
    ch <- result
}
```

[9] The saying in Go is: "Don't communicate by sharing memory, share memory by communicating." [36]

```
func Example() {
    var numbers [100000]int
    for i := range numbers {
        numbers[i] = i
    }
    ch := make(chan int, 2)
    go sum(numbers[:len(numbers)/2], ch)
    go sum(numbers[len(numbers)/2:], ch)
    fmt.Println(<-ch + <-ch)
    // Output: 4999950000
}
```

We've created the channel in this example via `make(chan string, 2)`, giving it a *capacity* of two messages. Channels with capacities are called **buffered channels**: if a channel holds fewer messages than its capacity, senders will not block. The type annotation on `sum`'s *ch* parameter, `chan<- int`, indicates the channel may only send. (There is, of course, a corresponding annotation for receive-only channels: `<-chan int`.)

In addition to reading a single item from a channel using `<-`, we can iterate over the channel with the `range` operator, consuming messages until the channel is closed. In the following example, a goroutine sends 100 messages through a channel, then closes the channel. A loop reads and prints each message, then exits cleanly as the channel was closed:

```
package main

import "fmt"

func generateMessages(ch chan string, n int) {
    for i := 0; i < n; i++ {
        ch <- "Hello"
    }
    close(ch)
}

func main() {
    messages := make(chan string, 5)
    go generateMessages(messages, 100)
    for message := range messages {
        fmt.Println(message)
    }
}
```

Reading from a closed channel does not block a goroutine. To detect whether a channel is closed, you can use the two-return-value form of `<-`; if closed the second value will be `false`, and the first will be the zero value of the channel's message type.

```
package main

import "fmt"
```

```
func Example() {
    ch := make(chan int)
    go func() {
        ch <- 100
        close(ch)
    }()
    value, ok := <- ch
    fmt.Println(value, ok)  // This will get something
    value, ok = <- ch
    fmt.Println(value, ok)  // This won't, but it won't block!
    // Output: 100 true
    // 0 false
}
```

Many concurrent programming scenarios involve goroutines that read or write multiple channels; reads and writes that should time out; or reads and writes that should back off if the other side is not immediately ready. Go's powerful `select` statement handles all of these concerns. It consists of zero or more *cases*, each of which is a channel read or channel write or (at most one) default case:

```
select {
case v1 := <- c1:
    statement_list_1
case v2, x2 := <- c2:
    statement_list_2
// more cases ...
case cn <- xn:
    statement_list_n
default:                        // The default case is optional
    statement_list_n+1
}
```

Go first evaluates, in order, each of the channel expressions and send-expressions. If none of the communications is ready to proceed and there is no default case, the statement blocks until a case can proceed, or forever if none ever become ready. If no case is ready and a default is present, the default statement list executes. If ready cases do exist, Go selects one at random to proceed, and executes the corresponding statement list.

The expression `time.After(d)`, for some `Duration`[10] value *d* produces a channel that blocks for the given duration, and when read from, yields the current time. This gives a convenient mechanism for timing out sends and receives—simply include a receive from this channel in a select statement. Either a communication will happen within the duration specified, or the channel will unblock the `select` when the delay expires. We'll see an example momentarily.

Now we're ready to build the Go version of the restaurant simulation we saw in the Java and Erlang chapters. Recall the scenario: three chefs serve twelve customers. Customers place their order in a bounded order queue of capacity 5. A customer unable to get an order placed in 7 seconds leaves for a shopping trip and returns later. Cooks prepare the order and

[10]The duration type is defined as `type Duration int64` in the package `time`, and represents the length of time between two instants in nanoseconds. Durations, therefore, have a range of approximately ±292 years.

deliver the finished meal directly to the customer. Customers leave after eating 10 meals. When all customers have finished, the simulation ends gracefully.

Let's start with the order type and the global order queue:

```
package main

type Order struct {
    food string
    customer string
    reply chan *Order
    preparedBy string
}

var Orders = make(chan *Order, 5)
```

Each order holds the menu item that was ordered, the name of the customer placing the order, and a channel through which to send the prepared order back to the customer. The global (to the package) variable `Orders` is the queue to which customers will place, and cooks will retrieve, orders. Our channels contain pointers to orders, allowing the orders themselves to be modified.

Cooks are goroutines that never terminate (call these **daemons** if you wish). Not explicitly terminating is not a problem; the entire application will end when the main goroutine finishes. A cook repeatedly retrieves an order from the global queue, prepares the meal, signs it in the `preparedBy` field, and sends the order back through the reply channel:

```
package main

func Cook(name string) {
    report(name, "starting work")
    for {
        order := <-Orders
        do(name, "cooking for " + order.customer, 12E9)
        order.preparedBy = name
        order.reply <- order
    }
}
```

Our utility function `report` logs a simple message, while the function `do` logs a message and sleeps, simulating an action such as cooking, eating, or shopping. Since multiple cooks and customers will log messages, we require a synchronized logging mechanism so that messages do not become interleaved. Fortunately, Go's `log` package contains a number of functions, including `Println`, that are protected against interleaving.

```
package main

import (
    "time"
    "log"
    "math/rand"
)
```

```
func report(name string, message string) {
    log.Println(name + " " + message)
}

func do(name string, message string, maxDelay time.Duration) {
    report(name, message)
    delay := int(maxDelay) / 2 + rand.Intn(int(maxDelay) / 2)
    time.Sleep(time.Duration(delay))
}
```

Before we look at customers, let's look at the main goroutine. We want each of the customers to complete their full lifecycle of ten finished meals before we exit the simulation. How should we do this?

Since we know the number of customers, we could require each customer goroutine to write to a channel at the end of its life. The main goroutine could end by receiving the desired number of messages from the channel before exiting. Alternatively, we can use a **wait group**. A `sync.WaitGroup` supports tree methods: (1) `Add(`*n*`)` to increase the wait group's counter, (2) `Done()`, decreasing the counter, and (3) `Wait()`, to block until the counter reaches zero. Our main goroutine simply creates each of cooks and customers, incrementing the wait count upon customer creation. The last line of the function just invokes `Wait`, relying on each customer to notify the group by calling `Done` when it finishes.

```
package main

import (
    "fmt"
    "sync"
)

const numberOfCooks, numberOfCustomers = 3, 15

func main() {
    var waitGroup sync.WaitGroup
    for i := 0; i < numberOfCustomers; i++ {
        waitGroup.Add(1)
        go Customer(fmt.Sprintf("customer-%v", i), &waitGroup)
    }
    for i := 0; i < numberOfCooks; i++ {
        go Cook(fmt.Sprintf("cook-%v", i))
    }
    waitGroup.Wait()
}
```

Now we can define the customer goroutine. Each customer is created with a name and the wait group to notify when finished. We begin by setting up the notification as a deferred operation; this common Go idiom of placing deferred operations for finalization and clean up at the beginning of a goroutine is good practice. Deferred operations are always executed, even if a function prematurely returns or panics; in our case we know customers will always notify the wait group no matter how they terminate.

The customer then loops until ten meals have been consumed. Each loop begins by placing a new order. We place the order in a `select` statement with a seven second time out. Each order will contain the customer's personal channel for receiving prepared meals. During each iteration the customer either eats a meal or gives up on placing the order. After eating all the meals, the customer reports being finished, and the deferred operation notifies the wait group.

```
package main

import (
    "sync"
    "time"
)

func Customer(name string, wg *sync.WaitGroup) {
    defer wg.Done()
    ch := make(chan *Order)
    for mealsEaten := 0; mealsEaten < 10; {
        select {
        case Orders <- &Order{food: "soup", customer: name, reply: ch}:
            meal := <-ch
            do(name, "eating " + meal.food + " prepeared by " +
                meal.preparedBy, 10E9)
            mealsEaten += 1
        case <-time.After(time.Second * 7):
            do(name, "waiting too long, going shopping", 5E9)
        }
    }
    report(name, "going home")
}
```

Admittedly, the fact that a customer will give up if unable to place an order in a given amount of time but will happily wait forever for a meal to be prepared may not be the best model of reality, but it keeps the example simpler than it would have been otherwise, since leaving while a meal is being prepared would force us to deal with throwing away prepared meals (i.e., draining channels).

We managed in this chapter to introduce many of the concurrency primitives available in Go, and looked at a few different usage patterns. Clearly, we only scratched the surface. A nice next step in your study of Go might be to write a server capable of dispatching goroutines to do work on behalf of clients. Web servers, chat servers, search servers, and load balancers are actually quite fun to write in Go.

One final note: while high-level communication between goroutines via channels will make up the majority of your concurrent programs' architectures, there are times when a channel is overkill and simple lock, condition variable, or atomic counter is best. Go's `sync` and `sync/atomic` packages will likely have what you need. A few end-of-chapter exercises will give you scenarios to employ a few of these objects.

11.8 REFLECTION

We'll finish our tour of Go with a brief look at some of its reflective capabilities. Go doesn't have macros like Clojure or Julia, but it can glean some information about the entities from which it is composed. Here, then, are a few highlights, beginning with computing both the syntax and the type of an expression at runtime. You can get this information in string form using the `%T` and `%#v` format specifiers:

```
package main

import "fmt"

func Example() {
    var c uint8 = 0x22
    values := []interface{}{c, [3]int{1,2,3}, "hello", '$', 10}
    for _, x := range values {
        fmt.Printf("%#v has type %T\n", x, x)
    }
    // Output:
    // 0x22 has type uint8
    // [3]int{1, 2, 3} has type [3]int
    // "hello" has type string
    // 36 has type int32
    // 10 has type int
}
```

The package `reflect`, however, defines the type `Type` to represent the runtime type of a value:

```
package main

import (
    "fmt"
    "reflect"
)

func Example() {
    var s reflect.Type = reflect.TypeOf("")
    b := reflect.TypeOf(true)
    m := reflect.TypeOf(map[string]bool{"a": true, "b": false})
    fmt.Println(m == reflect.MapOf(s, b))
    // Output: true
}
```

In addition, the package contains representations for expression values (`Value`), methods (`Method`), channel directions (`ChanDir`), struct fields (`StructField`) and other kinds of information about the program entities themselves. Structure field objects yield metadata such as the field type, index within the struct, and the field's `tag`. Tags annotate a struct's fields and can be used for any purpose. Perhaps the most common use case comes from the JSON package, with its facilities for encoding (marshalling) structs into JSON [64] strings. For example:

```
package main

import (
    "fmt"
    "encoding/json"
)

type Person struct {
    Id int `json:"id"`
    FirstName string `json:"first_name"`
    LastName string `json:"last_name"`
}

func Example() {
    if p, err := json.Marshal(Person{8, "Bea", "Bee"}); err == nil {
        fmt.Println(string(p))
    }
    // Output: {"id":8,"first_name":"Bea","last_name":"Bee"}
}
```

Internally, the implementation of `json.Marshal` reflects on the struct value passed to it, and iterates through its fields. You're invited to study the implementation of this method as an excellent example of using reflection in Go.

11.9 GO WRAP UP

In this chapter we were introduced to Go. We learned that:

- Go was designed for large-scale concurrent applications.
- Go is statically-typed with a fair amount of type inference (e.g., in the short variable declaration), though type annotations are required on function parameters and return types.
- Go defines distinct types for integers of various bit sizes and signedness, as well as distinct types for floating point values of different sizes.
- The string type is defined as a byte sequence, not a character sequence. To get code point sequences, you can cast the string to a rune slice.
- Variables are never really uninitialized in Go; if no initial value is provided in a declaration, the variable gets the zero-value of the type.
- Go features eight kinds of composite types: array, function, struct, map, pointer, slice, interface, and channel.
- The length of the array is part of its type, so slices, which can be of varying length, tend to be used more often. Slices are defined over a backing array.
- Functions can have more than one return value. This fact is often used to simulate option types and result types from other languages.
- Go's `for` statement includes functionality of the `while` statement in other languages. The `if` and `switch` statements can define local variables that can be used in these

statements' conditions. Go also contains the interesting `goto` statement, though with limited capabilities compared to the unrestricted `goto` much maligned throughout the years.

- Go is the first language in this text where pointers (references) are explicitly marked in the syntax. We can in Go refer to the reference and the referent separately. Pointers are used where sharing is required, and sometimes simply as a convenience to avoid copying arrays and structs.
- Go performs escape analysis during compilation to determine whether pointers to local variables or temporary expressions will outlive the function activation in which they were defined. If so, the Go runtime will ensure the corresponding referents will be heap-allocated.
- In Go, objects satisfy interfaces simply by virtue of defining the methods in the interface. There is no need for an explicit `implements` clause as in Java.
- Go does not have exceptions, but there is a panic and recover mechanism for situations where unexpected or semi-expected problems occur for which recovery is possible.
- Concurrency has a great deal in-language support. Both `go` and `select` are statements, not library calls, channels are instances of a built-in (parameterized) data type, and message send and receive have special syntax.
- Go's concurrent execution units are called goroutines, and many Go implementations support tens or hundreds of thousands of concurrently executing goroutines.
- Senders block on empty channels and receivers block on full channels. A channel with no capacity models synchronous communication where a sender and receiver must wait for each other. Channels with a non-zero capacity model asynchronous communication.
- Senders and receivers can both cancel sends and waits via a timeout mechanism, modeled in Go with a special channel.
- Senders and receivers can both cancel sends and waits if unable to communicate immediately, via a `default` case in the `select` statement.
- In addition to channel-based communication, Go supports wait groups and other synchronization mechanisms.
- Though it lacks the full power of Julia and Clojure macros, Go sports powerful reflection capabilities.

To continue your study of Go beyond the introductory material of this chapter, you may wish to find and research the following:

- **Language features not covered in this chapter**. Iota, type assertions, the `fallthrough` statement, and most of the standard library.
- **Open source projects using Go**. Studying, and contributing to, open source projects is an excellent way to improve your proficiency in any language. You may enjoy the following projects written in Go: Docker (`https://github.com/docker/docker`), Go (`https://github.com/golang/go`), Revel (`https://github.com/revel/revel`), and Cayley (`https://github.com/google/cayley`).
- **Reference manuals, tutorials, and books**. Go's home page is `https://golang.org/`. The docs page, `https://golang.org/doc/`, contains links to resources for

learning the language, information on best practices, and reference material for the language and its libraries. You can find a list of available books on Go at `https://github.com/golang/go/wiki/Books`, and links to notable video talks at `https://github.com/golang/go/wiki/GoTalks`.

EXERCISES

Now it's your turn. Continue exploring Go with the activities and research questions below, and feel free to branch out on your own.

11.1 Locate the Go Playground online and try out several of the standalone scripts from this chapter.

11.2 Install the standard Go distribution on your machine. Experiment running scripts (with `go run`) and tests (with `go test`). Explore other uses of the `go` command.

11.3 Find (online) and browse *A Tour of Go*, *Effective Go*, the official FAQ, and *The Go Programming Language Specification.*

11.4 The word count program early in this chapter does not check for reading errors. Make it more robust.

11.5 Write a Go function that returns the identity matrix of a given size, i.e., a call to `identity(3)` should produce `[3][3]float{{1,0,0}.{0,1,0},{0,0,1}}`.

11.6 Learn about authoring your own packages in Go. Create a package that includes (1) a type declaration (call the type `Predicate`) for functions that take a single argument of any type and return a boolean result, and (2) functions called `All` and `Any` that accept a predicate p and a slice s, returning whether $p(s_i)$ is true for all or any values of s, respectively.

11.7 Go has a `switch` statement that we did not cover. Research how it works (there are two flavors of this statement) and generate a working example. If you are familiar with switches in C, Java, and JavaScript, what do you find to be the most glaring difference between Go's switch and those of older languages?

11.8 Read Dijkstra's famous Goto letter [25]. Also look for and read the "rebuttal" article by Rubin. What do you make of Rubin's counterarguments? Do they even apply in modern languages?

11.9 Modify the binary tree example of this chapter to define binary *search* trees. Include a rich set of methods, including those to add and delete nodes, compute the height, and traverse the tree in various orders. Your traversal methods should write to a supplied channel.

11.10 See if you can find the language designers' rationale for making arrays and structs value objects requiring explicit pointer objects to reference them, while maps are always accessed through an implicit reference.

11.11 Perform experiments to discover how your Go implementation increases the capacity of a slice after `Append` operations overflow the existing capacity.

11.12 Research the literature—and the blogosphere!—regarding duck typing. Is there any consensus on whether Go is duck typed? What do you think?

11.13 How would you simulate Java's default interfaces in Go?

11.14 Write and execute a program that panics without recovery. What was the operating system's return code for the execution?

11.15 Find, in *The Go Language Specification* the complete list of predeclared identifiers.

11.16 The following example is a slight adaptation of code given by Rob Pike in [98]. Give a detailed explanation of its behavior.

```
package main

import "fmt"

func generate(first chan<- int) {
    for i := 2; ; i++ {
        first <- i
    }
}

func filter(in <-chan int, out chan<- int, prime int) {
    for {
        candidate := <-in
        if candidate % prime != 0 {
            out <- candidate
        }
    }
}

func main() {
    ch := make(chan int)
    go generate(ch)
    for i := 0; i < 1000; i++ {
        prime := <-ch
        fmt.Println(prime)
        nextCh := make(chan int)
        go filter(ch, nextCh, prime)
        ch = nextCh
    }
}
```

11.17 Reimplement the word count example from the first section of this chapter as a concurrent application. Read the input file with a goroutine and write the parsed and lowercased words to a channel. Process the words in the main goroutine.

11.18 Here's a common pattern:

```
select {
case service <- value:
    doSomething(value)
case <-quit:
    return
}
```

What is this pattern "saying?"

11.19 Perform a study to see how well your Go implementation takes advantage of your (hopefully) multiprocessor or multicore system. First, find the value of `runtime.NumCPU()` and make sure the value of `runtime.GOMAXPROCS` (don't forget to research the descriptions of these functions) is set correctly. Compare the runtimes of summing a massive array (1) split into $NumCPU$ pieces summed concurrently by goroutines, with (2) a single-goroutine sequential sum of the entire array. How close was the speedup to the number of CPUs? How do you explain your results?

11.20 Research the following types from the `sync` package: `Cond`, `Locker`, `Once`, `Mutex`, `RWMutex`, and `Pool`. Write scripts illustrating their use.

11.21 Research the functions from the `sync/atomic` package to get a feel for their capabilities. Look in best practices guides and other literature on the language to see which applications these are best suited for.

11.22 What is the difference between the format specifiers `%v` and `%#v`?

11.23 Write a function that prints the field names and types of a given struct. Would this be considered a "For educational purposes only" exercise? If so, what kind of real-world examples of Go reflection exist, beyond the JSON-marshalling example we encountered in the chapter?

CHAPTER 12

Swift

Swift is an innovative, open source language developed by Apple to make software safer, faster, and more fun to create.

First appeared 2014
Creator Apple Inc.
Notable versions 1.0 (2014) • 2 (2015) • 3 (2016)
Recognized for Being the modern alternative to Objective-C
Notable uses Native applications for macOS, iOS, watchOS, and tvOS
Tags Imperative, Safe
Six words or less The modern language for iOS

In the early 2010s, two operating systems dominated the mobile market: Android [40] and iOS [3]. While some mobile developers target only a device's web browser, many will opt to write *native* applications in a language and environment built to talk directly with the operating system. Android devices are generally programmed in Java. iOS devices favor Objective-C or the newer Swift.

Swift is a powerful, general purpose systems language that feels like a scripting language. While best known for its support for the Cocoa framework that provides services to iOS and its cousins macOS, watchOS, and tvOS, our focus in this chapter will be on Swift the language, not the APIs. Swift has much to offer the student of languages, especially as its initial developer, Chris Lattner, is famous for originating LLVM—a compiler framework used by many modern programming language implementations.

Swift's design exhibits an elegant, consistent integration of the best modern features around. We'll start with our usual three example scripts and present the basic elements of the language. We'll pay particular attention to the features of Swift that make it a *safe* language, such as guaranteed memory initialization, static typing, overflow checking, and excellent syntactic support for optionals. But our study will not be without new features. We'll see **external parameter names**, for one, and look at the role of **protocols** and **extensions** in the type system. Finally, we'll encounter **automatic reference counting**, a mechanism for memory management unlike the *tracing* garbage collection of the previous languages on our tour.

12.1 HELLO SWIFT

Let's begin with our traditional listing of small right triangle integers:

```
for c in 1...40 {
    for b in 1..<c {
        for a in 1..<b {
            if a * a + b * b == c * c {
                print("\(a), \(b), \(c)")
            }
        }
    }
}
```

Swift uses `...` for inclusive ranges and `..<` for ranges exclusive at the upper end. String interpolation uses `\()`. The function `print` comes from the standard library which need not be explicitly imported.

Our second traditional script, Heap's algorithm for string permutations, illustrates a fair number of Swift features and idioms:

```
import Foundation

func generatePermutations(of a: inout [Character], upTo n: Int) {
    if n == 0 {
        print(String(a))
    } else {
        for i in 0..<n {
            generatePermutations(of: &a, upTo: n-1)
            swap(&a[n % 2 == 0 ? 0 : i], &a[n])
        }
        generatePermutations(of: &a, upTo: n-1)
    }
}

if Process.arguments.count != 2 {
    fputs("Exactly one argument is required\n", __stderrp)
    exit(1)
}
let word = Process.arguments[1]
var charArray = Array(word.characters)
generatePermutations(of: &charArray, upTo: charArray.count-1)
```

Swift is statically typed. As in Go, we must explicitly specify parameter types, but local variable types can usually be inferred. We introduce mutable variables with `var` and immutable ones with `let`. For parameters, the modifier `inout` means that changes to the parameter are reflected in the passed argument, which we mark with the `&` prefix. In a function declaration, parameters can be given both a **local name** (used inside the function) and an **external name**, used to make readable function calls. Accessing standard error requires a bit of work: Swift is more often used for interactive applications, so it's no real surprise that support for command line applications looks a little "different."

Here's our word frequency script:

```
import Foundation

var counts = [String: Int]()

while let line = readLine()?.lowercased() {
    let range = line.startIndex ..< line.endIndex
    line.enumerateSubstrings(in: range, options: .byWords) {w,_,_,_ in
        guard let word = w else {return}
        counts[word] = (counts[word] ?? 0) + 1
    }
}

for (word, count) in (counts.sorted {$0.0 < $1.0}) {
    print("\(word) \(count)")
}
```

We begin this script by defining an initially empty dictionary to store words and their counts. As we've seen in other statically typed languages, we must mention the type of the empty dictionary explicitly since it can't be inferred.

The loop that gathers the words from standard input employs some excellent Swift idioms that almost make us forget we are using a safe, statically-typed language. Swift has optional types, and provides both the `?.` operator for safe dereferencing, and the **nil-coalescing operator** `??`, which evaluates and produces its second operand only if its first operand is `nil`. We iterate through the key-value pairs of the dictionary in key order by supplying a comparator as a **closure**. Our script uses two different closure syntaxes: {*params* `in` *body*} and the body-only syntax, in which parameters are named `$0`, `$1`, and so on. The comparator body illustrates element access for a tuple *t*: `t.0`, `t.1`, `t.2`, etc.

To extract words from a line, we use `enumerateSubstrings` from a string **extension** in the `Foundation` framework. The method takes a **range** (rather than a whole string), a value from an enumeration type, and a closure. Where the function's last argument is a closure, Swift lets us write it after the parenthesized arguments—a syntax reminiscent of Ruby. The callback to the substring enumerator receives an optional string, which we must unwrap before using. We've used a `guard` statement for this purpose; we'll explain its semantics, and alternatives to it, later in the chapter.

12.2 THE BASICS

Swift is a modern language, sporting default parameters, parameter packing and unpacking, static typing with type inference, first-class functions and closures, extensive pattern matching and binding capabilities, option types in lieu of null references, and generics. We'll hit the basics here very quickly, introducing new features and terms as they come up. We'll leave protocols, extensions, and discussions of safety features to future sections.

The standard library includes the signed integer types `Int8`, `Int16`, `Int32`, `Int64`, and `Int`; unsigned integer types `UInt8`, `UInt16`, `UInt32`, `UInt64`, and `UInt`; and floating-point types `Float` (32-bits) and `Double` (64-bits). The types `Bool`, `Character`, and `String` should be familiar, though unlike Go, Swift's character and string types understand actual characters rather than just code points. For example, the Swift string literal `"e\u{301}"` is formed from the two code points `U+0078` (LATIN SMALL LETTER E) and `U+0301` (COMBINING ACUTE ACCENT), which in sequence represent a *single* character. Swift also knows that Unicode has provided a **precomposed** character for this combination, namely `U+00E9` (LATIN SMALL LETTER E WITH ACUTE) and will equate the precomposed character with the two-code point sequence:

```
// e with acute accent
let s1 = "e\u{301}"                    // e + combining acute accent
let s2 = "\u{e9}"                      // the precomposed character
assert(s1.characters.count == 1)       // Swift combines to count
assert(s1 == s2)                       // == compares the characters

// x in a circle
let s = "x\u{20dd}"                    // x + combining circle
assert(s.characters.count == 1)        // one actual character
assert(s.utf16.count == 2)             // 0078 20DD
assert(s.utf8.count == 4)              // 78 E2 83 9D
```

Perhaps interestingly, none of these types are primitive nor built-in. They are defined in the standard library along with well over one hundred other types. All types in Swift belong to one of six kinds:

- *Structures*, with value semantics;
- *Classes*, with reference semantics;
- *Enumerations*, Swift's sum type, also with value semantics;
- *Protocols*, sets of requirements *adopted* by structures, classes, and enumerations;
- *Tuple types*, Swift's product type;
- *Function types*, containing the parameter and return types for functions.

Under **value semantics**, assignment and argument passing act as if they create copies, while under **reference semantics**, these act as if sharing takes place:

```
struct S {var x = 0}      // structs have value semantics
var s1 = S();             // create an *instance* of type S
var s2 = s1;              // copy, by value
s1.x = 5                  // mutate through s1
assert(s2.x == 0)         // s2 unchanged

class C {var x = 0}       // classes have value semantics
var c1 = C();             // create an *instance* of type C
var c2 = c1;              // reference copy
c1.x = 5                  // mutate through c1
assert(c2.x == 5)         // c2 sees change
```

The numeric types, `Bool`, `Character`, and `String` are all structures. So are:

- `[T]`, arrays of elements of type T;
- `Set<T>`, sets of elements of type T; and
- `[K:V]`, dictionaries with key type K and value type V.

Arrays, sets, and dictionaries, have a **count** method, equality operators (==) and (!=) that compare values, and iteration capabilities via the **for-in** statement. Let's explore some array operations:

```
var a = [10, 20, 30]
var b = [Int](repeating: 1, count: 4)
a[2] = 40
a.append(50)
a += b
assert(a.count == 8)
a[1...3] = [0, 2, 4]
a.remove(at: 3)
assert(a == [10, 0, 2, 1, 1, 1, 1])
```

and some set operations:

```
let a: Set<Int> = [10, 20, 30]
let b: Set<Int> = [20, 30, 40]
assert(a.union(b).count == 4)
assert(a.intersection(b) == Set([20, 30]))
assert(a.contains(10))
assert(!b.isEmpty)
```

and a dictionary, mapping the four constituent countries of the Kingdom of the Netherlands to their capitals:

```
let capitals = [
    "Netherlands": "Amsterdam",
    "Aruba": "Oranjestad",
    "Cura\u{e7}ao": "Willemstad",
    "Sint Maarten": "Philipsburg",
]

assert(capitals.count == 4)
assert(capitals.keys.contains("Aruba"))
assert(capitals.values.sorted().last == "Willemstad")

for (country, capital) in capitals {
    print("The capital of \(country) is \(capital).")
}
```

Declare constants with let and mutable variables with var. Unlike many languages we've seen, not only are let-defined set, array, or dictionary variables non-reassignable, but the values bound to the variable are immutable as well:

```
let a = [10, 20, 30]
// a[2] = 40     // is a compile time error!
// a.append(40) // is a compile time error!
```

Function definitions must indicate the types of all parameters and the return value if any. Parameters have both an **internal name** (to use within the function) and an **external name** (used in the call). By default, the internal and external names are the same; to distinguish the two names, list the external name first. When calling a function, you must specify the external name, unless it is the special name _:

```
func perimeter(base: Int, height: Int) -> Int {
    return 2 * (base + height)
}

func average(of x: Int, and y: Int) -> Double {
    return Double(x + y) / 2.0
}

func middle(_ x: Int, _ y: Int) -> Double {
    return Double(x + y) / 2.0
}

assert(perimeter(base: 4, height: 20) == 48)
assert(average(of: 5, and: 8) == 6.5)
assert(middle(5, 8) == 6.5)
```

Swift considers functions to be special cases of **closures**—code blocks that can be assigned, passed as parameters, and invoked. Swift closures (other than global functions) capture bindings from the context containing the closure's definition, and are thus lexical closures. Two forms of **closure expressions**, with type inference, implicit returns, and **shorthand argument names** ($0, $1, etc.) allow succinct code:

```
let a = [(3,1), (2,8), (3,20)]

assert(a.map({x, y in x*x + y*y}) == [10, 68, 409])
assert(a.map({$0*$0 + $1*$1}) == [10, 68, 409])
```

Classes and structs group properties (defined with `var` and `let`), methods (with `func`), subscripts, and initializers. Properties can be stored or computed:

```
import Foundation

struct Circle {
    var radius = 1.0
    var area: Double {
        get {return M_PI * radius * radius}
        set(a) {radius = sqrt(a/M_PI)}
    }
}

var c = Circle(radius: 10)
assert(c.radius == 10 && c.area == 100*M_PI)   // get property
c.area = 16*M_PI                               // set property
assert(c.radius == 4 && c.area == 16*M_PI)
```

A computed property can be made read-only by providing only a getter. When only a getter is present, you can omit the `get` keyword, writing for example, `var area: Double { return M_PI * radius * radius }`.

Classes, unlike structures, can inherit from each other creating superclass-subclass relationships. Both properties and methods can be overridden. Our common animal example illustrates how this looks in Swift:

```
class Animal {
    let name: String

    init(name: String) {
        self.name = name
    }

    func sound() -> String {
        preconditionFailure("This method must be overridden")
    }

    func speak() -> String {
        return "\(self.name) says \(self.sound())"
    }
}

class Cow: Animal {
    override init(name: String) {
        super.init(name: name)
    }

    override func sound() -> String {
        return "moooo"
    }
}

class Horse: Animal {
    override init(name: String) {
        super.init(name: name)
    }

    override func sound() -> String {
        return "neigh"
    }
}

class Sheep: Animal {
    override init(name: String) {
        super.init(name: name)
    }

    override func sound() -> String {
        return "baaaa"
    }
}
```

```
let h: Animal = Horse(name: "CJ")
assert(h.speak() == "CJ says neigh")
let c: Animal = Cow(name: "Bessie")
assert(c.speak() == "Bessie says moooo")
assert(Sheep(name:"Little Lamb").speak() == "Little Lamb says baaaa")
```

We'll introduce Swift's disjoint sum type, the **enumeration**, by way of example. We'll define a token as (1) the value `plus`, (2) the value `minus`, (3) a double, or (4) a string. Enumeration values must be tagged—the string `"x"` does not belong to our token type; `Token.identifier("x")` does. Processing enumerations often occurs via destructuring in switch statements:

```
enum Token {
    case plus
    case minus
    case number(Double)
    case identifier(String)

    func lexeme() -> String {
        switch self {
            case .plus: return "+"
            case .minus: return "-"
            case .number(let value): return String(value)
            case .identifier(let name): return name
        }
    }
}

let expression: [Token] = [.number(4.0), .plus, .identifier("x")]
let text = expression.map{$0.lexeme()}.joined(separator: " ")
assert(text == "4.0 + x")
```

Values of an enumeration type may be assigned **raw values**, from the enumeration's underlying **raw type**. This allows additional information to be associated with each enum value (not wholly unlike, but not the same as, structure field tags in Go). Raw types may be strings, numbers, or characters.

```
enum Suit: String {
    case spades, hearts, diamonds, clubs
}

enum Direction: Int {
    case north = 0, east = 90, south = 180, west = 270
}

assert(Suit.hearts.rawValue == "hearts")
assert(Direction.south.rawValue == 180)
assert(Direction.west == Direction(rawValue: 270))
```

Recursive enumerations, such as our starter binary tree below, must be marked **indirect**. A tree is either (1) empty, or (2) a node with a left subtree and a right subtree:

```
indirect enum Tree<T> {
    case empty
    case node(Tree, T, Tree)

    var size: Int {
        switch self {
            case .empty: return 0
            case .node(let left, _, let right):
                return left.size + 1 + right.size
        }
    }
}

let t: Tree<Int> = .node(.node(.empty, 3, .empty), 7, .empty)
assert(t.size == 2)
```

The T in `Tree<T>` is a parameter of the **generic type** Tree. We may parameterize structures, classes, and enumerations, and even functions. Here's a trivial generic function:

```
func twice<T>(_ x: T) -> [T] {
    return [x, x]
}

assert(twice(5) == [5, 5])
assert(twice("yak") == ["yak", "yak"])
```

In the following example, we apply the `>` operator to values of a parameterized type. As this operator is defined within the Comparable protocol[1], we must constrain the generic type parameter to those types that adopt Comparable:

```
func median<T:Comparable>(_ x: T, _ y: T, _ z: T) -> T {
    let (a, b) = (min(x, y), max(x, y))
    return b < z ? b : max(a, z)
}

for (x,y,z) in [(1,2,3),(1,3,2),(2,1,3),(2,3,1),(3,1,2),(3,2,1)] {
    assert(median(x, y, z) == 2)
}
```

Although the special syntax might deceive you, arrays and dictionaries are regular generic structure types defined in the standard library. The type expression `[T]` simply sugars `Array<T>` and `[K:V]` sugars `Dictionary<K:Hashable, V>`. Swift even sugars its optional type, `Optional<T>`, with `T?`. Because optionals appear so frequently in Swift, we've given them a special section, which we will now explore.

[1] Protocols are discussed in an upcoming section.

12.3 OPTIONALS

We've seen optionals in previous chapters, so let's cut right to some examples. To unwrap an optional, use an `if`, `while`, or `guard` statement with `let`-bindings in the condition, or one of several operators. For a simple illustration of `if-let`, let's convert strings to integers.

```
for s in ["42", "dog", "3xy", "-2E3", "-1"] {
    if let i = Int(s) {
        print("\"\(s)\" parsed as an integer is \(i)")
    } else {
        print("\"\(s)\" can't be parsed to an integer")
    }
}
```

The expression `Int(s)` produces an optional `Int`. If the optional wraps an actual integer, we bind the integer to the variable i of type `Int` and take the true branch. If `nil`, the false branch executes.

We encountered the `while-let` and `guard-else` forms in the word count example in the opening section of this chapter; each binds the unwrapped value, if present, to a new variable, or takes an alternative path through the code on `nil`. The else-branch of `guard` statements must transfer control out of the guard-statement's enclosing scope, via `break`, `continue`, `return`, `throw`, or calling a `noreturn` function. Guard statements are interesting for at least two reasons:

- The early exit pattern ("if something is wrong, abandon everything") avoids excessive indentation of the rest of the function within the false-branch of an `if-let`. The code stays focused on the "happy-path." [88]
- Unlike `let`-bindings in `if` and `while` conditions, `guard` condition bound variables are scoped until the end of the scope containing the guard statement.

In cases where you need the unwrapped value of an optional but have a default value in mind, use the (short-circuit) nil-coalescing operator:

```
let x: Int? = 5
let y: Int? = nil

assert(x ?? 3 == 5)
assert(y ?? 3 == 3)
assert(x ?? [1,2,3][100] == 5)  // short circuit
```

We can also avoid unwrap optional arrays during indexing, and optional functions in function calls:

```
var a: [Int]? = [10, 20, 30]
var b: [Int]? = nil
var f: ((Int)->Int)? = {$0 * 3}

assert(a?[2] == 30)
assert(b?[2] == nil)
assert(f?(100) == 300)
```

and optional instances in a similar fashion:

```
class Person {
    var name: String
    var supervisor: Person? = nil

    init(name: String, supervisor: Person? = nil) {
        self.name = name
        self.supervisor = supervisor
    }
}

var a = Person(name: "Alice")
var b = Person(name: "Bob", supervisor: a)
assert(a.supervisor?.name == nil)
assert(b.supervisor?.name == "Alice")
```

The `?.`, or `?[]`, and `?()` operators figure prominently in **option chaining** expressions, such as (the admittedly contrived) `previousWinner?().address?.city?.parks?[2]`.

Although Swift provides plenty of syntactic support for optionals, the programmer is not expected to use them for general error handling. Optionals model the possible *absence* of a result, but when operations can *fail*, one can (1) return an enum holding either a proper result or an error indication (recall Elm's `Result` type), or (2) throw an **error** instance:

```
enum BadLength: ErrorProtocol {
    case tooLong(by: Int)
    case tooShort
}

func third() throws -> () {
    throw BadLength.tooLong(by: 3)
}

func second() throws {
    try third()
}

func first() {
    do {
        try second()
    } catch BadLength.tooShort {
        print("Too short")
    } catch BadLength.tooLong(let howMuch) {
        print("Too long by \(howMuch)")
    } catch {
        print("Something else")
    }
}

first()
```

Swift errors represent recoverable situations; throwing and catching them serve as syntactic sugar for managing result objects. Errors must belong to a type conforming to the protocol[2] `ErrorProtocol`. Functions that may throw errors—known as **throwing functions**—must declare this fact by placing the keyword `throws` (see `second` and `third` in the example above) after its parameter list. Calls to throwing functions must be marked with `try`. Catch errors in a `do-catch` statement. Swift propagates any uncaught errors to the surrounding scope.

In addition to catching or propagating errors, you "convert an error to an optional" by invoking, for some throwing function f, `try? f()`, producing the (wrapped) result of the call, if any, or `nil` if the function throws.

12.4 OPERATORS

Just as Swift does not bake in its basic types but instead defines them in its standard library, so too does it define operators as regular functions within the library. The standard library defines a rich set of operators, including the prefix operators `!` (logical not), `~` (bitwise not), `+` (unary plus), and `-` (unary negation). The binary operators, from highest to lowest precedence, follow:

<table>
<tr><th>Operator(s)</th><th>Prec</th><th>Assoc</th><th>Description</th></tr>
<tr><td><< >></td><td>160</td><td>None</td><td>left shift, right shift</td></tr>
<tr><td>* / %
&* &</td><td>150</td><td>L</td><td>multiply, divide, remainder, multiply with overflow, bit-and</td></tr>
<tr><td>+ - &+
&- | ^</td><td>140</td><td>L</td><td>add, subtract, add with overflow, subtract with overflow, bit-or, bit-xor</td></tr>
<tr><td>..< ...</td><td>135</td><td>None</td><td>half-open range, closed range</td></tr>
<tr><td>is as as? as!</td><td>132</td><td>L</td><td>type check, cast, cast as optional, unwrap cast</td></tr>
<tr><td>??</td><td>131</td><td>R</td><td>nil coalesce</td></tr>
<tr><td>< <= > >=
= !=
=== !==
~=</td><td>130</td><td>None</td><td>less, less or equal, greater, greater or equal, equal, not equal, identical, not identical, pattern match</td></tr>
<tr><td>&&</td><td>120</td><td>L</td><td>(short-circuit) logical and</td></tr>
<tr><td>||</td><td>110</td><td>L</td><td>(short-circuit) logical or</td></tr>
<tr><td>?:</td><td>100</td><td>R</td><td>conditional</td></tr>
<tr><td>= *= /= %=
+= -= <<= >>=
&= |= ^=
&&= ||=</td><td>90</td><td>R</td><td>assignment</td></tr>
</table>

As in Elm, Swift operators are just functions, so you can define your own implementations of them on your own data types. We've done this below in our recurring vectors example, to which we've taken the liberty to add a prefix unary negation operator:

[2]We'll get to details of protocols shortly; for now, think of them as interfaces in Java or Go. And in case you are wondering whether there is any ambiguity in the enumeration declarations between specifying a raw type for the enum and adopting a protocol, don't worry. Raw types are restricted to certain structure types, not protocols.

```
import Foundation

struct Vector: CustomStringConvertible {

    let i: Double
    let j: Double

    func magnititude() -> Double {
        return sqrt(i * i + j * j)
    }

    var description: String {
        return "<\(i),\(j)>"
    }
}

func + (left: Vector, right: Vector) -> Vector {
    return Vector(i: left.i + right.i, j: left.j + right.j)
}

func * (left: Vector, right: Vector) -> Double {
    return left.i * right.i + left.j * right.j
}

prefix func - (v: Vector) -> Vector {
    return Vector(i: -v.i, j: -v.j)
}

let u = Vector(i: 3, j: 4)
let v = Vector(i: -5, j: 10)
assert(u.i == 3)
assert(u.j == 4)
assert(u.magnititude() == 5)
assert(String(u + v) == "<-2.0,14.0>")
assert(u * v == 25)
assert(String(-v) == "<5.0,-10.0>")
```

You can even define your own operators. There are quite a few rules surrounding which characters can appear in these **custom operators** (see [4] for full details). Among these rules: custom operator names must be made up of "symbol"-like characters, operators containing dots (`.`) must begin with a dot, a lone question mark is not an operator, and no postfix operator may begin with a question mark or exclamation mark.

Let's quickly run through the steps in creating a couple custom operators. We must *declare* the operators before defining them; for binary operators this means assigning precedence (in the range 0...255 inclusive) and associativity (`left`, `right`, or `none`). We've given our silly infix `~|*|~` a precedence of 145 so that it binds more tightly than addition (140) but less tightly than multiplication (150), just to show such a thing is possible:

```
infix operator ~|*|~ {precedence 145 associativity left}
postfix operator ^^ {}

func ~|*|~ (x: Int, y: Int) -> Int {
    return 2 * x + y
}

postfix func ^^ (x: Int) -> Int {
    return x - 2
}

assert(8^^ ~|*|~ 3 == 15)
```

12.5 PROTOCOLS

Now let's return to Swift's type system. We've seen five of the six kinds of types so far: structures (including numbers booleans, strings, arrays, and dictionaries), enumerations (including optionals), classes, tuples, and functions. The sixth, **protocols**, is Swift's analog of Java's interfaces and Ruby's mixins. A protocol specifies certain requirements that structures, enumerations and classes that wish to **adopt** the protocol must **conform** to.

For our introductory example, we define a protocol `Summarizable` for things with summaries, give a struct and enum that adopt it, and invoke the summary property through variables declared with the protocol type:

```
import Foundation

protocol Summarizable {
    var summary: String { get }
}

struct Circle: Summarizable {
    var radius = 1.0
    var summary: String {return "Circle with radius \(radius)"}
}

enum Direction: Int, Summarizable {
    case north, east, south, west
    var summary: String {return "Bearing \(90 * self.rawValue)"}
}

let a: [Summarizable] = [Circle(radius: 5), Direction.west]
assert(a[0].summary == "Circle with radius 5.0")
assert(a[1].summary == "Bearing 270")
```

The adopted protocols appear in the type declaration, comma separated, following any superclass specification (for classes) or raw type specification (for enums). The protocol definition itself defines a number of requirements for not only properties but initializers and functions (methods) as well.

The Swift standard library defines over 100 structures, enums, and classes, and several dozen protocols. Table 12.1 summarizes a handful of these structures and the protocols they adopt.

Struct	Conforms to Protocols
Array	ArrayLiteralConvertible, CollectionType, CustomDebugStringConvertible, CustomStringConvertible, MutableCollectionType
Bool	BooleanLiteralConvertible, BooleanType, CustomStringConvertible, Equatable, Hashable
Dictionary	CollectionType, CustomDebugStringConvertible, CustomStringConvertible, DictionaryLiteralConvertible
Double	AbsoluteValuable Comparable, CustomDebugStringConvertible, CustomStringConvertible, Equatable, FloatLiteralConvertible, FloatingPointType, Hashable, IntegerLiteralConvertible, Strideable
Int64	BitwiseOperationsType, CVarArgType, Comparable, CustomStringConvertible, Equatable, Hashable, RandomAccessIndexType, SignedIntegerType, SignedNumberType
Range	CollectionType, CustomDebugStringConvertible, CustomStringConvertible, Equatable
Set	ArrayLiteralConvertible, CollectionType, CustomDebugStringConvertible, CustomStringConvertible, Hashable
String	Comparable, CustomDebugStringConvertible, Equatable, ExtendedGraphemeClusterLiteralConvertible, Hashable, MirrorPathType, OutputStreamType, Streamable, StringInterpolationConvertible, StringLiteralConvertible, UnicodeScalarLiteralConvertible

Table 12.1 A selection of adoptions from the Swift Standard Library

Many of these standard protocols have names that indicate their intent very well. Adopt `Equatable` if it makes sense for your type to support value equality, adopt `Hashable` if you would like instances of your type to be used as dictionary keys, and adopt `Comparable` if your type has an underlying ordering. If you'd like values of your type to render in a certain way when printed, adopt (as we did in the previous section!) `CustomStringConvertible`. It's instructive to see how the Swift authors defined some of these protocols.

```
public protocol Equatable {
    @warn_unused_result
    func == (lhs: Self, rhs: Self) -> Bool
}
```

The type `Self` (note the capital "`S`") refers to the type adopting the protocol.[3]

If you adopt `Equatable` for your new type, you need only define `==`. You'll get its negation for free, as the library defines this function:

```
@warn_unused_result
public func != <T : Equatable>(lhs: T, rhs: T) -> Bool {
    return !(lhs == rhs)
}
```

[3]Anyone who's had to define `equals` methods in Java might find this feature very welcome.

`Comparable` requires you only to implement <, and ==. The protocol already adopts `Equatable`, and the operators (functions) <=, >, and >= are defined in terms of <.

`Hashable` is another single-requirement protocol:

```
public protocol Hashable : Equatable {
    var hashValue: Int { get }
}
```

Building the language's basic types and collections—and indeed its entire standard library—around protocols rather than classes, has earned Swift the title of the world's first *protocol-oriented language.* [1] Protocols support value types as well as reference types (Swift's classes). They are not limited to single inheritance. And, as we'll see in the next section, they interact nicely with Swift's approach to retroactive modeling: **extensions**.

12.6 EXTENSIONS

How can a language help a programmer safely extend a system? By safely, we generally mean, in part, "by some means other than editing existing, tested, source code." In some cases we need only create new packages or modules, or perhaps add new functions or types. What about adding methods to existing classes, or, in general, to new types? Can we do this without subclassing?

We can in Swift. Using **extensions** we can add new properties, methods, initializers, subscripts, and nested types to existing types. We can even make an existing type conform to a new protocol. When extending a type, the new functionality applies to all instances of the type, even those created before the extension was executed. Let's add some properties and methods to the `Int` type. We'll add an **instance property** `abs`, a **type property**[4] `allOnes`, and an **instance method** `factorial`:

```
extension Int {
    var abs: Int {return self >= 0 ? self : -self}

    func factorial() -> Int {
        return (1...self).reduce(1, combine: {$0 * $1})
    }

    static var allOnes: Int {return ~allZeros}
}

assert(8.abs == 8)
assert(Int.allOnes == -1)
assert(10.factorial() == 3628800)
```

Extensions solve the expression problem we first encountered with Clojure. We'll start with shapes as things which have an area. In Swift, we "start with a protocol, not a class". [1] We'll add a type implementing this protocol, too:

[4]Instance properties and methods apply to instances of the type, while type properties and methods apply to the type itself. Like Java, Swift uses the modifier `static` to refer to type properties and methods.

```
protocol Shape {
    func area() -> Double
}

struct Square: Shape {
    var side: Double
    func area() -> Double {return side * side}
}

let s = Square(side: 10)
assert(s.area() == 100)
```

Nothing surprising here. But now, let's add functionality. We'll add a perimeter method to squares, but through a new protocol, since other things—not only shapes!—may want boundaries. An extension allows us to adopt a new protocol and add a new method *without modifying existing source code.*

```
protocol Boundaried {
    func perimeter() -> Double
}

extension Square: Boundaried {
    func perimeter() -> Double {return side * 4}
}

assert(s.perimeter() == 40)
```

We can create new types that implement multiple protocols:

```
import Foundation
struct Circle: Shape, Boundaried {
    var radius: Double
    func area() -> Double {return M_PI * radius * radius}
    func perimeter() -> Double {return 2 * M_PI * radius}
}

let c = Circle(radius: 8)
assert(c.area() == 64 * M_PI)
assert(c.perimeter() == 16 * M_PI)
```

We can extend protocols with implemented methods, much like Java's default methods. All types adopting the extended protocol pick up the new methods with the given implementation:

```
extension Shape {
    var json: String { return "{\"area\": \(area())}" }
}

assert(s.json == "{\"area\": 100.0}")
```

Conforming classes may wish to provide their own implementation of a method from a protocol extension:

```
extension Square {
    var json: String {return "{\"kind\": \"square\", \"side\": \(side)}"}
}

assert(s.json == "{\"kind\": \"square\", \"side\": 10.0}")
```

12.7 SAFETY FEATURES

In the previous section, we saw ways in which Swift enables us to write *extensible* systems. We now turn to those elements of Swift's design that help us write *correct* software. A so-called **safe language** aids the programmer in writing correct code by providing as many guarantees and as few ambiguities as possible. Swift, for example, guarantees all variables have a value before use. Not only does the one line script

```
print(x)
```

fail to compile (`error: use of unresolved identifier 'x'`) but the two-line script

```
var x: Int
print(x)
```

also fails to compile (`variable 'x' used before being initialized`). Swift is a **definitive initialization** language: the compiler performs data flow analysis to make sure all paths leading to a use of a variable go through an initialization. Some languages, including C, give no initialization guarantees, leaving the value of an "uninitialized" variable to be "whatever happens to be in memory" at the time; the behavior, therefore, of such a C program is undefined. Go takes a different approach, **default implicit initialization**, giving uninitialized variables the zero value of the declared type. Lua takes the same approach as Go, initializing with `nil`. Still other languages force the programmer to include initializing expressions with every variable declaration. Swift's solution realizes that requiring an initial value may be unwieldy, and defaulting to a zero value (not to mention `nil`) might not be the most sensible choice, nor does it provide any checks against a programmer's possible unintentional omission of an initializer.

The convenience of specifying immutability of objects (in addition to preventing reassignment) through `let` declarations also limits the chance of some programmer error, as does the convenient syntax surrounding optionals—an attack against the Billion Dollar Mistake. Another safety feature shared with modern, statically-typed languages (but not with C), is the requirement that patterns in matching constructs must be exhaustive. If cases are left uncovered (as may happen when adding features to code), the compiler will reject the program and force to you address the missing cases.

Swift also eliminates the surprises that come from **modular arithmetic** of fixed-size integers. In Go, eight-bit integers, for example, have values between -128 and 127. Adding one to 127 "wraps around" to 128, and adding 100 to 100 wraps to -56. In Swift, however, arithmetic overflow crashes the program, hard. If you desire wraparound semantics, use the operators `&+`, `&-`, and `&*`:

```
let x: Int8 = 100
let y: Int8 = 100
assert(x &+ y == -56)   // x + y would crash
```

The last feature facilitating safe code that we'll mention here is Swift's `defer` statement. As in Go, this statement arranges to run code when the enclosing scope exits, for any reason, whether cleanly or due to an error.

12.8 AUTOMATIC REFERENCE COUNTING

Safety in programming practice is concerned not only with the lack of ambiguity, but also with the lack of surprises. Certain software systems often have time and space constraints, requiring events to happen at predictable instants or complete within a predictable duration. Software controlling vehicles or medical equipment, for example, should not be interrupted. Each of the languages of the previous chapters assumed the existence of a **tracing garbage collector**—a collector that fully determines all heap objects reachable from "active" (global and stack) variables (by tracing chains of references) and reclaims all nonreachable objects. Collection in these systems usually happens on a background thread, able to preempt the running program at any time.

Swift takes a different approach. Each heap-allocated object (class instance) is given a reference count, starting at 1 when first created. Each time a reference to the object is assigned to a new variable, or passed to a parameter, the reference count is incremented. The count is decremented when one of the referencing variables goes out of scope. When the count reaches zero, the deinitializer method is invoked and the object is reclaimed. This mechanism, **automatic reference counting**, or **ARC**, requires no garbage collection thread. Since allocation and deallocation occurs at identifiable points in execution, there is no need for a preemptive garbage collector. The following brief example shows an object being allocated and deallocated at predictable points in a script's execution:

```
var history: [String] = []

class C {
    init() {history.append("init")}
    deinit {history.append("deinit")}
}

func f(_ c: C) {                    // 3. count==2
    history.append("f")
    print(c)
    history.append("/f")
}                                   // 4. c out of scope, count == 1

func main() {
    history.append("main")
    let x = C()                     // 1. object created, count==1
    f(x)                            // 2. reference passed
    history.append("/main")
}                                   // 5. x out of scope, count == 0

main()
assert(history == ["main", "init", "f", "/f", "/main", "deinit"])
```

This example begins with a call to **main**, which allocates a local variable x in its stack frame. x is assigned a class instance, which in Swift is heap-allocated:

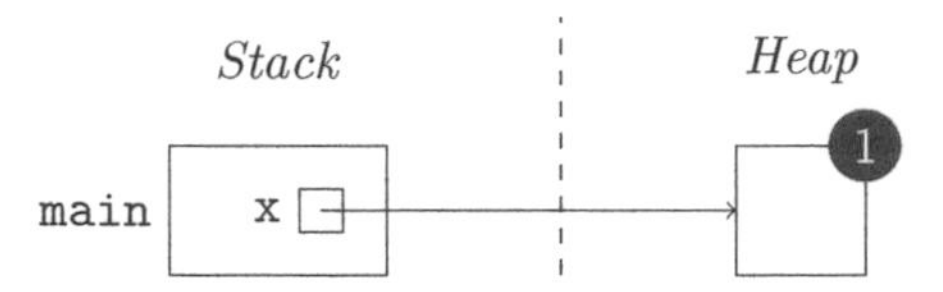

The instance at this point has a reference count of 1 (as indicated with the big red circle). Next, **main** calls **f**, passing x to the parameter c. Swift passes references to class instances, so our object now has a reference count of 2:

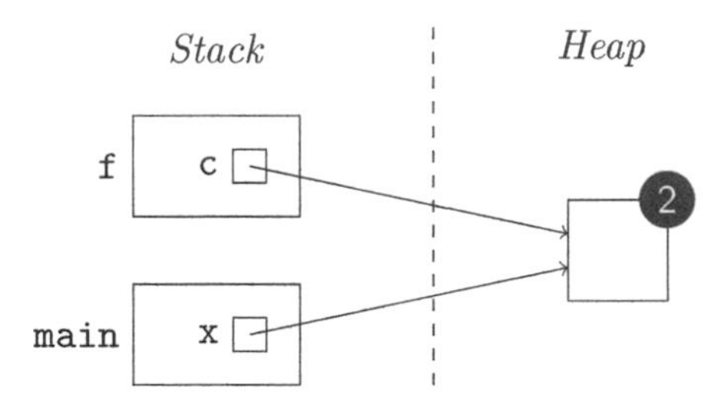

After **f** returns, its local variable c goes out of scope, removing one of the references to our object, reducing the reference count to 1:

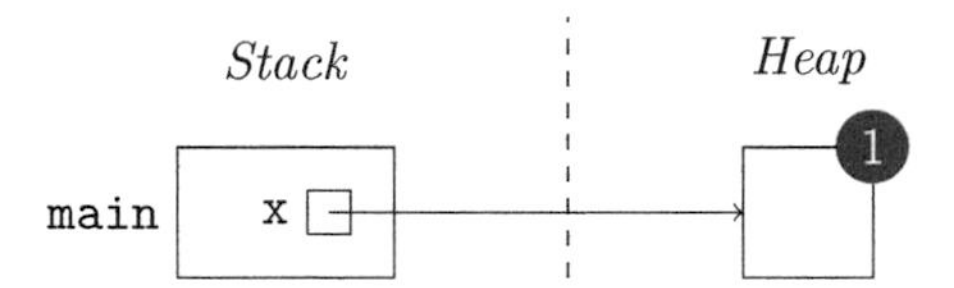

When **main** returns, the object count goes to zero and the object is deinitialized.

In return for avoiding the significant complexity and potential unpredictability of a tracing, thread-based garbage collector, Swift requires developers to understand the workings of ARC. The following example illustrates a memory-management pitfall, known as a **memory leak**, that programmers must know how to avoid:

```
class Person {
    var name: String
    var boss: Person?
    var assistant: Person?    // CAUSES A MEMORY LEAK

    init(name: String, boss: Person? = nil) {
        initCount += 1
        self.name = name
        self.boss = boss
    }
    deinit {
        deinitCount += 1
    }
}
```

```
func main() {
    let alice = Person(name: "Alice")
    let bob = Person(name: "Bob", boss: alice)
    alice.assistant = bob
}

var (initCount, deinitCount) = (0, 0)
main()
assert(initCount == 2 && deinitCount == 0)
```

This script allocates two objects on the heap that each reference each other, forming a **reference cycle**. Figure 12.1 shows a snapshot of program execution just before the function `main` exits. Each object has a reference count of 2. After main returns, the local variables will go out of scope, reducing the reference count of each object to 1. However, the two objects are left on the heap, holding on to memory that ARC can never reclaim.

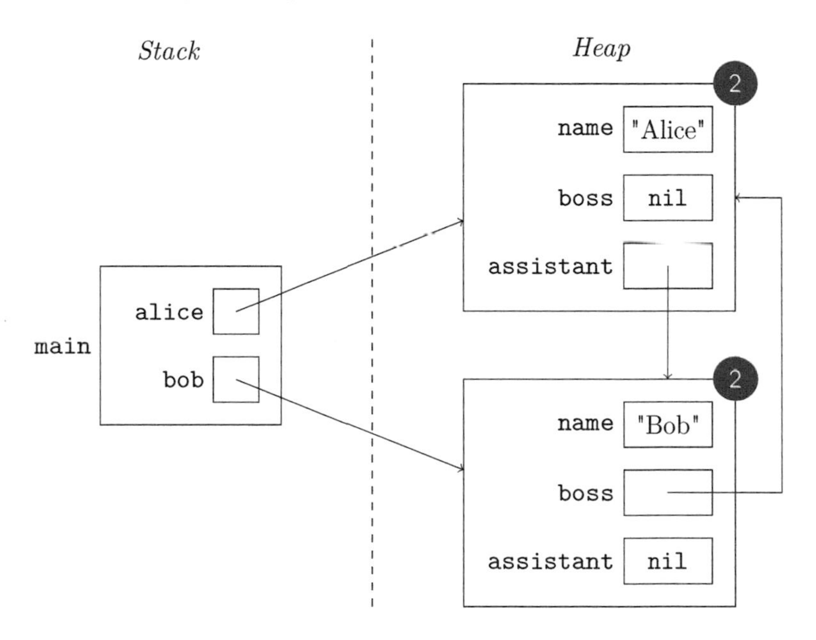

Figure 12.1 A reference cycle causing a memory leak under ARC

To prevent the memory leak, we will make one of the references (we'll choose the `assistant` property) to be a **weak reference**:

```
weak var assistant: Person?
```

The references we've seen so far have all been **strong references**. In Swift, a heap-allocated object can only be reclaimed when its strong reference count goes to zero. Figure 12.2 illustrates our fix to the memory leak in the previous example. Before `main` exits, the object representing Alice has strong reference count of 2, while only one strong reference refers to her assistant, Bob.

When `main` exits, the local variables go out of scope, decrementing the strong reference counts for both Alice and Bob. Alice's count goes to 1, and Bob's goes to 0. Therefore the Bob object can be reclaimed, decrementing the reference count for Alice to 0, at which time Alice is reclaimed as well.

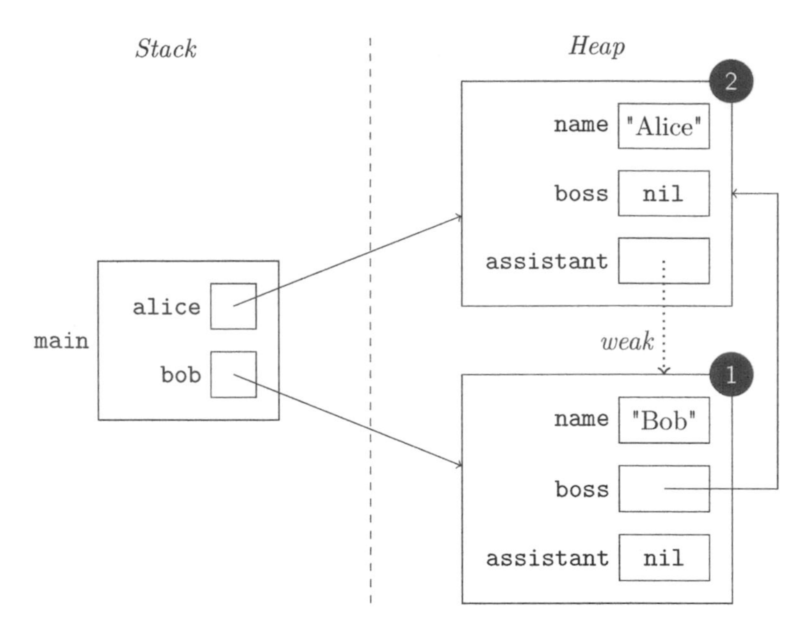

Figure 12.2 Weak references preventing a memory leak

Use weak references together with option-typed references to class instances. When cycles can occur among non-optional instances, you will use **unowned references**. We won't be discussing these here, but we do mention them in the end-of-chapter explorations.

We should mention in closing that although ARC does nearly all the work of reclaiming memory for you (as long as you manage your cycles), Swift is able to interoperate with code written in Objective-C or C, languages teeming with unsafe features. So...the language designers endowed Swift with the power of *unsafe* features. You can often recognize the features by name; the standard library gives you the types `UnsafePointer<T>` and `UnsafeMutablePointer<T>` among others, which serve as a standin for pointer types in C. Instances of these types are pointers to objects of type T and provide *no* automatic memory management whatsoever. *You* must both allocate and deallocate them yourself. They even lack the safety features of guaranteed initialization. You are responsible.

The following script illustrates the lifecycle of a typical unsafe pointer:

```
var p: UnsafeMutablePointer<Int>
// p is not allocated yet, using it here would be an error
p = UnsafeMutablePointer<Int>(allocatingCapacity: 1)
// allocated but not initialized, p.memory is an arbitarary integer
p.pointee = 8
// now p is initialized
assert(p.pointee == 8)
p.deinitialize(count: 1)
// at this point, the memory p points to is "garbage"
p.deallocateCapacity(1)
// now the memory is all freed up
```

An unsafe pointer can be in one of the following three states, as described in [5]:

- Memory is not allocated (for example, pointer is null, or memory has been deallocated previously).
- Memory is allocated, but [the] value has not been initialized.
- Memory is allocated and [the] value is initialized.

Obviously, one requires great discipline in using these unsafe features. Managing memory oneself is hard work, and indeed error-prone, but is generally expected of programmers who write device drivers, operating systems, and the virtual machines (on which our high-level languages run). Low-level languages, like assembly languages and the venerable C programming language, are entirely **unmanaged**: even strings and simple arrays require the programmer to perform manual allocation and deallocation of memory. To read from files the programmer must manually allocate memory chunks of just the right size.

But just as languages like C free most programmers from having to drop into assembly language, many modern languages free many systems programmers from dropping into C. C++ [133] features constructors and destructors in which the programmer can encapsulate memory allocation and deallocation where needed, though does not try to prevent memory leaks. The newer Rust [105] language is also unmanaged but uses concepts such as ownership and borrowing of pointers to eliminate nearly all memory leaks. Rust has a data type for automatically reference counted pointers, while Swift, as we've just seen, applies ARC throughout. These modern languages may lose a little bit of flexibility and a little bit of performance when compared to C, but they gain a great deal in terms of safety and expressiveness.

12.9 SWIFT WRAP UP

In this chapter we looked at Swift. We learned that:

- Apple first released Swift in 2014 with iOS applications in mind, though the language is modern, comprehensive, and general-purpose.
- Swift is known for its safety features. It checks types at compile time, ensures that all variables are initialized before use, manages memory with ARC, and provides `let` declarations to not only prohibit reassignment of a variable but to make the variable's object immutable. Integer arithmetic does *not* "wrap around" by default. Types, other than optional types, do not include `nil` values.
- Though statically typed, Swift uses type inference extensively, but not quite as extensively as possible—function parameters, for one, require explicit annotations.
- Swift's treatment of characters and strings and support for Unicode exceeds that of many other languages.
- Function parameters can have both local (used inside the function) and external (used in calls) names.
- Functions are considered a type of closure. There are two syntactic forms for anonymous closures, `{$0 * $0}` and `{x in x * x}`
- Rather than being wired into the language, the "basic types"—numbers, strings, characters, booleans—are defined as structures in the standard library. The same is true for operators, which are defined as plain old functions.

- All types in Swift belong to one of six *kinds*: structures, enumerations, classes, protocols, tuples, and functions. Structures and enumerations have value semantics (copied if necessary when assigned or passed as arguments); while classes have reference semantics.
- Arrays, sets, dictionaries, and optionals are defined as generic types in the standard library, though a good deal of syntactic sugar makes them pleasant to use.
- Classes can have superclasses and inherit properties and methods. Enumerations can specify a raw type. Classes, structs and enumerations can adopt protocols.
- Extensions allow us to add new properties, methods, initializers, subscripts, and nested types without modifying the source code of the existing type.
- It is possible to extend a protocol, but only with implemented methods.
- Swift's approach to garbage collection is to use automatic reference counting. The programmer does not have to insert code to free memory, but must code in such a way that heap-allocated memory will eventually become unreachable and thus freed. Weak and unowned pointers must be employed when objects appear in reference cycles.

To continue your study of Swift beyond the introductory material of this chapter, you may wish to find and research the following:

- **Language features not covered in this chapter**. Property observers, Type casting, Unowned references and implicitly unwrapped optional properties, Access levels, Associated types, Custom subscript functions, `Any` and `AnyObject`.
- **Open source projects using Swift**. Studying, and contributing to, open source projects is an excellent way to improve your proficiency in any language. You may enjoy the following projects written in Swift: Alamofire (`https://github.com/Alamofire/Alamofire`), Dollar.swift (`https://github.com/ankurp/Dollar.swift`), Surge (`https://github.com/mattt/Surge`), and Cartography (`https://github.com/robb/Cartography`).
- **Reference manuals, tutorials, and books**. Swift's home page is `https://swift.org/`. The docs page contains links to a tour of Swift, the Language Guide, and the Language Reference. The curated Awesome Swift (`https://github.com/matteocrippa/awesome-swift`) links to hundreds of Swift resources on the web. Several companies and independent authors, notably Ray Wenderlich, have provided online training resources; these are easy to find via a web search. As Swift is a young and evolving language, expect the number of available books and other resources to keep increasing.

EXERCISES

Now it's your turn. Continue exploring Swift with the activities and research questions below, and feel free to branch out on your own.

12.1 Locate the Swift home page, and browse its tutorials and references.

12.2 Practice with the Swift REPL.

12.3 Run the first three scripts from this chapter as command line applications.

12.4 Install the Swift Playground and experiment with pre-built applications. Spend time working through the examples in the Swift Book [6].

12.5 Many of the examples in this chapter began with `import Foundation`. What is `Foundation` exactly, and what capabilities does importing it bring to our code?

12.6 Other imports commonly seen in Swift applications include `Darwin` and `Cocoa`. What capabilities are gained by importing these?

12.7 What is the type of the expression `9..<12`?

12.8 How is the function `print` from the standard library declared? Give the names, types, and default values of each of its parameters.

12.9 What happens if you try to invoke a function with an `inout Int` parameter with the argument `&5`? Is an error generated? If so, at compile time or run time? Or is a warning issued? If no error occurs, how does Swift handle a change to the corresponding parameter?

12.10 Make a list of the possible values for the `options` parameter of `enumerateSubstrings`.

12.11 In our Swift word count example, we extracted words not with regular expressions but with the `.byWords` option of `enumerateSubstrings`. This was not because Swift lacks regular expressions, but because word-based enumeration is the *right* way to proceed: many natuaral languages do not use whitespace to separate words! Run the script at `https://github.com/rtoal/ple/blob/master/swift/enumerate_words_example.swift`. How many words does Swift find?

12.12 Find out how to do culture-dependent word separation in the programming languages of previous chapters. You many need to find third-party libraries for this functionality. Implement the word counting script in these languages using what you find.

12.13 Can an expression of type `Int` be assigned to a variable of type `Int64` without an explicit cast? What are the rules in Swift, if any, regarding which assignments can be made without explicit casts?

12.14 For the string literal `"\u{1f1e8}\u{1f1f7}"`: (a) How many characters does this string have? (b) Express the string as a sequence of UTF-16 code points. (c) Express the string as a sequence of UTF-8 code points. (d) What does this string represent?

12.15 How does one define a function to force every argument in a call to be named? How does one allow every argument in a call not to be named?

12.16 In this chapter, we populated our classes and structs only with properties, methods, initializers, and deinitializers. Give an example that defines a subscript.

12.17 Find out if there is any way for a running Swift application to recover from a precondition failure.

12.18 Research the ways in which Swift errors are similar to, and different from, exceptions in Python.

12.19 For the indirect tree enumeration in this chapter, see if it is possible to assign an object of type tree as a member of the tree's own children. Does the compiler stop you from doing so? Do you get a run time error? Were you able to build a cyclic object? Should this even be possible, given that enumerations have value semantics?

12.20 We mentioned, but did not explain, the `noreturn` keyword. What is this used for?

12.21 What happens if an error is thrown but never caught?

12.22 Evaluate the expression `37 & 95 + 14` in both Python and Swift. Explain in detail why the answers are different.

12.23 The Swift Language Reference describes cases in which the spacing, or lack of spacing, around certain operators changes the meaning of the program. See if you can make a comprehensive list of these cases. Do you find it problematic or confusing? Or pragmatic? Why?

12.24 Do some research on Swift custom operators. Which characters are allowed in operators? Why are prefix and postfix operators not assigned a numeric precedence? If a term consists of both a prefix and a postfix operator, which gets applied first?

12.25 Rewrite the animals example from this chapter—the one we used to illustrate classes—to use protocols instead.

12.26 Use extensions to augment the animals example to vocalize the sounds.

12.27 Create an example that shows the difference between weak and unowned references.

CHAPTER 13

Additional Languages

The preceding twelve chapters comprised a tour of several modern languages as a vehicle for introducing concepts and practices important to both professional developers and students of computer science. We make no apologies for leaving out your personal favorite language, or languages you feel we *should* have covered. However, we recognize that any selection of a dozen languages will miss several interesting topics. So in this chapter, we're extending our tour with a few lightning stops, visiting a mix of historically important, influential, experimental, and popular languages not covered in the main tour.

13.1 THE CLASSICS

Our first few lightning stops take us through some of the classics. These "early" languages were often the first of their kind, and their historical importance and influence on subsequent languages cannot be overstated. Some of the languages in this group survive today in modern forms, while some have simply morphed into their successors.

Fortran

Fortran debuted in the mid-1950s as an "automatic coding system"—automatic in the sense that programmers did not have to write their own machine code (considered at the time to be *the* code): the system created it from a specification (i.e., the source). It debuted at a time when the need for high-level languages was still being questioned, so the designers made every effort to ensure the compiler and the generated code was as efficient as possible.

First appeared 1957
Designed by John Backus at IBM
Notable versions I • II • IV • 66 • 77 • 90 • 95 • 2003 • 2008 • 2015
Recognized for Being the first high-level general purpose programming language
Notable uses Numeric computation and scientific computing
Tags Imperative

The first Fortran, Fortran I, looks very different from modern versions: code was not indented, recursion was prohibited, and conditionals and while loops were made by referencing line numbers. Here's a short Fortran I program, taken from [43]:

```
C       PROGRAM FOR FINDING THE LARGEST VALUE
C   X        ATTAINED BY A SET OF NUMBERS
        DIMENSION A(999)
        FREQUENCY 30(2,1,10), 5(100)
        READ 1, N, (A(I), I= 1,N)
    1   FORMAT (I3/(12F6.2))
        BIGA= A(1)
    5   DO 20 I= 2,N
   30   IF (BIGA-A(I)) 10,20,20
   10   BIGA= A(I)
   20   CONTINUE
        PRINT 2, N, BIGA
    2   FORMAT (22H1THE LARGEST OF THESE I3, 12H NUMBERS IS F7.2)
        STOP 77777
```

The instruction on the line labeled "30" is a **computed goto**. It evaluates a conditional expression and transfers control to one of (up to) three instructions according to whether the condition is negative, zero, or positive.

Fortran has evolved a great deal throughout the years. Line numbers, **fixed formatting** (did you notice that all instructions started in column 7 above?), and the computed goto are part of the language's past. Today's Fortran sports many modern features including support for object-oriented and parallel computing. The language remains widely used in the areas of numerical and scientific computing where high performance is a major concern. Modern dialects of Fortran include support for parallel computing through distribution of data across processors.

COBOL

Another language from the 1950s, COBOL, emerged to support business and finance applications. It is best known for its English-like syntax (with `MOVE`...`TO` for assignment and operators with names including `ADD`, `MULTIPLY`, `GREATER THAN`), a large number of reserved words, and its use of the new (at the time) record type. COBOL had a run as one of the most widely used languages in the world. While less popular than in the past, COBOL-based applications can still be found today powering many business software systems.

First appeared 1959
Designed by The CODASYL Consortium
Notable versions 60 • 68 • 74 • 85 • 2002 • 2014
Recognized for Readability, Verbosity, Batch processing of transaction data
Notable uses Financial processing and business applications
Tags Imperative, Procedural

COBOL was designed and developed by CODASYL, a consortium of government and industry members. Perhaps the most famous technical consultant to the group was Grace Hopper, the inventor of FLOW-MATIC (1995), the first English-like data processing language, and the A-0 System (1951), credited as being the first compiler for an electronic computer. [107] Hopper's FLOW-MATIC was a major influence on COBOL. Hopper herself played an active role in programming languages in her career, including a period as director of the U.S. Navy's Programming Languages Group. [134]

Lisp

The 1950s also brought us Lisp, a language widely praised [44] for its elegance, expressiveness, beauty, and homoiconicity, and sometimes teased for its unusual parentheses-laden syntax. It has been called "the greatest single programming language ever designed." [113]

First appeared 1958
Designed by John McCarthy at MIT
Notable dialects 1.5 • InterLisp (1967) • CommonLisp • Scheme • Clojure
Recognized for Metaprogramming
Notable uses Artificial intelligence research
Tags Expressive, Dynamic, Homoiconic

Much of what makes Lisp Lisp was covered in our chapter on Clojure, and you can read more about its essential character and the reverence in which it is sometimes held in [46] and [45]. The latter reference is required reading for those interested in programming language history, as it relates the story of Steve Russell's realization that Lisp was not just a notation for writing functions, as McCarthy himself believed, but a true programming language that could be implemented.

Russell in fact *did* write the first Lisp interpreter. It didn't take too long to write, as McCarthy's language specification was very brief. Alan Kay noted, "the big revelation to me when I was in graduate school—when I finally understood that the half page of code on the bottom of page 13 of the Lisp 1.5 manual was Lisp in itself. These were Maxwell's Equations of Software!" A few functions, including `eval` and `apply`, was all that was needed to define a complete programming language. For an in-depth analysis of the half page of code and the rationale for Kay's realization, see [89].

Algol

The language Algol (short for Algorithmic Language), perhaps best known as the direct ancestor of many of today's most popular languages (including C, C++, C#, Java, and JavaScript), began as the product of a joint committee of European and American Computer Scientists in 1958. The publication of *The Algol 60 Report* introduced the world to now-ubiquitous method for specifying syntax, Backus-Naur Form (BNF). Among Algol's innovations were block structure and lexical scope, characteristics frequently taken for granted today.

First appeared 1958
Designed by A Committee of European and American Computer Scientists
Notable versions 58 • 60 • W • 68
Recognized for Block structure, nested functions, lexical scope
Notable uses Computer science research
Tags Procedural, Imperative, Structured

The second Algol, Algol 60, was not only influential, but good. Sir Tony Hoare, on the impact of Algol 60, remarked, "Here is a language so far ahead of its time that it was not only an improvement on its predecessors but also on nearly all its successors." [58] In fact,

Algol 68, the language designed to succeed Algol 60, gained a degree of infamy as a large and complex language, and acquired quite a number of famous detractors. [27]

Simula

In the early 1960s, Ole-Johan Dahl and Kristen Nygaard created an extension of Algol 60 which came to be known as Simula. Simula was originally developed as a vehicle for writing simulations, a design goal its authors ultimately satisfied with classes, objects, inheritance, and coroutines. [138] Simula 67 was the first of a number of languages, including Mesa, CLU, Modula, and Euclid, which highlighted the importance of encapsulation in the construction of software systems.

First appeared 1965
Designed by Ole-Johan Dahl and Kristen Nygaard
Notable versions 1 (1962) • 67
Recognized for Objects, classes, inheritance, and subtyping
Notable uses Simulations, VLSI designs
Tags Object-oriented

Simula is now considered to be the first object oriented programming language. Its influence on its better-known immediate object-oriented descendants, Smalltalk and C++, is difficult to overstate.

Smalltalk

Many of the innovations pioneered by Simula were more fully developed, refined, and expanded by a group at the Xerox Palo Alto Research Center (PARC), led by Alan Kay, Dan Ingalls, and Adele Goldberg in the early 1970s. Smalltalk was conceived and grew up in the same laboratory as did the mouse-driven user interface, Ethernet, and the laser printer. [94]

First appeared 1972
Designed by Alan Kay
Notable versions 71 • 72 • 74 • 76 • 80 (1980)
Recognized for GUI Workstations, Metaprogramming, Object-orientation
Notable uses Education, Financial analysis and modeling
Tags Object-oriented, Dynamic, Reflective

Although a great admirer of LISP, Kay's motivation to create Smalltalk stemmed in part from realizing that

> ...the surface beauty of LISP was marred by some of its key parts having to be introduced as "special forms" rather than as its supposed universal building block of functions. The actual beauty of LISP came more from the *promise* of its metastructures than its actual model. I spent a fair amount of time thinking about how objects could be characterized as universal computers without having to have any exceptions in the central metaphor. [71]

The notion of objects as "universal computers" arose from Kay's application of the notion, credited to Bob Barton by Kay, that the key to successful decomposition of a system is to make the parts resemble the whole. Decomposing a computer into dissimilar data structures and algorithms violates this idea; decomposing it into little computers (bundles of data and code now known as **objects**) does not. Smalltalk, therefore, eventually came to follow six principles, recorded in [71]:

1. Everything is an object.
2. Objects communicate by sending and receiving **messages** (in terms of objects).
3. Objects have their own **memory** (in terms of objects).
4. Every object is an instance of a **class** (which must be an object).
5. The class holds the shared **behavior** for its instances (in the form of objects in a program list).
6. To eval a program list, control is passed to the first object and the remainder is treated as its message.

Here's a brief glance at Smalltalk code—our animals script—written in the GNU Smalltalk dialect, illustrating classes, objects, and messages:

```
Object subclass: Animal [
    | name |
    Animal class >> named: aString [^self new name: aString]
    name: aString [name := aString]
    speak [^name, ' says ', self sound]
]

Animal subclass: Horse [
    sound [^'neigh']
]

Animal subclass: Sheep [
    sound [^'baaaa']
]

Animal subclass: Cow [
    sound [^'moooo']
]

h := Horse named: 'CJ'.
h speak displayNl.
c := Cow named: 'Bessie'.
c speak displayNl.
(Sheep named: 'Little Lamb') speak displayNl.
```

13.2 SYSTEMS LANGUAGES

It's often helpful to make a distinction between **application software** and **system software**. The former solves problems of interest to humans and is device and platform independent; the latter is concerned with directly managing the resources of a computer system (memory, files, processes) and may be device dependent. An **application programming language** provides rich data modeling constructs and assumes the existence of a large runtime system to perform memory management without programmer intervention. A **systems language** provides programmers with abstractions for memory (such as pointers) and neither provides nor requires much (if any) runtime support. We require systems programming languages to write the virtual machines, garbage collectors, and runtime systems supporting our modern, highly expressive applications languages.

C

C is the quintessential systems programming language. Though dating from 1972, it remains today one of the most popular programming languages in the world.

First appeared 1972
Designed by Dennis Ritchie at Bell Labs
Notable versions K & R (1972) • ISO (1989/90) • 99 (1999) • 11 (2011)
Recognized for Systems programming
Notable uses Unix and Unix-like operating systems
Tags Imperative, Low-level, System, Non-managed

C is the language of choice for operating systems, device drivers, and embedded devices. Most operators and statements map fairly closely to machine instructions. Its macro system and its means to drop into assembly language ensure the language keeps up with hardware advances. The language itself assumes no garbage collector and no "smart pointers" of any kind. Programmers must be acutely aware of all memory-related operations. Even strings require allocation of the proper amount of memory to hold them, and all deallocation must be performed by the programmer at the proper times. Our anagrams example illustrates some of these limitations:

```
#include <stdlib.h>
#include <stdio.h>
#include <string.h>

void swap(char *i, char *j) {
    char saved = *i;
    *i = *j;
    *j = saved;
}

void generate_permutations(char* a, int n) {
    if (n == 0) {
        printf("%s\n", a);
    } else {
        for (int i = 0; i < n; i++) {
            generate_permutations(a, n-1);
            swap(&a[n % 2 == 0 ? 0 : i], &a[n]);
        }
        generate_permutations(a, n-1);
    }
}
```

```
int main(int argc, const char* argv[]) {

    if (argc != 2) {
        fprintf(stderr, "Exactly one argument is required\n");
        return 1;
    }

    size_t len = strlen(argv[1]);
    char *word = malloc(len + 1);
    word = strncpy(word, argv[1], len);

    generate_permutations(word, len-1);

    free(word);
    return 0;
}
```

A string in C is a pointer to a block of memory. The "characters" of the string are the bytes[1] of this memory block, up to, but not including, the first zero byte. Command line arguments are given as pointers into read only memory, so we must allocate (on the heap, with `malloc`) a copy of our string that we will mutate as we generate our permutations. We need to allocate one byte more than the length of the string in order to copy over the zero byte that terminates the string. No garbage collector will reclaim this memory, so we must call `free` ourselves.

As we saw in Go, we can make pointers to objects using `&` and pass these pointers as arguments to other functions, as we've done in our `swap` function. But there's a difference. C will not, as will Go, perform escape analysis on pointers to local variables and expressions and choose heap allocation for escaped objects. Passing pointers to local variables into contexts that outlive the function owning the locals is asking for trouble! A compiler may emit a warning, but won't stop you from running such problematic, insecure code. At runtime, the function's frame can, and probably will, go away and leave the pointer **dangling**. C is permissive by design.

C++

Object-orientation was the domain of non-mainstream and research languages until C++ came along. Bolting this new paradigm onto an existing popular language for low-level systems programming brought objects to the world like no other language before it.

First appeared 1983
Creator Bjarne Stroustrup
Notable versions 98 (1998) • 03 (2003) • 11 (2011) • 14 (2014)
Recognized for Extending C with classes and much more
Notable uses Simulation, Game engines, General purpose computing
Tags Systems, Procedural, Static, Class-based, Non-managed
Six words or less Near superset of C for OOP

[1] The type `char` has a size of one byte. If needed, the type `wchar_t` exists to hold characters (code points) instead of just bytes.

In the 1970s, Bjarne Stroustrup implemented a simulator to study communication between programs in a distributed environment. The first version, written in Simula, was a "pleasure to write" but performance problems (80% of the run time was in the garbage collector) nearly doomed the project, so Stroustrup rewrote the simulator in BCPL (a predecessor of C). However, "the experience of coding and debugging the simulator in BCPL was horrible." [115] This experience led him to design a new language, a superset of C with Simula's classes, but without the need to garbage collect objects, without run-time type checking, and without the impossibly low-level near-typelessness of BCPL. The language went by the name C with Classes and evolved, by 1983, into C++.

At its core, C++ looks very much like a superset of the C systems programming language. Like Go and Swift, complex objects can be allocated in static or stack memory (when efficiency constraints come into play) or on the heap via explicit pointers (when dynamic allocation cannot be avoided). Unlike Go and Swift, C++ has *no* built-in garbage collector, so programmers must be acutely aware of all memory-sensitive operations. The language does provide various features, like **smart pointers**, that help programmers avoid certain memory bugs.

C++ is a venerable, long-lived, and still massively popular language used in enterprise applications, games, and performance-critical environments. It has evolved well: the so-called *Classic C++* has grown into *Modern C++*, a kind of language rebirth that began with C++ 11, and has continued through C++ 14 and C++ 17. Many of the programming patterns in Modern C++ replace the "old ways" of doing things. C++ programmers now enjoy extensive type inference, range-based `for`-loops, atomics, and anonymous functions (even supporting generics). The combination of expressive constructs and system-level memory operations have kept the language relevant and widely-used.

Here's a quick peek:

```
#include <iostream>
#include <map>

int main() {
  std::map<std::string, std::string> capitals = {
    {"Netherlands", "Amsterdam"},
    {"Aruba", "Oranjestad"},
    {"Cura\u00e7ao", "Willemstad"},
    {"Sint Maarten", "Philipsburg"},
  };
  for (auto& c: capitals) {
    std::cout << c.first << "'s capital is " << c.second << '\n';
  }
  return 0;
}
```

Rust

Rust is "a systems programming language that runs blazingly fast, prevents segfaults, and guarantees thread safety." [105]

First appeared 2012
Created at Mozilla
Notable versions 0.4 (2012) • 1.0 (2015) • 1.10 (2016)
Recognized for Safe systems programming
Notable uses Servers
Tags Systems, Safe
Six words or less Systems programming without memory errors

Graydon Hoare began working on Rust in 2006 in his spare time as a manager at Mozilla. Hoare's goal for the language was to mitigate many of the memory, type safety, and other preventable developer errors common in many systems languages like C++ without compromising on speed. Mozilla took interest and supported the development of the language. Rust is today a general-purpose, multiparadigm language, supporting both functional and object-oriented programming styles, and featuring traits, threads, channels, and mechanisms for wrapping unsafe code.

While a non-managed language, like C++, Rust eliminates memory leaks, dangling pointers, and even most data races *at compile time*, by defining and enforcing strict rules on the ownership of references. You'll want to check out the Rust Book [106] for complete details, but the basic idea is this: each resource has a single **owner**, and when the owner ceases to exist, the resource is deallocated. If ownership of a resource with reference semantics (such as the vector object in the following example) is transferred, the original owner can no longer access the resource:

```
fn main() {
    let a = vec![0, 1, 2];  // owned by variable a
    let b = a;              // now owned by variable b

    // println!("{:?}", a); will not compile!
    println!("{:?}", b);    // This works, b is the owner
}                           // vector deallocated, b out of scope
```

More common than an outright transfer, or move, is a **borrow**, where ownership is temporarily given. The lifetime of the borrow generally coincides with the scope of the pointer doing the borrowing:

```
fn main() {
    let mut a = 3;
    f(&mut a);
    assert!(a == 5);
}

fn f(b: &mut u8) {  // borrow the 8-bit unigned value
    *b = 5;
}                   // transfer ownership back
```

Multiple immutable borrows (defined using `&`) may be in force at a time, but mutable borrows (using `&mut`) must be exclusive of *any* other borrow. In addition to preventing certain data races, these rules prevent iterator invalidation and dangling pointers [106]. Memory safety is enforced by the compiler; there are no run-time checks to slow down execution. Rust programmers pay for these remarkable guarantees with time spent learning the rules and "fighting the borrow checker" [106, Sec. 4.9] during programming. Given the potential enormous costs of memory and concurrency errors in systems programming, the research and development of Rust continues to contribute greatly to our understanding of language-based approaches to preventing these problems from occurring.

13.3 THE ENTERPRISE

An important domain targeted by some languages is the so-called **enterprise**, a term roughly meaning "a large collection of applications, custom and third-party libraries, and scripts for managing the mission-critical operations of a business or organization." In the business world, enterprise software manages data stores, processes bulk data from partner companies (often called *feeds*), services customers' browsers and mobile applications, and generates massive amounts of usage data, or analytics. In other domains, enterprise software may refer to the millions of lines of code running sophisticated aircraft or spaceships. Regardless of the domain, building systems with hundreds of thousands, or millions of lines of code requires language support for writing correct, reliable, robust, scalable, efficient code that is easy to maintain and extend.

CLU

The first language to combine abstract data types and encapsulation, CLU, was authored by a group led by Barbara Liskov beginning in 1973. Its pioneering approach to information hiding has been adapted by hundreds of subsequent languages.

First appeared 1974
Designed by Barbara Liskov and her students at MIT
Recognized for Abstract data types, iterators, clusters
Notable uses Research into data abstraction
Tags Object-based, procedural

Although not widely implemented, CLU was an enormously influential language. It provided "linguistic support for data abstraction [and] was the first implemented language to do so. In addition, it contain[ed] a number of other interesting and influential features, including its exception handling mechanism, its iterators, and its parameterized types." [75] Its construct for modularity, the **cluster** (which gives the language its name), extended Simula's `class` by encapsulating, or hiding, all of the objects and functions not explicitly exported. The object and type systems were essentially copied to a tee by Python: values of all variables are references to entities, and assignment and parameter passing therefore copies references. Liskov herself coined the term **pass-by-sharing** for this scenario, a term popularized by Python. In addition, Python's universal supertype, `object`, is exactly CLU's `Any`.

CLU's design and implementation came from years of research on programming abstractions and structuring mechanisms, and was guided by a number of principles including "minimality, simplicity, expressive power, uniformity, safety, and performance [75, p. 8], goals adopted by subsequent languages. Liskov writes:

> CLU was geared toward developing production code. It was intended to be a tool for "programming in the large," for building big systems (e.g., several hundred thousand lines) that require many programmers to work on them. As the work on

> CLU went on, I developed a programming methodology for such systems. CLU favors program readability and understandability over ease of writing, since we believed that these were more important for our intended users. [75, p. 6]

Ada

Ada, named after Ada Lovelace, often credited as the first computer programmer, [100] was the winning entry in a competition sponsored by the United States Department of Defense to create a standard language to supersede the then hundreds of programming languages being used in the Department.

First appeared 1983
Designed by Jean Ichbiah and a team at CII Honeywell Bull
Notable versions 83 • 95 • 2012
Recognized for Code safety and maintainability
Notable uses Avionics, air-traffic control, satellites, defense systems
Tags Structured, Imperative, Static, Concurrent

Ada has an international standard [67], and happens to be the world's first internationally standardized object-oriented language. [11] Ada is a large, general-purpose language with *several* design goals. Among these are:

- *Support for megaprogramming.* Ada was designed to support systems with millions of lines of code through various modular constructs split into interfaces and implementations, and can perform type-checking across separately compiled modules.
- *Support for real-time, concurrent, embedded systems.* Ada's representation clauses, tasks, and protected objects allow for hardware support with high-level constructs.
- *A laser focus on safety.* Ada prohibits virtually all type coercions, provides many constructs that hide all non-exported entities, and requires that generic types be explicitly declared. Ada has a reputation as a verbose language due to its favoring of explicitness over implicitness nearly everywhere.

Here's a small Ada application—just to get the flavor of its syntax. A main process fires off square-this-number and print-four-dots messages to seven worker tasks, each of whom loop until no more messages can possibly be sent, thanks to the `or terminate` clause in the `select` statement. Unlike Go and Erlang, tasks are started implicitly! Communication is completely synchronous: the sender and receiver wait for each other. The `accept` statements receive messages. The portion in the `do` block is executed by the receiver while the sender remains blocked; this **rendezvous** area is generally used by the receiver to perform a thread-safe copy of the message to local storage and freeing the sender as soon as possible.

```
with Ada.Text_IO; use Ada.Text_IO;

procedure Task_Demo is

   task type Worker is
      entry Square(X: Integer);
      entry Ping;
   end Worker;
```

```
    task body Worker is
      Y: Integer;
    begin
      loop
        select
          accept Square(X: Integer) do Y := X; end;
          Put_Line(Integer'Image(Y) & "^2 = " & Integer'Image(Y * Y));
        or
          accept Ping;
          Put("....");
        or
          terminate;
        end select;
      end loop;
    end Worker;

    Workers: array(0 .. 6) of Worker;

  begin
    for I in Integer range 1..100 loop
      if I mod 2 = 0 then
        Workers(I rem 7).Square(I);
      else
        Workers(I rem 7).Ping;
      end if;
    end loop;
  end Task_Demo;
```

Though not widely used outside its niche in safety-critical environments, Ada continues to evolve today. It has both influenced and has been influenced by a number of languages for whom safety is a critical motivating design factor.

C#

C# is the flagship enterprise language for the .NET framework [84], a massively popular software platform created by Microsoft.

First appeared 2000
Principal Designer Anders Hejlsberg
Notable versions 2.0 (2005) • 3.0 (2007) • 4.0 (2010) • 6.0 (2015)
Recognized for .NET integration, delegates, properties, extension methods, LINQ
Notable uses Enterprise applications
Tags Static, Object-oriented, Multi-paradigm, Safe

C# is one of several dozen languages to enjoy an implementation on the Common Language Infrastructure, or CLI [66], and is the most widely used language for writing applications targeted to Microsoft's .NET platform. Although early versions of C# were notably similar to Java, C# has, through its evolution, kept ahead of Java by supporting features that Java only later adopted, has yet to adopt, or may never adopt. These include nullable types, reified generic types,[2] partial methods, properties, functional programming support (e.g., lambda expressions), extension methods, anonymous types, and string interpolation.

C# sports many features useful in an enterprise language, including safety features such as static type checking, bounds checking, isolation of unsafe memory operations, and a prohibition against certain forms of error-prone shadowing. It prohibits compilation of code in which local variables are not assigned, but guarantees initialization of certain other variables, including instance variables. In addition, programmers find many conveniences such as properties (sugaring getters and setters, as in Swift), namespaces, list comprehensions, operator overloading, and metaprogramming capabilities.

Let's take a brief look at a short C# script, namely, our common word frequency application:

```
using System;
using System.Text.RegularExpressions;
using System.Collections.Generic;

namespace Examples
{
    static class TraditionalWordCounter
    {
        static void Main(string[] args)
        {
            var counts = new SortedDictionary<string, int>();
            var wordRegex = new Regex(@"[\p{L}']+");
            string line;
            while ((line = Console.ReadLine()?.ToLower()) != null)
            {
                foreach (Match match in wordRegex.Matches(line))
                {
                    int count;
                    counts.TryGetValue(match.Value, out count);
                    counts[match.Value] = count + 1;
                }
            }
            foreach (var pair in counts)
            {
                Console.WriteLine("{0} {1}", pair.Key, pair.Value);
            }
        }
    }
}
```

C#'s type inference capability is similar to that of Swift; here `counts` and `wordRegex` have types inferred by the constructor expression to which they are initialized, while `pair` receives the inferred type `KeyValuePair<string, int>`.

Here's the same script, using a more fluent coding style—and a custom generator:

[2] A reified generic maintains type information about the instantiated type at run time. Contrast this approach with Java's *type erasure*, which removes this knowledge at run time.

```
using System;
using System.Collections.Generic;
using System.Text.RegularExpressions;
using System.IO;
using System.Linq;

namespace Examples
{
    static class WordCounter
    {
        static void Main(string[] args)
        {
            Regex wordRegex = new Regex(@"[\p{L}']+");
            StandardInputLines()
            .SelectMany(line => wordRegex.Matches(line).Cast<Match>())
            .Select(x => x.Value.ToLower())
            .GroupBy(x => x)
            .OrderBy(x => x.Key)
            .ToList()
            .ForEach(p => Console.WriteLine("{0} {1}", p.Key, p.Count()));
        }
        static IEnumerable<string> StandardInputLines()
        {
            string line;
            while ((line = Console.ReadLine()) != null)
            {
                yield return line;
            }
        }
    }
}
```

The `Select` (think "map"), `SelectMany` (think "flatMap"), and similar methods come from the `System.Linq` namespace, which provides an extension class for immutable arrays defining the usual "streaming" methods. Linq stands for "language-integrated query." One can write Linq-style code using the methods directly (as we've done here) or employ a **query expression syntax**, for which details are a quick web search away.

Scala

C# and Java, which we toured in Chapter 7, dominate the enterprise language space today, due in no small part to the massive ecosystems supporting their platforms. But recent years have seen an explosion on the number of languages that interoperate with these leaders. You can, for example, "write Java" in Ceylon, Kotlin, Clojure, and Scala.

First appeared 2003
Creator Martin Odersky
Notable versions 2.0 (2006) • 2.8 (2010) • 2.12 (2016)
Recognized for Blending the functional and object paradigms
Notable uses Infrastructure, APIs
Tags Functional, Object-oriented
Six words or less Modern, scalable language for the JVM

Scala (the *Sca*lable *la*nguage) was created by Martin Odersky at École Polytechnique Fédérale de Lausanne in the early 2000s as a functional, statically-typed, object-oriented language for the Java platform. Scala programs are composed of classes which can be freely mixed with Java classes. Though Scala is interoperable with Java and its threaded concurrency model, its preference for immutability and safety drove it to adopt the actor model as its main concurrency feature, though like Clojure, it also features software transactional memory.

Scala programs are *much* less verbose than their Java equivalents, thanks to type inference, string interpolation, and the ability to just write scripts without an enclosing class. Consider our recurring animals example:

```
abstract class Animal(name: String) {
  def speak = s"$name says $sound"
  def sound: String
}

class Cow(name: String) extends Animal(name) {
  override def sound() = "moooo"
}

class Horse(name: String) extends Animal(name) {
  override def sound() = "neigh"
}

class Sheep(name: String) extends Animal(name) {
  override def sound() = "baaaa"
}

var h = new Horse("CJ")
assert(h.speak == "CJ says neigh")
var c = new Cow("Bessie")
assert(c.speak == "Bessie says moooo")
assert(new Sheep("Little Lamb").speak == "Little Lamb says baaaa")
```

The field names of each class are found in the parameter list of the class declaration, and constructors are automatically generated.

Naturally, Scala's ability to interoperate with Java gives it access to the entire Java platform making it a player in the enterprise space.

13.4 SCRIPTING

In a 1998 article [92], John Ousterhout identified a class of languages he termed **scripting languages**, remarking:

> Scripting languages assume that a collection of useful components already exist in other languages. They are intended not for writing applications from scratch but rather for combining components. For example, Tcl and Visual Basic can be used to arrange collections of user interface controls on the screen, and Unix shell scripts are used to assemble filter programs into pipelines. Scripting languages ...are rarely used for complex algorithms and data structures, which are usually provided by the components.

Today, the notion of what defines a scripting language is much more fluid, at times encompassing any language with sufficiently dynamic, high-level, expressive features. We're interested here in two classic exemplars: Bash and Perl.

Bash

Bash is one of the best-known of the Unix shell languages.

First appeared 1989
Original author Brian Fox
Notable versions 3 (2004) • 4 (2009) • 4.2 (2011)
Recognized for Shell scripting, Text processing
Notable uses The primary shell for the popular GNU and macOS operating systems
Tags Scripting

Today, operating system shells do much more than simply invoke and combine programs. Even the earliest of the Unix shells, the **Bourne Shell**, colloquially known as **sh**, provided variables, rudimentary control flow, and command substitution. Bash, intended as a replacement and extension of sh, added additional looping and conditional constructs, unbounded arrays, new forms of command substitutions, and support for in-process arithmetic.

Here's a trivial Bash script, defining a function that turns a hexadecimal string (e.g. ADC2DB8B) into its dotted quad representation (e.g., 173.194.219.139). It illustrates a couple string processing tricks we'll leave to you to explore on your own.

```
#!/usr/bin/env bash
function hex2dottedquad {
    counter=0
    while [ $counter -le 6 ]; do
        hex=${1:$counter:2}
        let decimal=0x$hex
        ip+=.$decimal
        let counter+=2
    done
    echo ${ip:1:${#ip}}
}
```

```
hex2dottedquad $1
```

Perl

According to its designer, Larry Wall, Perl has two slogans: "There's more than one way to do it," (or TMTOWTDI) and "Easy things should be easy and hard things should be possible." [124] Perl rose to fame in the early days of the World Wide Web due to its extraordinarily powerful text processing capabilities.

First appeared 1987
Designed by Larry Wall
Notable versions 1.0 (1987) • 3 (1989) • 4.036 (1993) • 5.004 (1997) • 5.24 (2016)
Recognized for Powerful text processing, popularizing regular expressions
Notable uses Early web backends (CGI), Networking, Bioinformatics
Tags Imperative, Shell, Object-oriented

Perl continues to hold on to a fair share of popularity in the scripting world, where it competes with Python, Ruby, Tcl, and others. It is, to some old timers, the first language that comes to mind when hearing the phrase *regular expressions*. Perl's extensive regular expression support inspired the popular PCRE library.[3]

Perl has a rather unique culture in the programming language space. The language is very flexible and supports a great many programming styles. Wall's training as a linguist led to many of the language's highly expressive—and sometimes quirky—features, including various quoting mechanisms, a flip-flop operator, a spaceship operator, the diamond, and the tilde-tilde. Wall has even come up with a self-deprecating backronym for the language: *Pathologically Eclectic Rubbish Lister.*

13.5 THE ML FAMILY

While working on the theorem proving system LCF (short for Logic of Computable Functions), Robin Milner and colleagues in the early 1970s created a language for writing scripts to create proofs. Since LCF was itself a language for representing formulas in an underlying logic, the theorem proving language was called the *Meta Language*, or **ML** for short. The original ML language soon evolved into a general-purpose programming language. Like Lisp, it featured first-class functions and pattern matching, and was expression-oriented and garbage-collected. Unlike Lisp, ML was statically typed, and was the first language to implement the Hindley-Milner type system and its associated type inference mechanism.

A number of ML dialects soon appeared, with the most popular family branches being led by Standard ML and OCaml. Several closely-related languages such as Elm, Haskell, and F# include the characteristic features of ML mentioned above that we consider them, at least for this book, to be part of the (extended) ML family.

[3] PCRE is an acronym for Perl Compatible Regular Expressions.

Standard ML

In April of 1983, Milner authored *A Proposal for Standard ML* to advocate for a language that would consolidate the strengths of the multiple ML dialects created to that time. The new language appeared shortly after.

First appeared 1984
Principal Designers Robin Milner, Mads Tofte, Robert Harper
Notable versions 90 (1990) • 97 (1997)
Recognized for Pattern matching, Hindley-Milner typing, sophisticated modules
Notable uses Theorem proving
Tags Functional

The most notable feature of Standard ML is its module system.

- **Structures** define collections of types and values (simple, structured, and function).
- **Signatures** provide interfaces into structures; they are often used to provide a restricted view of the structure to clients. Structures can be parameterized.
- **Functors** can compute new structures, giving an interesting computational view into generic instantiation, among its other capabilities.

Here's an example of a signature and a matching structure—perhaps you recognize it as a persistent stack data structure. We'll leave functors to your self-study.

```
signature STACK =
sig
  type 'a stack
  val empty: 'a stack
  val push: 'a * 'a stack -> 'a stack
  val pop: 'a stack -> 'a * 'a stack
  val size: 'a stack -> int;
end;

structure Stack :> STACK =
struct
   type 'a stack = 'a list
   val empty = nil
   fun push (elem, stack) = elem::stack
   fun pop (first::rest) = (first, rest)
     | pop _ = raise Domain
   fun size (s) = length s;
end;

val s = Stack.empty;
val s = Stack.push(10, s);
val s = Stack.push(3, s);
val s = Stack.push(6, s);
val s = Stack.push(12, s);
val (x, s) = Stack.pop(s);

Stack.size(s) = 3 orelse raise Fail "failure";
```

OCaml

While Milner's efforts to consolidate the disparate ML dialects into a single standard had some success, one family branch—that of INRIA's CAML—did not join the party. CAML soon evolved into OCaml, also known as Objective Caml.

First appeared 1996
Architect Xavier Leroy
Notable versions 3.00 (2001) • 3.10 (2008) • 4.03 (2016)
Recognized for Performance, Expressiveness, Safety
Notable uses Compilers, Financial Trading
Tags Functional, Object-oriented

OCaml and Standard ML are very similar languages, inheriting all of the pattern-matching, type inference, and other powerful features of their ancestor, but differ greatly in syntax. OCaml does have better support for what we today call object-orientation, and has turned out to be better known than its "standard" sibling. It has notably become a convenient language for authoring compilers—both Facebook's Hack and early versions of Mozilla's Rust were written in OCaml. Modern languages such as Elm, F#, Scala, and Rust have listed OCaml as one of their primary influences.

Haskell

Haskell arose from an effort to build an open standard bringing together various ideas and contributions from some twenty or so closely-related functional languages that grew up in the 1980s. It has since become the most popular pure functional language in use today, with many fans both in academia and industry.

First appeared 1990
Notable versions 1.0 (1990) • 98 (1998) • 2010
Recognized for Pure functions, non-strict semantics, type classes, monads
Notable uses Academia, Software validation and verification, Modeling, Compilers
Tags Non-strict, Lazy, Functional

Haskell sports the usual ML-style features such as higher-order functions, static Hindley-Milner typing, tuple types, pattern matching, and modules. But Haskell is quite different in at least three ways from its relatives, and, indeed, from nearly every language we've encountered so far: Haskell has a **non-strict semantics**, categorizes types via **type classes**, and meets enough of the requirements to be known as a **pure** functional language. We'll discuss these language attributes in the remainder of this section.

A non-strict semantics means that expressions can produce a valid value even when subexpressions have no value.[4] For example, if we define the function *zero* as:

```
zero x = 0
```

then the Haskell expression `zero (1 `mod` 0)` has the value 0, while evaluating the equivalent expression `zero (1 % 0)` in Elm, whose semantics is strict, produces a division by zero

[4] Loosely, an expression without a value is one whose evaluation would enter an infinite loop or fail. In formal semantics, the meaning assigned to a "valueless" expression is $\perp$, pronounced "bottom." A semantics is **strict** iff the meaning of $f(\perp)$ is $\perp$ for all functions, operators, and constructors f.

error. A non-strict semantics allows us to write "infinite structures" as seen in the function definition gracing the language's home page:

```
primes = filterPrime [2..]
  where filterPrime (p:xs) =
          p : filterPrime [x | x <- xs, x `mod` p /= 0]

main = print(take 10 primes)
```

Here we define `primes` as the (infinite) list of all values x greater than or equal to 2, such that x is not divisible by any prime number p. The list comprehension and the ability to define a local, recursive function in a `where` clause permit concise code. The application of `take` forces the infinite list to produce its initial values.

Haskell's built-in types include `Int` (bounded integers), `Integer` (unbounded), `Char` (all Unicode characters), `Bool` (defined with two zero-argument constructors `False` and `True`), `Float` (32-bit floating-point) and `Double` (64-bit). The type `[a]` is the list of elements of type a; `(a,b)` is the type of tuples whose first element is of type a and the second from type b; and `a->b` is the type of functions from type a to type b. The type `String` aliases `[Char]`. Define new types with `data`, define type aliases with `type`, and define functions by cases on the new type's constructors:

```
data PrimaryColor = Red | Green | Blue
data Shape = Circle Float | Rectangle Float Float
data Response a = Ok a | Error String
data Tree a = Leaf a | Internal a [Tree a]

type Point = (Float, Float)
type Polygon = [Point]

size :: Tree a -> Int
size (Leaf _) = 1
size (Internal _ children) = 1 + foldr (+) 0 (map size children)
```

Unlike Elm, which deals with the sometimes tricky problem of overloading functions and operators (e.g., `*` across numeric types, `==` across types that support notions of equality, `<` across comparable types, and `++` across appendable types) with fixed, special type variables, Haskell has a full-blown system of **type classes**.

How does Haskell infer the type of the function f defined below?

```
f x y z = (x == y) && (z < z)
```

You might answer `a -> a -> b -> Bool`, but this is too general. Not every type in Haskell admits equality testing (functions do not!), and not every type has an underlying ordering via `<`. Types admitting `==` are members of the type class `Eq`, and types understanding `<` belong to the type class `Ord`. Haskell thus denotes the inferred type as:

```
(Eq a, Ord b) => a -> a -> b -> Bool
```

Note the similarity with Swift's

```
func f<T:Equatable, U:Comparable>(x: T, y: T, z: U) -> Bool {
    return x == y && z < z
}
```

While Swift types *adopt* protocols, Haskell types *are instances of* type classes. Classes may inherit from multiple parents. For example, Integral inherits from Enum and Real; Real inherits from Num and Ord; and Num inherits from Eq and Show. The complete list of predefined types and type classes resides at [96, Ch. 9]. And by the way...Haskell programmers can *define their own type classes.*

Finally, Haskell functions are *pure*: they operate only on their arguments and are side-effect-free. They produce the same result for the same arguments every time—simplifying testing and formal verification, and creating opportunities for optimization. [69] But the real world isn't pure: we need to do I/O, generate random numbers, manage external state, etc. Haskell isolates **impure** actions, such as I/O, by rolling them into objects that are later executed to perform effects. Let's get a feel for how this is done by looking at the I/O system.

The type IO a represents an action that, when executed, transforms the global state via I/O and returns a value of type *a*. So getLine has type IO String, getChar has type IO Char, and putStrLn has type String -> IO (). The latter maps a string into an action that produces a real-world-state with a string written to standard output, returning nothing of interest (represented by the empty tuple, ()). To send the result of one action to another, use the operator >>= of type IO a -> (a -> IO b) -> IO b:[5]

```
main =
  getLine >>= \s -> putStrLn $ "Hello, " ++ s
```

In practice, you'll employ syntactic sugar for I/O operations (e.g., <- and do) rather than >>=. Usage can be found in any tutorial; for an in-depth technical look at Haskell I/O see [50] and [51]. (Speaking of syntax, the $ operator is equivalent to Elm's <|.)

Haskell is an incredibly rich language with much to offer both programmers and those interested in the study of programming languages. For an excellent history of the language, including its founding, basic principles, and numerous research contributions, see [62].

F#

F# is a modern ML for the .NET world.

First appeared 2005
Designers Don Syme, Microsoft
Notable versions 1.0 (2005) • 2.0 (2010) • 3.0 (2012) • 4.0 (2016)
Recognized for Being a modern ML
Notable uses Machine learning, Business Intelligence, Analytics, Gaming
Tags Static, Functional, Multi-paradigm

Microsoft's F# first appeared in 2005 and was developed as an open-source, cross-platform CLI language. Microsoft bills it as a "functional-first" language. While the similarities to ML languages such as OCaml and Elm are unmistakable, F# contains an impressive list of features and support for imperative programming, asynchronous workflows and agents, object-orientation, and parallel code.

[5] The existence of the "chaining" operation >>= on values, together with the "wrapping" operation return, which maps a value of type a to one of type IO a, means that IO is a **monad**, and that I/O in Haskell is said to be **monadic**. The language features several other monads, including Maybe. [96, Ch. 21]

Idris

Our last ML-influenced language is one to watch: Idris. Idris debuted in 2009 as a language featuring **dependent types**. Dependent types allow many aspects of a program's behavior to be specified within the type system itself. Common examples of dependent types include "the type of all vectors of length n", "the type of all pairs of integers in which the second element is greater than the first," and "the type of all functions returning the value seven."

First appeared 2009
Designed by Edwin Brady
Notable versions 0.10 (2016) • 0.11 (2016)
Recognized for Dependent Types
Notable uses Research, Mechanical Theorem Proving
Tags Pure, Functional

A type system with dependent types allows computations to be performed within the type checker, *at compile-time*. The programmer can then set guarantees that a program, when executed, will meet certain specifications. In fact, Idris borrows a number of features from interactive proof assistants such as Coq and Agda. While Idris can be used for theorem proving, its development has focused on general purpose computation, and it is actively being developed as such today.

Syntactically, Idris looks a great deal like Haskell, with some superficial differences beyond those forms required to express dependent typing. On the semantic side, Idris is, like Haskell, a pure functional language. However, Idris has additional machinery for managing side effects (networking, file I/O, exceptions, user interaction), using **algebraic effects** in addition to monads. Idris's creator, Edwin Brady, describes the theory and implementation of algebraic effects, and the means by which they solve various shortcomings of Haskell's monadic approach, in [13].

13.6 CONCURRENCY MATTERS

Most modern languages feature some concurrency support, but experienced developers working on mission-critical applications require a deep understanding of the fundamental principles of concurrent and distributed systems. This knowledge comes not only from day-to-day practice but also from early seminal work such as Hoare's CSP and Milner's CCS, and the early experimental and research languages occam (featuring named channels, as adopted by Go), SR, and Linda (sporting a unique means of process coordination through a "tuple space").

Naturally, research in concurrent and distributed systems remains extremely active. We've spent time in previous chapters on a few principles of concurrency (especially in our tours of Clojure, Erlang, and Go), and should note that a few languages in *this* chapter (Rust, Ada, Scala, and C# for instance) feature many capabilities for concurrent and distributed computing. Still, a few languages created in recent years are worth a very brief mention.

ParaSail

ParaSail, or *Parallel Specification and Implementation Language*, is "a new parallel programming language designed to support the development of inherently safe and secure, highly parallel applications that can be mapped to multicore, manycore, heterogeneous, or distributed architectures." [116] The language has a familiar syntax and traditional object-oriented features such as classes, interfaces, and parameterized modules. Safety features are designed into the language itself and enforced by the compiler; there are no pointers, global variables, parameter aliasing, exceptions, or explicit locking.

First appeared 2009
Designed by Tucker Taft
Notable versions 4.7 (2013) • 5.2 (2014) • 6.3 (2015)
Recognized for Parallel processing, Compile-time enforcement of safe parallelism
Notable uses Parallel processing for multicore processors
Tags Safe, Static, Parallel

Chapel

Chapel, developed at Cray Inc. and originating as part of DARPA's High Productivity Computing Systems (HPCS) program, aims to improve the programmability of parallel computers. It is strongly influenced by classic languages such as (High-Performance) Fortran, C, and C++ and their associated libraries, but has the look and feel of more modern, expressive languages such as Python. An important design goal has been to make the language expressive enough to formulate all parallel tasks with out the need to drop into a library. Architecturally, writes Brad Chamberlain, Chapel's principal engineer, "Chapel supports distinct concepts for describing parallelism ('These things should run concurrently') from locality ('This should be placed here; that should be placed over there'). This is in sharp contrast to conventional approaches that either conflate the two concepts or ignore locality altogether." [18]

First appeared 2009
Designed at Cray Inc.
Notable versions 1.7 (2013) • 1.13 (2016)
Recognized for Separation of parallelism and locality
Notable uses High Performance Computing
Tags Parallel, Portable, Scalable

Elxir

Elixir was created by José Valim in 2012. Roughly, it's a modern Erlang, running on the Erlang virtual machine (EVM), with all of Erlang's process capabilities, but with a syntax influenced by Ruby. It adds to Erlang powerful metaprogramming facilities, with hygienic macros (operating on abstract syntax as in Julia and Clojure), protocols, and reducers. You'll also find working with strings to be much nicer than in Erlang (recall those lists of integers), and enjoy the `|>` operator we saw in Elm. The Ruby-style syntax makes the authoring of domain-specific languages within Elxir quite pleasant.

First appeared 2012
Designed by José Valim
Notable versions 1.0 (2014) • 1.3.0 (2016)
Recognized for Concurrency, Immutability
Notable uses Web development through Phoenix framework
Tags Functional, Concurrent, Process-Oriented

13.7 THE WEB

Given the ubiquity of the World Wide Web, it's not at all surprising that programming languages specifically targeted to the web as a platform would arise. The web is a **client-server** system, where multiple clients interact asynchronously with a stateless server using a protocol with a small, limited number of verbs, namely GET, POST, PUT, PATCH, DELETE, OPTIONS, HEAD, and TRACE.

While just about any language can be—and has been—used on the server side of the web (e.g., Python, Ruby, Java, and many others come to mind), one, PHP, owes its very existence to its creator's desire for a server-side web application scripting program. We'll look at PHP, and a successor language, Hack. On the client-side, there is just one choice: JavaScript. And hundreds of others! These other, compile-to-JavaScript, languages include ClojureScript (a Clojure dialect), PureScript (which looks suspiciously like Haskell), and CoffeeScript (from Chapter 2). You can find a list of these languages at [9]. We'll give a mention in this section to TypeScript and Dart.

PHP

PHP is a server-side scripting language developed by Rasmus Lerdorf, who originally built it to maintain his **p**ersonal **h**ome **p**age.

First appeared 1994
Designed by Rasmus Lerdorf
Notable versions 4.0.0 (2000) • 5.0.0 (2004) • 7.0.0 (2015)
Recognized for Server-side web applications
Notable uses E-commerce, Powering many of the world's web servers
Tags Imperative, Weakly typed, Web scripting

PHP now stands for **P**HP **H**ypertext **P**rocessor, and is today an expressive, general-purpose programming language with plenty of features, including closures, null-coalescing, exception handling, generators, and variadic functions.

PHP does, however, remain dogged by some less-than-stellar design choices[6] in its somewhat organic evolution. Ample good-natured criticisms of these design choices are documented in [29] and [126] among others. The language has, over the years, shed some of the early problematic features, but even with its remaining issues is able to stay high on the list of the world's most popular languages.

[6] For example: inconsistent spelling: `gettype` vs. `get_class`; the left-associative conditional: `TRUE ? 1 : FALSE ? 2 : 3` is 2, and strangely named methods: `DateTimeImmutable::modify()`.

Hack

How does one deal with the many problems (real or perceived) of PHP? "Using a different language" comes to mind, but so does evolving the language in subtle, but radical, ways. Facebook took this approach in creating Hack.

First appeared 2014
Developer Facebook
Recognized for Being PHP with static typing and modern features
Notable uses Facebook codebase migration from PHP
Tags Imperative, Gradually typed, Web scripting

Facebook released Hack in 2014. It can be thought of as a fix of PHP. Facebook says that Hack "reconciles the fast development cycle of PHP with the discipline provided by static typing, while adding many features commonly found in other modern programming languages." [32] You'll find type annotations, generics, nullable types, anonymous functions, type aliasing, and support for asynchronous computation. Learning resources and thorough documentation can be found at [33].

TypeScript

TypeScript is a typesafe superset of JavaScript. [85]

First appeared 2012
Developer Microsoft
Notable versions 0.8 (2012) • 0.9 (2013) • 1.0 (2014) • 1.8 (2016)
Recognized for Type annotations, Static checking
Notable uses Large web clients
Tags Object-oriented, Structured, Imperative, Functional

One of the many compile-to-JavaScript languages for web client development, Microsoft's TypeScript, is a superset of JavaScript—if you know JavaScript, you can start using TypeScript right away—that allows type annotations to open up possibilities for static analysis of source code. In addition, TypeScript also features classes, inheritance, generics, and modules, backing up its tagline "JavaScript that scales."

Dart

Dart is a another alternative to JavaScript. Unlike TypeScript, Dart is no JavaScript superset, nor even a dialect, but rather a distinct language for client-side web applications.

First appeared 2011
Developer Google
Notable versions 1.12.0 (2015) • 1.16.0 (2016)
Recognized for Optional typing, Isolates, SIMD support
Notable uses Adobe, AdSense, Google Fiber
Tags Object-oriented, Imperative, Functional

Dart was created by Google to create scalable web applications. Its flavor of object-orientation, with interfaces, abstract classes, and generics are familiar to fans of Java and

C#. While supporting client-side web applications and interoperability with JavaScript, Dart can manipulate files, sockets, and other I/O channels not available to typical browser-based runtimes.

Dart supports **optional typing**. You can provide type annotations on any subset of your variables, parameters, and functions, and these will be verified at run time in *checked mode*. In *production mode*, annotations will neither be checked nor enforced.

Dart enables concurrency through **isolates**, a multithreaded message passing system that compiles to the HTML5 Web Worker API [127] (or to the browser's event loop if web workers are unavailable). The language also defines the SIMD types `Float32x4`, `Int32x4`, `Float32x4List` and `Int32x4List` supporting hardware-level parallelism crucial for high-performance graphics, and other computationally intensive tasks.

Here is our recurring word count example in Dart:

```
import 'dart:io';
import 'dart:convert';
import 'dart:collection';

final words = new SplayTreeMap<String, int>();

void main() {
  stdin.transform(UTF8.decoder)
    .transform(const LineSplitter())
    .listen(addLines, onDone: printWords);
}

void addLines(String data) {
  data.toLowerCase().split(new RegExp(r"[^a-z'\w]+")).forEach((word) {
    if (word.length > 0) {
      words[word] = words.putIfAbsent(word, () => 0) + 1;
    }
  });
}

void printWords() {
  words.forEach((k, v) => print('$k $v'));
}
```

13.8 CRYSTALLIZATIONS OF STYLE

Some languages are strong exemplars of a particular style or paradigm, often featuring small, elegant specifications that take up only a few pages. Nearly every computation in such a language looks about the same. Everything is an object, or an array. Each computation step generally has the same form: manipulate a stack (Factor), send a message to an object (Io), apply a function to an array (K). We've seen one such language a few pages back, Smalltalk, exemplifying the object paradigm. In this section, we'll encounter languages that epitomize the array, logic, prototype, and concatenative paradigms.

APL

Kenneth Iverson's work on a mathematical notation to manipulate arrays culminated in a 1962 publication at Harvard called *A Programming Language*. By 1964 at IBM, Iverson adapted this notation into a programming language, APL.

First appeared 1962
Designers Kenneth Iverson
Recognized for Arrays, conciseness, symbolic alphabet
Notable uses Finance, Scientific research
Tags Array-based, Dynamic

Few languages can match APL for conciseness. APL programs are expressions in which operators are (successively) applied to arrays. APL features many dozens of operators, which often have different meanings when used in a unary (monadic)[7] versus a binary (dyadic) context. We can get a feel for array processing in the famous expression for producing an array of all prime numbers less than or equal to R:

$$(\sim R \in R \circ . \times R) \,/\, R \leftarrow 1 \downarrow \iota R$$

For concreteness, assume R is 8. In APL, all operators associate to the right, so we first evaluate $\iota 8$ which produces `[1 2 3 4 5 6 7 8]`. Next we apply the binary drop ($\downarrow$) operator to remove one element from the front, resulting in `[2 3 4 5 6 7 8]`. We then assign ($\leftarrow$) this array to R. On the left side of the compress operator (/), we have $R \circ . \times R$ which builds a two-dimensional array containing the outer product, or multiplication table, of R with itself:

4	6	8	10	12	14	16
6	9	12	15	18	21	24
8	12	16	20	24	28	32
10	15	20	25	30	35	40
12	18	24	30	36	42	48
14	21	28	35	42	49	56
16	24	32	40	48	56	64

Next we apply $\in$ to R and this table, producing the array containing 1 for each element of R present in the table and 0 where not present, resulting in `[0 0 1 0 1 0 1]`.[8] Applying the $\sim$ (not) operator over each element yields `[1 1 0 1 0 1 0]`. Now we apply the compression operator which selects those elements of the right operand having a 1 in the left:

$$\begin{array}{ccccccc} 1 & 1 & 0 & 1 & 0 & 1 & 0 \\ \downarrow & \downarrow & & \downarrow & & \downarrow & \\ 2 & 3 & 4 & 5 & 6 & 7 & 8 \\ \downarrow & \downarrow & & \downarrow & & \downarrow & \\ 2 & 3 & & 5 & & 7 & \end{array}$$

The prime numbers up to 8 are thus: `[2 3 5 7]`.

[7]Don't confuse APL's use of the term *monadic* with Haskell's; the two meanings are wholly distinct.
[8]Read this result as saying that of the values 2..8. only 4, 6, and 8 were found in the table.

Prolog

Prolog is the brainchild of a collaboration between Alain Colmerauer and Phillipe Roussel of the University of Aix-Marseille and Robert Kowalski of the University of Edinburgh in the late 1960s and early 1970s. As one of the first logic programming languages, Prolog has been extremely influential in the domains of theorem proving, expert systems, and natural language processing. It also greatly influenced the development of Erlang.

First appeared 1972
Designed by Alain Colmerauer
Recognized for Logic programming, Natural Language Processing
Notable uses Artificial Intelligence (notably IBM's Watson), Computational Linguistics
Tags Declarative, Logic

A Prolog consists of a list of facts and inference rules. Programming is generally declarative; the engine figures out how to complete queries based on the given facts and rules. Here's a very simple Prolog "database:"

```
plays(kira,waterpolo).
plays(colette,diving).
plays(emily,waterpolo).
plays(olivia,diving).
plays(michelle,piano).
club(colette,rosebowl).
club(kira,rosebowl).
club(emily,rosebowl).

diver(X) :- plays(X,diving).
teammate(X,Y) :- plays(X,Z), plays(Y,Z), club(X,C), club(Y,C), X\=Y.
```

We can now run several different computations, or **queries**, against this database. For example:

- `?- plays(olivia,S)` binds the variable S to the atom `diving`.
- `?- plays(P,waterpolo)` produces multiple successive bindings for P, for each water polo player.
- `?- diver(colette)` produces `yes`, while `?- diver(alex)` produces `no`. There may be a diver named Alex, but Prolog cannot prove this fact from the given database. In Prolog, `yes` means "I can prove this" and `no` means "I cannot prove this."
- `?- teammate(oliva,colette)` produces `no`. We don't know Olivia's club.
- `?- teammate(X, Y)` finds all pairs of teammates. Our first result is X = `kira`, Y = `emily`, and the second X = `emily`, Y = `kira`.
- `?- plays(X, waterpolo), club(X, rosebowl)` binds X successively to each of the known Rose Bowl water polo players.

Prolog searches through the universe of possible facts to find query results, using backtracking internally. Occasionally, the programmer needs knowledge of this process in order to apply linguistic devices such as the **cut** (`!`) to drastically improve efficiency. Exploration of these devices are, alas, beyond the scope of this text.

K

K "is the executable notation at the heart of a high performance programming platform. It is designed for analyzing massive amounts of real-time and historical data—ideal for financial modelling." [131]

First appeared 1993
Designers Arthur Whitney
Notable versions N/A
Recognized for Array operations, Expressive syntax
Notable uses Financial products
Tags Array-based, Dynamic

K is an array-based language in the APL family (which also includes the tersely-named J and Q). Its designer, Arthur Whitney, had extensive APL experience at several companies, most notably at Morgan Stanley, where he worked on financial and trading applications and developed another APL-family language, A+. In 1993, Whitney left Morgan Stanley to form Kx Systems to commercialize the newer language, K.

Like APL, K's fundamental data type is the array, and nearly all operations apply to, or are easily extended to—arrays. Unlike APL, K programs are written in the ASCII character set. Let's take a look at the factorial function, defined as follows:

```
factorial:{*/1+!:x}
```

Working from right to left, as K does:

```
factorial 10
  = {*/1+!x} 10                -- call a function
  = */1+!10                    -- parameters are x, y, and z
  = */1+0 1 2 3 4 5 6 7 8 9   -- !n makes the array of 0..n-1
  = */1 2 3 4 5 6 7 8 9 10    -- scalar+array makes an array
  = 3628800                    -- op/array reduces (folds) the op
```

In K terminology, +, * and * are called **verbs**, as they apply to values, while / is an **adverb**, as one of its arguments is a verb. We may recognice the slash as the higher-order fold, or reduce operation, but K calls it OVER. Other adverbs include the apostrophe, called EACH, and \ called SCAN, similar to OVER but accumulates partial results. Let's run some examples in the REPL:

```
  factorial:{*/1+!x}
{*/1+!x}
  factorial' 3 8 0 4
6 40320 1 24
  +/3 8 0 4
15
  +\3 8 0 4
3 11 11 15
```

K is a proprietary language defined in [73], but an open source version of the language, Kona, is available at `https://github.com/kevinlawler/kona/`. The Kona wiki has a number of example functions, a few of which we've duplicated here. Enjoy your self-study of K, and have fun puzzling out the inner workings of the following concise function definitions.

```
leapyear:{(+/~x!'4 100 400)!2}

pascal:{x{+':0,x,0}\1}

gcd:{:[~x;y;_f[y;x!y]]}

sort:{x@<x}

collatz:{(1<){:[x!2;1+3*x;_ x%2]}\x}
```

Io

Steve Dekorte's Io is another simple, "pure" language, whose programs consist entirely of objects sending messages. Dekorte created Io in 2002 "to refocus attention on expressiveness by exploring higher level dynamic programming features with greater levels of runtime flexibility and simplified programming syntax and semantics." [114]

First appeared 2002
Designers Steve Dekorte
Recognized for Simplicity of design
Notable uses Exploratory software development
Tags Prototypal

Like JavaScript and Lua, Io is a prototype-based language. In Io, all computation, including assignment, conditionals, and loops, proceeds by sending messages. All objects are run-time accessible, as is the source code tree (yes, metaprogramming opportunities abound). The language provides actors, futures, and coroutines for concurrency. Io has much to offer and we can't do the language justice in this short section, but we'll leave you with a look at our right triangle measurement script:

```
for(c, 1, 40,
  for(b, 1, c-1,
    for(a, 1, b-1,
      if(a*a + b*b == c*c,
        (list(a, b, c) join(", ")) println,
        nil))))
```

Factor

Although this book's main tour managed to visit a number of programming paradigms, we've up to now missed **concatenative programming**. Concatenative programing languages build subroutines for evaluation by composing functions. Concatenative programming is generally differentiated from applicative programming in which functions are applied to arguments. Factor is a modern concatenative language—PostScript is another—and like many, but not all, concatenative languages, it is also **stack-based**.

First appeared 2003 **Designers** Slava Pestov **Notable versions** 0.97 (2014) **Recognized for** Flexibility, Large standard library **Notable uses** Optimization technique research **Tags** Stack-based, Concatenative

In Factor, as in other stack-based languages, each operation takes its implicit operands from a stack. Basic arithmetic has a postfix flavor:

```
9 7 * 2 + 13 neg /
```

Operands generate pushes. Operators pop values, compute, then push results:

1. Push 9, $Stack = \langle 9 \rangle$
2. Push 7, $Stack = \langle 9\ 7 \rangle$
3. Pop operands, push product, $Stack = \langle 63 \rangle$
4. Push 2, $Stack = \langle 63\ 2 \rangle$
5. Add, $Stack = \langle 65 \rangle$
6. Push 13, $Stack = \langle 65\ 13 \rangle$
7. Negate, $Stack = \langle 63\ \ -13 \rangle$
8. Divide, $Stack = \langle -5 \rangle$

The stack can hold code blocks, too, as in:

```
3 [ "Hello, world" print ] times
```

Developers and students of computer science will do well to study stack-based execution. There is a long history of virtual machines using stack execution, including the very popular Java Virtual Machine (JVM).

13.9 ESOTERIC LANGUAGES

No tour of programming languages should omit the ever-growing family of **esoteric languages**, with members such as Ook!, Malbolge, Befunge, LOLCODE, Shakespeare, Chef, Chicken, Whitespace, and GOTO++. You won't see any of these powering any real-world applications; instead, their designers work outside the box to produce artistic, crazy, funny, themed (we'll see LOLCODE momentarily), weird, or insightful creations that may push the boundaries of minimalistic computing models, look at computing in a whole new way, or simply amuse.

Brainf**k

Brainf**k is one of the better known of the esoteric languages and one of the earliest. Urban Müller's creation grew out of an effort to design a Turing-complete[9] programming language implementable with the smallest possible complier (296 bytes at the time, with smaller compilers created since).

[9]Formally, as powerful as a Turing Machine; informally, capable of doing anything a Python program can do assuming infinite memory.

First appeared 1993
Designer Urban Müller
Recognized for Minimalism
Notable uses Research into small compilers
Tags Esoteric, Imperative

A Brainf**k program is a sequence of one-character commands operating on a single infinite array of bytes (indexed from zero) and two byte streams, standard input and standard output. The *current* memory cell is initially the leftmost. The eight commands are:

`,`	Read the byte from stdin into the current cell
`.`	Write the byte in the current cell to stdout
`<`	Move the current cell pointer left
`>`	Move the current cell pointer right
`+`	Increment the byte in the current cell
`-`	Decrement the byte in the current cell
`[`	If the byte in the current cell is zero, jump ahead past the matching `]`
`]`	If the byte in the current cell is nonzero, jump backwards to just beyond the matching `[`

The following program:

```
+++++++[>+++++[>++>+++<<-]<-]>>++.>.
```

prints the string `Hi` to standard output. As all cells have an initial value of zero, the program first increases the first cell to 7, then the second to 5, then the next two cells to 2 and 3, with the fourth cell as our current cell:

```
7  5  2  3
         ^
```

We're inside a doubly-nested "loop" here. Our next instruction says to go back two cells and decrement the value. If not yet zero, we'll continue our inner loop of incrementing the third and fourth cells by 2 and 3 respectively. After this is done five times, our memory becomes:

```
7  0  10  15
   ^
```

We're back in an "outer loop" now, and we're supposed to go back one cell to the left and decrement...we're being told to do what we just did seven times. This takes us to the following state:

```
0  0  70  105
^
```

We now need to move two cells to the right, bump the value by 2 to 72, the code point for LATIN CAPITAL LETTER H and print it with the `.` command. Moving one cell to the right we find, happily, 105, the code point of LATIN SMALL LETTER I, which we print. And we're done.

You might want to now work through the classic Hello world program, as found on [132]:

```
++++++++[>++++[>++>+++>+++>+<<<<-]>+>+>->>+[<]<-]>>.>
---.+++++++..+++.>>.<-.<.+++.------.--------.>>+.>++.
```

Finally, do consider writing your own Brainf**k compiler. If you get stuck, consult the direct mappings from Brainf**k commands to fragments of C code online at [132].

Befunge

While an average programmer can write a Brainf**k compiler without much trouble, writing a compiler for Befunge is, by design, notoriously difficult. In this two-dimensional language, self-modifying code happens.

First appeared 1993
Designer Chris Pressey
Notable versions 93 • 98
Recognized for Being multi-dimensional
Notable uses Challenge to would-be compiler writers
Tags Funge, Stack-based, Self-modifying

Befunge programs are written in a two dimensional grid of instructions. Some instructions manipulate the **data stack** in the usual way (as in Factor), some do input and output (to/from the stack), and some instructions—`<`, `^`, `>`, and `v` change the direction in which the program is read. Even better, or worse, is the `p` instruction: it pops from the stack three values y, x, and v, then replaces the character at position (x, y) *in the code grid* with v. So the grid contains both instructions and data, allowing you to change code on the fly.[10]

Here's a factorial function in Befunge (can you see the loop?):

```
&>:1-:v v *_$.@
 ^    _$>\:^
```

The full instruction set and semantics for Befunge are available at [17], and you'll find writing an interpreter to be quite possible. Compilers *can* and have been written, utilizing some very interesting techniques you may wish to look into.

LOLCODE

A year after the massive popularization of the internet meme LOLcats, Adam Lindsay developed LOLCODE, a procedural language that adapted LOLcats' "kitty pidgin" speech to its own programming vocabulary.

First appeared 2007
Designer Adam Lindsay
Notable versions 1.2 (2007)
Recognized for Kitty pidgin vocabulary
Notable uses Lulz
Tags Imperative, Themed

LOLCODE features all uppercase keywords such as `HAI`, `KTHXBYE`, and `O RLY?`. Underneath

[10]Not unlike traditional computers, as you may know.

its amusing syntax, LOLCODE is actually a regular, imperative (though small) programming language, with loops, conditionals, and functions. Here's our recurring example script for computing Pythagorean triples:

```
HAI 1.2
  IM IN YR FIRST UPPIN YR Z TIL BOTH SAEM Z AN 41
    IM IN YR SECOND UPPIN YR Y TIL BOTH SAEM Y AN Z
      IM IN YR THIRD UPPIN YR X TIL BOTH SAEM X AN Y
        I HAS A W ITZ SUM OF PRODUKT OF X AN X AN PRODUKT OF Y AN Y
        BOTH SAEM W AN PRODUKT OF Z AN Z, O RLY?
          YA RLY, VISIBLE X ", " Y ", " Z
        OIC
      IM OUTTA YR THIRD
    IM OUTTA YR SECOND
  IM OUTTA YR FIRST
KTHXBYE
```

Cats don't write very well, so you won't see much punctuation in LOLCODE; we've only used the comma, which does nothing more than abbreviate a newline. And fixed-arity operators don't need parentheses (try proving this!). A few symbols do appear within YARN literals (YARNs are strings, get it?): `:)` represents the newline, `:>` is the tab character, and so on. The theme continues with well-chosen token names for breaks (`GTFO`), switches (`WTF?`), cases within switches (`OMG`), returns (`FOUND YR`), and the else branch of an `O RLY?` statement (`NO WAI`).

13.10 ASSEMBLY LANGUAGES

An **assembly language** describes a machine's execution, often with commands that map one-to-one to **machine instructions**. Even if you use only high-level languages, something about machines—machines of the past, present, and future—may be relevant to you. Uncle Bob tells us why: [79]

> Did you know, for example, that the `++` operator in C, C++, Java, and C# derived from the fact that there were special registers in early computers that automatically incremented their contents every time you accessed them? The compiler tried to make use of these auto-increment registers whenever you used a `++` operator.
>
> ...our languages are projections of the hardware. Some languages are more abstract projections than others; but the most popular languages, C, Pascal, C++, Delphi, Java, and C# are very closely related to the hardware. Even languages like Smalltalk, Python, and Ruby, while more abstract than the others, still show their hardware roots.

Machines can be quite interesting. While the average programmer with little exposure to assembly language may dread the number of instructions required to implement a `for`-loop, the fact is machines often contain relatively powerful instructions, combining comparisons and jumps, locking, exchanging regions of memory, and operating on dozens of data items in a single clock pulse.

We'll cover two short examples in the NASM assembly language, which targets the x86-64 architecture. [65] The first is a function computing the maximum of three integers. We have here only a function, not a complete program. After assembling and linking the code, the function can be called from any language that can be compiled to the same executable file format that the assembler produces.

```
; A 64-bit function that returns the maximum value of its
; three 64-bit integer arguments. The function has signature:
;
;    int64_t maxofthree(int64_t x, int64_t y, int64_t z)
;
; Note that the parameters have already been passed in rdi,
; rsi, and rdx.  We just have to return the value in rax.

        global   _maxofthree
        section .text
_maxofthree:
        mov      rax, rdi            ; result (rax) initially holds x
        cmp      rax, rsi            ; is x less than y?
        cmovl    rax, rsi            ; if so, set result to y
        cmp      rax, rdx            ; is max(x,y) less than z?
        cmovl    rax, rdx            ; if so, set result to z
        ret                          ; the max will be in rax
```

Here we see that functions accepting integers accept arguments in various registers, the first three of which are `rdi`, `rsi`, and `rdx`. The instructions for performing comparisons and conditional moves are explained in the comments.

And now let's look at one last example. The x86 architecture features hardware registers that can store, and operate on, multiple units of data at once, making **SIMD** (single-instruction, multiple-data) processing possible. The so-called XMM registers are 128-bits wide; the YMM registers are 256-bits wide; ZMM registers are 512-bits wide. Here's a very brief illustration: we'll add two 4-element arrays of floating-point numbers using the XMM registers:

```
; void add_four_floats(float x[4], float y[4])
; x[i] += y[i] for i in range(0..4)

        global    _add_four_floats
        section   .text

_add_four_floats:
        movdqa    xmm0, [rdi]       ; all four values of x
        movdqa    xmm1, [rsi]       ; all four values of y
        addps     xmm0, xmm1        ; do all four sums in one shot
        movdqa    [rdi], xmm0
        ret
```

A good compiler for a high-level language should be able to recognize when SIMD instructions can be used. And you may often see software libraries with functions that map directly to these instructions. But sometimes, you may wish to break out an assembler and experience the feeling of writing assembly language directly, feeling at one with the machine.

Or leave that to the specialists.

The question that opened the preface of this book, "Why would you want more than machine language?" was asked by none other than John von Neumann. History has shown that high-level languages matter. But somehow, it does seem rather fitting to finish our tour looking at...of all things, an assembly language. As long as our programs run on hardware, we'll need to be aware of the connections between our high-level languages and the machine and assembly languages that make the hardware run.

Von Neumann's name now labels one of the most prolific electronic computer architectures of the last century—one in which programs and data share memory, a program counter ticks away running sequential processes on each core, an arithmetic-logic unit (ALU) processes instructions, and a bus shuttles data to and from memory. But hardware engineers have been building alternative architectures for quite some time! Dataflow and reduction machines have been around for a while, and to some extent, functional and array programming languages really do show very little connection to the von Neumann architecture at all. Today we see research in a number of brain-inspired computer architectures, such as IBM's TrueNorth [83], and other neuromorphic [87] systems. Will we see inklings of these architectures showing up in future programming languages? Or will new architectures, no matter how novel, be programmed in the languages of today, employing little more than new programming paradigms?

We hope you've enjoyed the tour.

CHAPTER 14

Afterword

Hello again! If you've made it this far, you've been introduced to programming language practice through chapter-length overviews of twelve languages and brief glances at three dozen more. The intent was not to provide a comprehensive tutorial for any of the languages, but rather to use the overviews as a vehicle to introduce important *concepts* of programming language design and practice. Our treatment was non-mathematical and introductory; for an in-depth study and coverage of the theoretical foundations that will enable you to design the languages of the future, we recommend the works by Turbak and Gifford [119], Pierce [97], and Harper [49].

Let's take a look at where we've been and where you can go next.

14.1 WHERE WE'VE BEEN

The primary takeaway from this book is: The study of multiple languages frees us from getting stuck on particular approaches to solving particular problems, making us question how things can be different, and...better.

14.1.1 Functions

Take the well-known function, for example. What could be simpler to understand than passing arguments to a function and waiting for the result to be computed? Quite a lot, it turns out! Knowing even a handful of languages enables us to ask many reasonable questions:

1. Must a function be called with a specific number of arguments? What if we pass too few arguments? Too many? Can we roll up many arguments into a single parameter?
2. Can we (or must we, or must we not) specify the names of the parameters in the call?
3. Can we arrange that certain parameters accept only arguments of specific types?
4. Can we change the value of a parameter within the function? If so, does this action also change the variable holding the argument (if any)? If it does, is the change made immediately or only at the point the function returns?

5. Is a parameter of the called function a distinct variable from its corresponding argument, or just an alias for the argument?
6. If a complex expression is used as an argument, is it evaluated once before the call? Or evaluated fresh every time the parameter is used inside the call?
7. If all arguments are evaluated before the call, in what order are they evaluated? Left to right? Right to left? Arbitrary order? In parallel?
8. How do we arrange for the function to return something? Through a special variable? A `return` statement? Implicitly as the last expression? Some other way?
9. Can a function return zero values? Must it return one? Can it return many?
10. Are there any restrictions on the types of things a function is allowed to return?
11. How do we distinguish between the function itself as a value, and the value resulting from a call of the function?
12. Can the function itself (not just its result) be assigned to a variable? Passed as an argument? Returned as a value from another function?
13. Can a function call itself? If so, are its variables shared among all invocations or does each invocation have a distinct set of variables?
14. Can a function access variables defined outside of the function? If so, must they be explicitly imported or are they available by default?
15. Can "local" variables (defined inside the function) be seen by outside code, or are local variables completely private to the function?
16. If a function can access "non-local" variables, does it use variable bindings in effect in its *caller*, or those in effect in the textually surrounding environment?
17. If a function can access non-local variables and is passed somewhere where those (non-local) variable names have different bindings, which bindings are used when the function is eventually called?
18. Is it possible to test two functions for equality? If so, what does it mean for two functions to be equal?[1]
19. Is a function like a regular object, one that we can attach properties to?
20. Do functions have to have names?

14.1.2 Types

We can ask similar questions about types. For a given language:

1. Is there or is there not a distinction made between primitive and reference types?
2. Is there exactly one numeric type or multiple numeric types? Are integer types divided across signed and unsigned types? Fixed size and unbounded? Modular and saturated? Are there separate types for 8, 16, 32, 64, and 128-bit fixed-size integers? Are floating-point types so divided? Is there an unbounded floating-point type?
3. Can a programmer find ratio or decimal types within the language? In the standard library?
4. Are characters differentiated from strings of length one?
5. Is the length of a string, or of any kind of collection, part of the type?

[1]This is a pretty deep question. What would *you* say the answer is? Same function object? Same source code? Same input-output behavior? Wait, did you notice something unusual about that last question?

6. Is there a boolean type?
7. Are the basic types like numbers, booleans, and strings, wired so strongly into the language that reserved words define them? Or do they simply live in a standard library like all other user-defined types?
8. Can the programmer obtain the type of an expression via a function call, method call, or `typeof` operator? If so, is the response just a string naming the type, or a first-class type object?
9. Are types distinguished from classes?
10. Can types have supertypes? If so, more than one? If so, how is the inherited behavior determined in the presence of name conflicts?
11. Can we add new types?
12. Can we conveniently make both sum and product types? Are there tagged or labeled versions of each (i.e., regular vs. tagged unions, tuples vs. records)?
13. Are there explicit pointer types? If so, can we do arithmetic on instances of these types?
14. Do function objects have types? Is the return type part of the function type? What about the types of the parameters? Is the order of parameters significant in determining the function type? What about the modes of the parameters?
15. Can we define parameterized types? If so, are they parameterized only by other types, or by arbitrary values?
16. What are the covariance, contravariance, and invariance rules for parameterized types?
17. What kind of type expressions are allowed in parameterization? Can we go beyond Java's `<? super T>` and `<? extends T>` mechanisms?
18. Are types first-class? Are type variables allowed? If so, can they be...assigned?
19. If types are objects, are all types instances of a single type (perhaps called `Type`)? Do type classes exist in the language? If so, can a type belong to more than one type class? Can a programmer define new type classes?
20. What kind of expressions are allowed in dependent type specifications?

14.1.3 Expressions

What questions have we learned to ask about expressions?

1. Are expressions evaluated when they are first encountered, or only when needed?
2. Is the type of every expression known at compile time (static typing) or not known until run time (dynamic)?
3. If the language detects that an operator is given operands of a type that it does not expect, does an error occur (strong typing) or is there an attempt made to convert the operands to operands of an acceptable type (weak)?
4. How about syntax? Are all operators prefix or is there a mix of prefix, infix, and even postfix operators?
5. Can we overload operators? Can we define new operators? If so, how do we specify precedence and associativity? Are all new operators given the same precedence, or can we choose our own?
6. Can we change the precedence of "built-in" operators?

7. Are operators built into the language, or are they just functions from a standard library?
8. Do variable declarations have to be annotated with a type, or can they be inferred? What about function parameters?
9. Can the type of a variable change?
10. Can a variable be marked as being mutable or immutable? If it can be marked immutable, does this mean that the variable may not be rebound, or that the object bound to the variable may not be mutated?
11. Are variables bound? Or assigned? Can they be bound or assigned once, or multiple times?
12. Can variables be bound via destructuring? In a pattern match? Are they bound in parallel or sequentially?
13. How do we know, when looking at an expression, to *which* entities do each of the variables refer? In other words, how do we determine the scope of our bindings?
14. Are there lexical mechanisms for distinguishing local variables from global (or other non-local) variables? Or is this resolved by scope computation within a compiler?
15. Are variables allowed to shadow variables of the same name in outer scopes? If not, are name collisions prohibited or do they simply create new locals or refer to globals depending on context?
16. Which constructs define scope? Function bodies only? Certain compound statements but not others? Is there a block construct? `let`-expressions?
17. When a variable is updated, what happens to the "old" value?
18. Which "magic" expressions does the language provide? Anything similar to JavaScript's `this`?
19. Do type expressions live in a different world than other (value) expressions?
20. Does the language require that expressions be reduced in a particular fashion? For example, must `square(inc(8))` be reduced to `square(9)` or to `inc(8)*inc(8)`?

14.1.4 Control Flow

We've seen a few ways to express control flow.

1. Are statements, expressions, and declarations distinct syntactic and semantic entities, or is everything—or perhaps *nearly* everything—an expression?
2. How do we express the ordering of evaluation or execution? Can we combine statements sequentially (s_1`;`s_2), in parallel (s_1`||`s_2), or even via some kind of nondeterministic choice mechanism (s_1`|`s_2)? Or are we allowed to express sequencing only via composition of function calls?
3. In a multiway selection, must guards be evaluated in any order or a specific order? Do we evaluate only enough guards until we find one that is true, or must we evaluate them all?
4. In the case of a nondeterministic choice construct, what guarantees does the language place on its implementors to ensure, or to not ensure, fairness among choices?
5. Are short circuiting boolean operators (such as `&&` and `||`, or their generalizations, `all` and `any`) available?

6. Is there a looping construct or must we rely on tail recursion?
7. Can we conveniently express looping mechanisms for all of the usual patterns: forever, n times, through a range of numbers (with an optional step), through a collection, or while a condition is true?
8. Are there iterator objects? Or can iteration be implicit? Can we do both? What happens when a collection is changed during iteration?
9. Can we break out a loop? Continue to the next iteration? Start the current iteration over? Start the entire loop over?
10. Are exceptions relied upon for control flow, or are result objects (tagged unions with success and failure) or option objects favored?
11. Does the language provide a `goto` statement? If so, what restrictions, if any, are placed on its use?
12. Does the programmer have access to a timer, either for sleeping or delaying execution?

14.1.5 Concurrency

We saw quite a few approaches to concurrency. As we review, think of how each of the languages we saw answer the questions below.

1. How does one decompose a program for concurrent execution?
2. Are there linguistic constructs for coroutines, threads, and processes, or are these "just functions"?
3. Can we define concurrent execution at different levels of granularity? Expression-level? Function-level? Program-level?
4. Can mutable memory be shared? If so, are there constructs other than explicit locks that help to prevent race conditions? Is there support for transactional memory?
5. Is lock-free synchronization (e.g. atomic integers) available?
6. Are tasks spawned implicitly, at the time their enclosing block begins execution, or must they be explicitly started?
7. Does the runtime unceremoniously kill tasks when the enclosing block, or the spawning task, terminates? Or does the enclosing/spawning task wait for all internal tasks to terminate?
8. Are tasks spawned with or without a link to the task doing the spawning?
9. If an asynchronous task is launched, how is a result obtained? Via a callback, promise, or some other mechanism?
10. Is message passing between processes done by mentioning processes by name (or process identifier), or through specific channels, or are messages put into a global pool where they can be processed by any willing process?
11. If channels are used to wire processes for communication, must the wiring be fixed at compile time or can we rewire during execution?
12. Is message passing only synchronous or only asynchronous? Can the programmer choose?
13. Can synchronization be timed (backed off if the other party does not connect within a given duration)? Do we have a conditional synchronization mechanism (to connect *only* if the other party is ready and waiting)?

14. How, if at all, do we detect that a process has crashed?
15. What happens if a message is sent to a terminated process? Is an error generated or is the message ignored?
16. Do introspection mechanisms exist to get the state of a process (alive, dead, waiting, etc.)? If so, can we ensure that the state is still valid by the time we are ready to act upon it?

14.1.6 Modularity

Large-scale, complex software systems must by nature be hierarchical. Often, a programming language may provide direct support for subsystems and subsubsystems, rather than leaving it to builder programs to cobble together the system. Whether they are called "modules," "packages," or "clusters," language designers can work with these in many ways.

1. What kind of top-level constructs are defined by the language? Scripts and modules? Scripts only?
2. Can top-level modules be compiled separately? That is, are modules standalone entities that can be used in multiple scripts? Must a module's specification and implementation be separate?
3. How does one arrange to hide entities within a module from the outside world? Are they marked private? Or are entities private by default, requiring explicit export?
4. Can access permissions be changed at runtime?
5. Are the types of entities retained when imported? (Don't be surprised at this question. In early languages, they were not!)
6. Can modules be nested within functions? Within other modules?
7. How does a module import entities from other modules? Implicitly? Via an explicit import statement? If a module in nested within an outer scope, are the outer entities automatically imported?
8. Do entities in a nested module shadow entities of the same name?
9. Are the modules (or packages) first-class?
10. Can modules be parameterized, so that new modules are created upon instantiation?

14.1.7 Metaprogramming

If you've ever found yourself in need of writing multiple constructs that all look about the same, you'd have certainly appreciated a language with good metaprogramming facilities. Metaprogramming, the art of writing programs that write other programs, or modify or extend the program that is currently running, is definitely one of the more fascinating areas of programming language design and pragmatics. It relies, of course, on reflection and introspection. What options do language designers have here?

1. Can a program ask which classes have been loaded? Which variables are global? Local?
2. Can a program ask for the currently executing function, and perhaps, its caller?

3. Given a function, can the program ask which parameters, and of which type, the function would accept as arguments? Can it ask for the return type of the function?
4. Given a class or module, can the program ask for its members, constructors, methods, initializers, etc? Its supertypes and subtypes?
5. Can a variable be read or set via its name (a string) only?
6. Can a function or method be invoked by its name (a string) only?
7. Can new variables or types or functions be created at run time, given only their name and some indication of how they should look?
8. Does the language have a macro system? If so, do macros generate strings of code which must then be scanned, parsed, and compiled? Or do they generate abstract syntax trees?
9. Is it possible to unquote within a macro?
10. Are macros hygienic? What syntactic devices are provided to explicitly capture in-scope variables, if any?

We've just laid out only a tiny sampling of the questions we can ask about programming languages. We haven't even tried to list all of the questions considered in this book, let alone those that have answers in languages not covered in this book. The fun questions, though, are the ones that have never been conceived. Perhaps you are now in a good position to think of some!

14.2 WHERE TO GO NEXT

We do hope you are able to use this book's exploration of various programming language concepts and practice to further your academic and professional development along multiple directions. We'll offer three roads for you to take from here:

- **Professional Practice**. We hope you will discover, and be able to employ, a wide variety of programming styles. Style tends to be informed by language, but not tied to any particular language. Cristina Lopes provides an excellent and comprehensive account of 37 programming styles, and their implementation, in [76].
- **Formal Syntax and Semantics**. Our overviews have intentionally been informal. But research in the field of programming languages is ongoing and intense. We hope many readers will go on to discover and work on formal (mathematically rigorous and unambiguous) mechanisms for precisely defining languages. Your ability to read and write formal syntactic (e.g., EBNF, ABNF) and semantic (e.g., Denotational, Operational, Natural, Axiomatic) descriptions may generate new insights into linguistic constructs that are difficult to see at a high level.
- **Programming Language Implementation**. Oh, to write your own compiler or interpreter! And to make it efficient! Few, if any, activities deepen your understanding of a language more than that of implementing it. The implementation effort itself exposes you to a number of new questions: how do you, for example, efficiently manage thousands of lightweight concurrent processes on a machine that, well, doesn't look very parallel? Implementation also increases your understanding of virtual and real machines, and the need to manage resources wisely.

It has been noted than since the 1950s, computers (hardware, that is) have improved *several orders of magnitude* in speed, storage capacity, size, and weight. [79] Where are the orders of magnitude improvements in software? How much better are today's languages from the early versions of Fortran and Lisp? Are they anywhere near orders of magnitude better? What kind of revolutionary thinking would it take to design and implement a language that is? This seems a fitting question on which to end. We are looking forward to what you will create.

APPENDIX A

Numbers

If there are topics truly shared by *every* language that deserve special treatment, they would be the representations of numeric and textual data. Modern languages use two of the most influential standards of our time: *IEEE-754* [63] for floating-point quantities and *Unicode* [121] for text. Rather than working the technical details of these topics into the language overviews, we've consolidated the important things to know about each into two appendices.

A.1 INTEGERS

The **integers** are ..., $-3, -2, -1, 0, 1, 2, 3, \ldots$. Virtually all programming languages distinguish **fixnums** (integers stored in a fixed number of bits) and **bignums** or **bigints** (of arbitrary size). Fixnums may come in a variety of sizes, with values either wholly nonnegative ("unsigned") or both negative and nonnegative ("signed"). Table A.1 shows typical signed fixnum types, and Table A.2 typical unsigned fixnum types.

Bits	Typical Names	Range
8	`i8`, `Int8`, `int8_t`, `byte`	−128...127
16	`i16`, `Int16`, `int16_t`, `short`	−32768...32767
32	`i32`, `Int32`, `int32_t`, `int`	−2147483648...2147483647
64	`i64`, `Int64`, `int64_t`, `long`	−9223372036854775808... 9223372036854775807

Table A.1 Fixed-size signed integers

Type names for the unbounded integers vary quite a bit. Python uses `long`, Ruby uses `Bignum`, and Clojure calls the type `BigInt`. Generally, you can expect seamless casting between the fixed and unbounded types in dynamic languages, and a demand for explicit type conversions in static languages.

Bits	Typical Names	Range
8	`u8`, `UInt8`, `uint8_t`, `unsigned byte`	0...255
16	`u16`, `UInt16`, `uint16_t`, `unsigned short`	0...65535
32	`u32`, `UInt32`, `uint32_t`, `unsigned int`	0...4294967295
64	`u64`, `UInt64`, `uint64_t`, `unsigned long`	0...18446744073709551615

Table A.2 Fixed-size unsigned integers

A.2 FLOATING POINT NUMBERS

A **floating-point number** represents both integral and nonintegral values in a fixed-size storage location, generally, but not necessarily, 32 or 64 bits. Types of the former size have been called `single precision`, `float`, `Float32`, or `f32`. The latter size traditionally holds types with names such as `double precision`, `double`, `Float64`, or `f64`. JavaScript calls the 64-bit floating-point type simply `number`, though certain operations (`&`, `|`, `~`, `<<`, and `>>`) can be invoked to give the programer access to 32-bit integers.

Many languages specify that their floating-point types map directly to the IEEE 754 specification [63], in which, for the 64-bit format, assigns the following role to each bit:

- Bit 63 is the sign bit, s
- Bits 62–52 comprise the exponent, e
- Bits 51–0 comprise the mantissa, or fraction, f

The numeric value is computed from the bits as follows (see Table A.3):

- If $1 \leq e \leq 2047$, the value is $-1^s \cdot (1 + f \cdot 2^{-52}) \cdot 2^{e-1023}$ ("the normals")
- If $e = 0$, the value is $-1^s \cdot (f \cdot 2^{-52}) \cdot 2^{-1022}$ ("the denormals")
- If $e = 2048$, the value is $-1^s \cdot \infty$ if $f = 0$ and `NaN` otherwise.

The two infinities result from operations producing numbers whose magnitude exceed the largest finite number ($2^{1024} - 2^{971}$), approximately 1.7977×10^{308}, or division of a non-zero finite value by zero, e.g. `5.0/0.0`. The value `NaN` results from operations with meaningless or ambiguous results, such as 0.0/0.0 or $\infty - \infty$. The spacing between representable numbers becomes more pronounced as magnitudes increase. This is appropriate: the *relative* difference between 0.0003 and 0.003 is enormous (the latter is *ten times* the former), while the relative difference between 7,198,993.0003 and 7,198,993.003 is (probably) insignificant, at least if we were measuring distances, weights, or prices.

The inexactness of floating-point representations often leads to surprising results, either because the result is not representable, e.g. $10^{20} + 500 = 10^{20}$ (representable numbers are spaced by 16,384 at those magnitudes), but also due to **roundoff error**: $0.1 + 0.2 = 0.30000000000000004$, because

- The closest bit pattern to 0.1 is `3fb999999999999a`
- The closest bit pattern to 0.2 is `3fc999999999999a`
- The sum of these is `3fd3333333333334`
- But the closest bit pattern to 0.3 is `3fd3333333333333`

Hex	Values	Description
`0000000000000000`	$+0$	Positive Zero

`0000000000000001`
...
`000fffffffffffff`
	2^{-1074}
	...
	$2^{-1022} - 2^{-1074}$
		Positive Denormals

`0010000000000000`
...
`7fefffffffffffff`
	2^{-1022}
	...
	$2^{1024} - 2^{971}$
		Positive Normals

`7ff0000000000000`	$+\infty$	Positive Infinity

`7ff0000000000001`
...
`7fffffffffffffff`
	NaN
	...
		NaNs

`8000000000000000`	-0	Negative Zero

`8000000000000001`
...
`800fffffffffffff`
	-2^{-1074}
	...
	$-2^{-1022} + 2^{-1074}$
		Negative Denormals

`8010000000000000`
...
`ffefffffffffffff`
	-2^{-1022}
	...
	$-2^{1024} + 2^{971}$
		Negative Normals

`fff0000000000000`	$-\infty$	Negative Infinity

`fff0000000000001`
...
`ffffffffffffffff`
	NaN
	...
		NaNs

Table A.3 64-bit IEEE 754 floating-point representations

A.3 RATIOS AND DECIMALS

Ratios, also known as **rationals**, avoid roundoff errors from division by storing both the numerator and denominator. So rather than being surprised at the sum of the floating-point values 0.1 and 0.2, we could instead add the rationals one-tenth and one-fifth to receive the sum three-tenths.

The Julia language has a special operator (//) to construct rationals:

```
julia> 1/10 + 1/5
0.30000000000000004
julia> 1//10 + 1//5
3//10
julia> typeof(ans)
Rational{Int64}
```

Ruby has a `Rational` class; rational numbers can be constructed from strings or from a numerator and denominator:

```
>> 0.1+0.2
=> 0.30000000000000004
>> Rational('1/10')+Rational(2,10)
=> (3/10)
```

```
>> _.class
=> Rational
```

Clojure overloads the / operator to produce ratios where necessary:

```
user=> (+ 0.1 0.2)
0.30000000000000004
user=> (/ 3 10)
3/10
user=> (type *1)
clojure.lang.Ratio
```

Financial applications generally benefit from a type of rational number in which the denominator is a power of 10, known as a **decimal type**. A decimal representation physically stores the base-10 digits, leading to exact computation, provided enough digits are stored. (Languages supporting decimal types may allow the programmer to explicitly set the number of stored digits.) Python, for one, has a decimal type in its standard library:

```
>>> from decimal import Decimal
>>> Decimal('0.1') + Decimal('0.2')
Decimal('0.3')
```

APPENDIX B

Text

In this appendix, we provide a few details behind text representation and manipulation, concentrating on the Unicode character set.

B.1 CHARACTERS, GLYPHS, AND GRAPHEME CLUSTERS

A **character** is a named symbol, e.g., CYRILLIC CAPITAL LETTER TSE or MUSICAL SYMBOL F CLEF. A **glyph** is a picture of a character. Different characters may share the same glyph; for example, ∅ can represent LATIN CAPITAL LETTER O WITH STROKE, DIAMETER SIGN, or EMPTY SET. Similarly, a single character can be represented by many glyphs: LATIN CAPITAL LETTER A can be rendered as A, A, *A*, 𝔸, or 𝒜.

A **character set** is a particular selection of characters, each assigned a unique **code point**. The very popular **Unicode** character set includes several hundred thousand characters, with a code point range of 0...1114111 (although 62 of the code points are explicitly disallowed). It is customary to describe the code points in hexadecimal, prefixed with `U+`. Here is a sampling of Unicode characters:

U+002C	COMMA
U+071d	SYRIAC LETTER YUDH
U+13EA	CHEROKEE LETTER WE
U+20B4	HRYVNIA SIGN
U+2248	ALMOST EQUAL TO
U+30FE	KATAKANA VOICED ITERATION MARK
U+AAB9	TAI VIET VOWEL UEA
U+1201F	CUNEIFORM SIGN AK TIMES SHITA PLUS GISH
U+1F0CE	PLAYING CARD KING OF DIAMONDS
U+1F353	STRAWBERRY

A character with an assigned code point is called an **encoded character**, but it may take more than one encoded character to make up a "user-perceived" character; we call the encoded character sequence a **grapheme cluster** [23], which will often be rendered with its own glyph. Examples include:

Encoded Character Sequence	Grapheme
U+0052 LATIN CAPITAL LETTER R U+030A COMBINING RING ABOVE	Latin capital R with ring above
U+0BA8 TAMIL LETTER NA U+0BBF TAMIL VOWEL SIGN I	Tamil ni
U+1F6B4 BICYCLIST U+1F3FE EMOJI MODIFIER FITZPATRICK TYPE-5	Bicyclist Type-5
U+1F1F1 REGIONAL INDICATOR SYMBOL LETTER L U+1F1E7 REGIONAL INDICATOR SYMBOL LETTER B	Flag for Lebanon

In some cases, a single Unicode character exists for an equivalent grapheme cluster. For example, the code point sequence `U+0065` (LATIN SMALL LETTER E) `U+0303` (COMBINING DIAERESIS) represents the same text as the single code point `U+00EB` (LATIN SMALL LETTER E WITH DIAERESIS), namely ë. Your favorite programming language may or may not be able to compute such equivalences natively.

Many languages include functions or methods to produce both the character for a given code point and the code point of a given character. Characters can usually be specified by code point within a string literal; the following string, valid in JavaScript, CoffeeScript, Ruby, Julia, Java, Scala, and many other languages, includes a happy face character:

```
"Have a nice day! \u263a"
```

In addition to this four-hex-digits-following-`\u`, you typically see two characters following `\x`, eight following `\U`, and anywhere between one and eight characters in `\u{...}`.

B.2 CHARACTER PROPERTIES

A unicode character may seem like a simple entity but in reality has a large number of properties beyond its name and code point. These properties include its block, category, age, script, and a large number of normalization properties. You can find more information about these properties at [130], but we do wish to highlight two: block and category.

The code point space is partitioned into over 200 different **blocks**. We show a few (randomly selected) here; for the full list, see [120]:

Code Point Range	Block
0000..007F	Basic Latin
0080..00FF	Latin-1 Supplement
0300..036F	Combining Diacritical Marks
0840..085F	Mandaic
0A80..0AFF	Gujarati
1CC0..1CCF	Sundanese Supplement
2190..21FF	Arrows
2200..22FF	Mathematical Operators
2300..23FF	Miscellaneous Technical
3000..303F	CJK Symbols and Punctuation
3040..309F	Hiragana
4E00..9FFF	CJK Unified Ideographs
A800..A82F	Syloti Nagri
10280..1029F	Lycian

13000..1342F	Egyptian Hieroglyphs
1F000..1F02F	Mahjong Tiles
1F600..1F64F	Emoticons
1F650..1F67F	Ornamental Dingbats
1F680..1F6FF	Transport and Map Symbols
F0000..FFFFF	Supplementary Private Use Area-A
100000..10FFFF	Supplementary Private Use Area-B

Characters are grouped into **categories**, helping to designate characters as letters, numbers, symbols, etc. Here is the complete list of categories as of Unicode 9.0:

Category	Description
Lu	Letter, Uppercase
Ll	Letter, Lowercase
Lt	Letter, Titlecase
Lm	Letter, Modifier
Lo	Letter, Other
Nd	Number, Decimal Digit
Nl	Number, Letter
No	Number, Other
Sm	Symbol, Math
Sc	Symbol, Currency
Sk	Symbol, Modifier
So	Symbol, Other
Pc	Punctuation, Connector
Pd	Punctuation, Dash
Ps	Punctuation, Open
Pe	Punctuation, Close
Pf	Punctuation, Final quote
Pi	Punctuation, Initial quote
Po	Punctuation, Other
Zl	Separator, Line
Zp	Separator, Paragraph
Zs	Separator, Space
Cc	Other, Control
Cf	Other, Format
Co	Other, Private Use
Cs	Other, Surrogate
Cn	Other, Not Assigned
Mc	Mark, Spacing Combining
Mn	Mark, Nonspacing
Me	Mark, Enclosing

Many languages will feature standard library functions to determine whether a character belongs to a specific category. Furthermore, many regular expression engines allow the pattern `\p{`Xx`}`, to match any character in the category Xx. These functions and matching operations likely use an internal Unicode database.

B.3 CHARACTER ENCODING

A **character encoding** specifies the representation of code points in bytes. For Unicode, the UTF-32 encoding presents the code point directly in 4 bytes. UTF-32 is a **fixed-length encoding** since every character is represented in the same number of bytes. The **variable-length encoding** UTF-16 encodes the characters `U+0000` through `U+FFFF` in two bytes, and `U+10000` through `U+10FFFF` in four bytes. The latter group is encoded by subtracting 0x10000 from the code point and spreading the resulting 20-bit value $x_{19}x_{18}...x_1x_0$ into a 32-bit word as follows:

$$\texttt{110110}x_{19}x_{18}x_{17}x_{16}x_{15}x_{14}x_{13}x_{12}x_{11}x_{10}\texttt{110111}x_9x_8x_7x_6x_5x_4x_3x_2x_1x_0$$

For example, the character `U+1F6A1` AERIAL TRAMWAY has the UTF-16 encoding `D83D DEA1`.

Perhaps the most popular encoding scheme is the variable-length UTF-8. We spread the code point bits $x_{n-1}x_{n-2}...x_0$ across 1–4 bytes as follows:

Code points	Encoding
0...7f	$\texttt{0}x_7...x_0$
80...7ff	$\texttt{110}x_{10}...x_6\ \texttt{10}x_5...x_0$
800...ffff	$\texttt{1110}x_{15}...x_{12}\ \texttt{10}x_{11}..x_6\ \texttt{10}x_5...x_0$
10000...1fffff	$\texttt{11110}x_{20}...x_{18}\ \texttt{10}x_{17}...x_{12}\ \texttt{10}x_{11}..x_6\ \texttt{10}x_5...x_0$

Many languages with a string type will provide methods to get both the number of *characters* and the number of *bytes* in the string. For example, using UTF-8, the string "$\forall x.A \vee \neg A$" has a character length of 7 (code points `2200`, `78`, `2E`, `41`, `2228`, `AC`, `41`), but a byte length of 12 (`E2 88 80 78 2E 41 E2 88 A8 C2 AC 41`).

Character counting varies widely between languages. Some languages explicitly count encoded characters; while some understand and count grapheme clusters as a single character. Some may even count UTF-16 code units. For example, the Lebanese flag emoji is:

- A single grapheme cluster, or character,
- composed of two encoded characters (`U+1F1F1 U+1F1E7`),
- encoded in four UTF-16 code units, and
- encoded in 8 bytes in UTF-8.

JavaScript's `length` method on strings counts in UTF-16:

```
// One character that JavaScript counts as 4
const assert = require('assert');
assert("\u{1f1f1}\u{1f1e7}".length == 4);
```

Swift, in contrast, does understand grapheme clusters, and can provide UTF-8 and UTF-16 views into a string as well:

```
// Swift understands grapheme clusters and encodings
let s = "\u{1f1f1}\u{1f1e7}"
assert(s.characters.count == 1)
assert(s.utf16.count == 4)
assert(s.utf8.count == 8)
```

Glossary

Although this book has taken a language-by-language approach to the study of programming languages, the real benefit to the study of languages as a discipline comes from learning the concepts that transcend multiple languages. Here is a summary of some of the more important ones. You won't find any examples here—just bare-bones definitions.

Abstract Class A class that cannot have any instances.

Abstract Method A method of an abstract class that has no implementation.

Activation Record See *frame*.

Actor A concurrent process that communicates by sending and receiving messages.

Agent An entity that acts on behalf of another entity.

Alias One of the names of a multiply-named entity. An entity to which multiple names are bound is said to be *aliased*.

Annotation A form of metadata attached to an entity.

Argument An expression passed to a (parameterized) entity during a call, instantiation, or other invocation.

Assignable A mutable storage location. Used in place of the term *variable* when emphasizing mutability.

Assignment The act of associating a value with a (mutable) variable.

Asynchronous A type of invocation in which the caller does not wait for the callee to complete.

Atom A named primitive entity whose value is, essentially, itself. Also known as a *symbol*.

Atomic Object An object that can be operated upon safely in a multithreaded environment without locks. Often the object holds a numeric value on which "compound" operations such as compare-and-set or get-and-add can be done atomically.

Automatic Reference Counting (ARC) A mechanism in which the compiler inserts code to release and retain blocks of memory so the programmer does not have to.

Binding An association of a name with an entity.

Borrow The *use* of a resource without taking full ownership.

Callback A function f both passed to a function g and invoked within g.

Character 1. A named, abstract symbol. 2. The type of characters.

Class An entity from which objects are created. Objects instantiated from a class have the structure and behavior defined by the class, though in some languages instances are allowed to be customized.

Class Variable A property that belongs to a class, rather than to the instances of the class.

Closure A block or function with free variables that takes values from the environment in which it is defined.

Code Point The unique, nonnegative integer value assigned to a character in a character set.

Coercion An implicitly applied type cast.

Comprehension A composite value defined by means of an expression that generates the components.

Concrete Class A class that is allowed to have instances.

Concurrency The modeling and coordination of independent and distinct computing activities whose execution spans may overlap in time.

Continuation Roughly, a representation of "the rest of the program."

Coroutine A line of execution that, under explicit programmer control, can yield to other coroutines and be resumed. Coroutines share a processing unit in the sense that only one coroutine is executing at a time.

Covariance The existence of a subtype relationship between `T<U>` and `T<V>`, where T is a parameterized type and U is a subtype of V.

Curried Function A function of one parameter that returns a function of one parameter, in contrast to a function that takes two parameters (or a pair of objects).

Dangling Pointer An in-scope pointer variable whose referenced memory has been freed.

Deadlock A situation in which two or more competing threads are blocked forever, each waiting for the other to finish.

Decimal Type A datatype that stores the decimal digits of numeric quantities.

Decorator An object providing additional behavior to another object.

Deferred Execution A block of code executed at some point in the future, generally in response to a triggering event.

Deep Copy A copy which recurses through an an object graph, copying all referenced values.

Dependent Type A type that depends on a value or constraint, such as the type of trees with height 8, or the type of all sorted lists.

Dereference The evaluation of the the object referenced by a (pointer) variable.

Destructor Code (generally a function) executed when an object is destroyed.

Dictionary A set of key-value pairs, with unique keys. Also known as a *map*.

Discriminated Union See *tagged union*.

Dynamic Language A language in which tasks such as modifying code and adding types can be done at run time.

Dynamic Typing The state of having types of some expressions not known until run time.

Eager Evaluation The evaluation of all of a function's arguments (or an operator's operand) before application of the function (or operator). Also known as *strict evaluation.*

Encapsulation The bundling of data elements into a single component, often in a way that the constituent elements cannot be directly accessed.

Encoding 1. The representation, in bytes, of an entity. 2. A specification of how certain entities should be encoded in bytes.

Enumeration A complete ordering of a given set of elements. The term may refer either to (1) a type whose constituent values are listed (an *enumerated type*) or (2) a sequence of values emitted by a generator.

Event An occurrence that can be detected and handled.

Exception An entity that is *thrown*, or *raised* to abort the current normal control flow and transfer control to the point at which the exception is *caught.*

Falsy Treated as false in a boolean context (e.g, when evaluating conditions).

First-class Able to be assigned to variables, passed as parameters, and returned from routines.

Fixnum An integer type represented with a fixed, finite, number of bits.

Frame A structure in the execution model of a program representing a particular activation of a routine. A frame records the parameters and local variables of the activation, as well as additional bookkeeping information. Also known as an *activation record.*

Function A possibly parameterized, and possibly named, callable block of code.

Functional Programming A programming paradigm in which computation proceeds entirely by evaluating side-effect-free expressions, generally involving function calls.

Future See *promise.*

Garbage Collection The reclamation of storage holding objects that are no longer needed.

Generic Able to be specified just once and "work in about the same way" in many different contexts.

Guard A boolean expression used to allow (when true) or disallow (when false) access to an entity.

Heap Storage The portion of memory in which new objects are allocated (and deallocated) at run time.

Higher-Order Function A function that takes functions as arguments or returns functions.

Homoiconicity The property of a language in which code is represented as a data structure and both have the same syntactic forms.

Hygienic Macro A macro whose expansion will not capture identifiers.

Immutable The quality of an entity whose state may not change.

Inheritance The property of a class or object acquiring the same structure or behavior of another entity, which can then be optionally specialized.

Instance An object (when referred to in reference to its class or interface).

Interface A programming structure that enforces certain behaviors on an object.

Interoperability The capability of two or more programming languages to interact and operate on shared data or invoke shared functions.

Interpolation The process of evaluating a string or object literal to yield a result in which placeholders are replaced with computed values.

Iterator An object that provides the capability to access the elements of a container in some order.

Keyword Argument An argument that is matched to a corresponding parameter using the parameter's name, rather than by its position in a call.

Lazy Evaluation An evaluation strategy for expressions in which evaluation of subexpressions is delayed until an actual value is needed.

Lifetime The time between an objects' creation and destruction.

Literal A representation of a value in source code.

Lvalue A storage location into which values can be assigned.

Macro A rule or pattern that specifies how an input should be mapped to a replacement output. Often used in the context of rewriting source code prior to compilation.

Mailbox A container for data into which one process writes data for another process to read (generally in an asynchronous fashion).

Match An object storing information regarding the successful location of a pattern in some data.

Memoization An optimization technique for repeatedly-called functions that caches results, and returns cached results when the same inputs occur.

Metaclass A class whose instances are classes.

Method A function intimately associated with an object.

Module A language construct wrapping a collection of (usually encapsulated) related entities, providing an interface to other modules of a system.

Monad Roughly, a structure providing a mechanism for chaining computations over different structural types. In practical use, a monad provides a function for computing a wrapped value from a raw value (e.g. $\alpha \rightarrow M\alpha$) and a way to combine a wrapped value with a function on wrapped values to a new wrapped value (e.g., $M\alpha \rightarrow (\alpha \rightarrow M\beta) \rightarrow M\beta$).

Monitor An object or module providing mutually exclusive access among its clients to achieve thread safety.

Object 1. An entity in a program intended to be stored in memory. 2. (OOP) An instance of class.

Operator A function, generally, though not necessarily, named with non-alphanumeric symbols that computes well-known (often algebraic) and widely applicable operations.

Operator Overloading The allowance for a single named operator to have different implementations depending on their arguments. For example, the name * can overload integer multiplication, floating-point multiplication, string repetition, vector dot product, or vector scalar product.

Option Type A type whose values are wrappers that either contain or do not contain a wrapped value.

Parameter A variable to which arguments are passed.

Pass-by-Name An argument passing mechanism in which an argument is not evaluated prior to a call, but rather within the function using (essentially) the text of the argument expression.

Pass-by-Reference An argument passing mechanism in which a reference to the argument is passed to the parameter, so that mutations through the parameter are immediately applied to the argument.

Pass-by-Sharing An argument passing mechanism in which only (a copy of) the object identifier of the argument is passed.

Pass-by-Value An argument passing mechanism in which an argument is evaluated and then (a copy is) subsequently passed to the parameter.

Pass-by-Value Result An argument passing mechanism in which an argument is evaluated before the call, passed by value to the corresponding parameter, and, after execution of the callee, any changes made to the parameter are copied to the corresponding argument.

Persistent Data Structure A data structure that retains the prior version of itself when modified.

Pointer An object that references a location in memory.

Polymorphism The allowance for a single name to be bound to multiple entities at the same time, generally with the understanding that any invocation of the name will refer to the desired entity based on context.

Promise Roughly, a placeholder for an (initially) unfinished computation. A promise is immediately returned when an asynchronous computation is launched to be queried later or used in a callback. Also known as a *future*, though in some contexts, futures and promises are distinguished.

Prototype 1. An object from which other objects are derived. 2. An object to which property requests are forwarded to by an object that does not own a property of a given name. 3. In C, a function signature.

Proxy An object that provides an interface to another object. The proxy takes requests on behalf of the proxied object and may modify or deny these requests before forwarding them to the proxied object.

Reactive Programming A programming style centered on streams of data that are managed asynchronously.

Record An object whose components are labeled values.

Reference An object that refers to another object, called its referent. Often used to allow sharing of data. A reference may be *dereferenced* to obtain the referent, but does not act on behalf of the reference as does a proxy.

Regex See *Regular Expression.*

Regular Expression A pattern that describes sets of strings according to a large number of widely-accepted mechanisms, and sometimes with programming-language specific mechanisms.

Reserved Word A word in the syntax of a programming language that cannot be used as an identifier.

Runtime The period of time in which a program is being executed.

Rvalue A value that can be assigned. So named because traditional assignment statements are written with the destination storage location on the left (hence the term l-value) and the data to be assigned on the right (hence the term r-value).

Scope The region of source code in which a binding refers to a given entity.

Scripting Language 1. A language designed for running jobs, i.e., coordinating (or scripting) the activities of other programs. 2. A language low on "ceremony" but rich in expressive features, designed to allow its programmers to be productive.

Second-class An entity that is not allowed to be represented by a variable in an expression.

Semaphore A variable or object used to permit only a specific number of threads to access a shared resource at a time.

Shallow Copy A component-by-component copy of the values of the original object, in which references are not followed but are themselves copied.

Sigil A symbol affixed to a variable name that shows the variable's scope.

Signal A value that changes over time but that is processed as a stream.

Signature A specification of a function, method, or module that is completely free of implementation details, providing the minimum amount of information to use the object.

Single Inheritance The restriction of a type or class to have only a single supertype or superclass.

Software Transactional Memory A mechanism for controlling updates in a concurrent environment in which all updates within a marked transaction either happen (get committed) or they are rolled back.

Splat A term used primarily in CoffeeScript referring both to a function parameter that is to be unpacked and to a function argument spread. (See *spread.*)

Spread An operator that expands a single expression into multiple expressions, within the context of a collection literal, argument list, or similar construct.

Stack Storage Location in memory used to store frames.

Starvation A scenario in which a process or thread cannot proceed because it is continually denied requests to resources it requires.

Static Language A language in which the vast majority of metadata (types of variables, which functions are called, etc.) can be inferred by examining source code only and not requiring execution to determine.

Static Storage The region of memory used to store data whose lifetime coincides with the entire program.

Static Typing The inference of and checking of types prior to execution.

Stream A sequence of objects made available over time.

Strict Evaluation An evaluation strategy in which all operands are evaluated before being given to an operator.

Subclass A class derived from another class, inheriting many of the original class's properties and optionally adding its own properties and behaviors.

Superclass A class that has been subclassed.

Symbol A named primitive entity whose value is, essentially, itself. Also known as an *atom*.

Synchronous A form of communication in which the caller must wait for the callee to be ready to accept the call, or in some cases, must wait for the callee to finish the requested service.

Syntactic Salt Syntactic forms that tend to make programs verbose, or difficult to follow, even though they may have been added to the language with good intentions.

Syntactic Sugar A syntactic form that improves upon an equivalent primitive form by being more concise, expressive, powerful, or elegant.

Systems Language A programming language used for writing system software.

Tagged Union A union type whose members are each tagged with a label indicating the underlying type from which they came. Also known as a *disjoint union*, *discriminated union*, *disjoint sum*, or *discriminated sum*.

Thread A sequential flow of control, generally preemptible and generally able to share memory with other threads.

Trait A set of methods, generally expected to be mixed in to a class to enhance its behavior or to increase its capabilities.

Transient The property of any element in a system that is not persisted or serialized.

Transpiler A complier that changes the source code of a program written in one language to the equivalent source code in another programming language.

Truthy Treated as true in a boolean context (e.g., when evaluating conditions).

Type A set of values with common behavior enforced by the programming language.

Type Compatibility The property of a relationship between an expression e and type t in which e can be used in any context expecting a value of type t.

Type Inference Automatic deduction of the type of an expression.

Type Variable A variable that ranges over types.

Union Type A type whose set of values is the union of zero or more other types.

Value A unit of data.

Variable 1. An identifier bound to a value. 2. A storage location containing a value that can change.

Volatile Variable A variable that must be made available to multiple threads and thus cannot be implemented with multiple divergent copies per-thread.

Wildcard A character in a pattern meant to be replaced.

Bibliography

[1] Dave Abrahams. Protocol-oriented Programming in Swift. https://developer.apple.com/videos/play/wwdc2015/408/, 2015. Presented at Apple WWDC 2015.

[2] The Apache Software Foundation. Apache Commons. https://commons.apache.org/.

[3] Apple, Inc. iOS. http://www.apple.com/ios/.

[4] Apple, Inc. Swift Standard Library Operators Reference. https://developer.apple.com/reference/swift/1851035-swift_standard_library_operators.

[5] Apple, Inc. Swift Standard Library Reference: UnsafeMutablePointer Structure Reference. https://developer.apple.com/reference/swift/unsafemutablepointer.

[6] Apple, Inc. The Swift Programming Language.

[7] Joe Armstrong. *Programming Erlang: Software for a Concurrent World.* Pragmatic Bookshelf, 2007.

[8] Joe Armstrong. Big changes to Erlang. http://joearms.github.io/2014/02/01/big-changes-to-erlang.html, 2014.

[9] Jeremy Ashkenas. List of languages that compile to JS. https://github.com/jashkenas/coffeescript/wiki/list-of-languages-that-compile-to-js.

[10] Jeremy Ashkenas et al. CoffeeScript. http://coffeescript.org.

[11] Ada Resource Association. Introducing Ada 95: First Internationally Standardized, Object-Oriented Language. http://www.adaic.org/ada-resources/standards/ada-95-documents/95intro/.

[12] Gary Bernhardt. WAT. https://www.destroyallsoftware.com/talks/wat, 2012.

[13] Edwin Brady. Programming and reasoning with algebraic effects and dependent types. *SIGPLAN Not.*, 48(9):133–144, September 2013.

[14] Trevor Burnham. *CoffeeScript: Accelerated JavaScript Development.* Pragmatic Bookshelf, 2nd edition, 2015.

[15] Mike Cantelon, Marc Harter, TJ Holowaychuk, and Nathan Rajlich. *Node.Js in Action.* Manning Publications Co., Greenwich, CT, USA, 1st edition, 2013.

[16] Luca Cardelli and Peter Wegner. On understanding types, data abstraction, and polymorphism. *ACM Computing Surveys*, 17(4):471–523, December 1985.

[17] Cat's Eye Technologies. Befunge-93. http://catseye.tc/node/Befunge-93.html.

[18] Brad Chamberlain. Chapel: Productive Parallel Programming. http://www.cray.com/blog/chapel-productive-parallel-programming/, 2013.

[19] Nick Coghlan. Python 3 Q & A. http://python-notes.curiousefficiency.org/en/latest/python3/questions_and_answers.html.

[20] Douglas Crockford. *JavaScript: The Good Parts.* O'Reilly, 2008.

[21] Bjarne Däcker, Joe Armstrong, Mike Williams, and Robert Virding. Erlang: The Movie. https://www.youtube.com/watch?v=xrIjfIjssLE, 1990.

[22] Luis Damas. *Type Assignment in Programming Languages.* PhD thesis, University of Edinburgh, 1984.

[23] Mark Davis. Unicode Text Segmentation. Technical Report 29, Unicode Consortium, 2015.

[24] Steve Dekorte et al. Io: A Programming Language. http://iolanguage.org/.

[25] Edsger W. Dijkstra. Letters to the editor: Go to statement considered harmful. *Communications of the ACM*, 11(3):147–148, March 1968.

[26] Edsger W. Dijkstra. The humble programmer. *Communications of the ACM*, 15(10):859–866, October 1972.

[27] Edsger W. Dijkstra, Fraser Duncan, Jan V. Garwick, C. A. R. Hoare, Brian Randell, G. Seegmueller, Władisław M. Turski, and Michael Woodger. News item — minority report. *Algol Bulletin*, 31:7–7, March 1970.

[28] ECMA International. Standard ECMA-262. http://www.ecma-international.org/publications/standards/Ecma-262.htm.

[29] Melissa Elliott. PHP Manual Masterpieces. http://phpmanualmasterpieces.tumblr.com/.

[30] Epic Games, Inc. Unreal Engine Technology. https://www.unrealengine.com/.

[31] Ericsson. *Erlang Reference Manual User's Guide*, chapter Guard Sequences. Ericsson AB, 2003–2016.

[32] Facebook. Hack. http://hacklang.org/.

[33] Facebook. Hack Documentation. https://docs.hhvm.com/hack/.

[34] Stuart Feldman. A conversation with Alan Kay. *ACM Queue*, 2(9):20–30, December 2004.

[35] Andrew Gerrand. Defer, Panic, and Recover. https://blog.golang.org/defer-panic-and-recover.

[36] Andrew Gerrand. Share Memory By Communicating. https://blog.golang.org/share-memory-by-communicating, 2010.

[37] The Go Authors. The Go Programming Language. https://golang.org/.

[38] The Go Authors. The Go Programming Language Frequently Asked Questions. https://golang.org/doc/faq.

[39] The Go Authors. The Go Programming Language Specification. https://golang.org/ref/spec.

[40] Google. Android. https://www.android.com/.

[41] Google. Guava: Google Core Libraries for Java. https://github.com/google/guava.

[42] James Gosling and Henry McGilton. *The Java Language Environment: A White Paper.* Sun Microsystems, 1995.

[43] Paul Graham. Fortran I. `http://www.paulgraham.com/history.html`.

[44] Paul Graham. Lisp Quotes. `http://www.paulgraham.com/quotes.html`.

[45] Paul Graham. Revenge of the Nerds. `http://www.paulgraham.com/icad.html`.

[46] Paul Graham. The Roots of Lisp. `http://lib.store.yahoo.net/lib/paulgraham/jmc.ps`, 2002.

[47] James Hague. Puzzle Languages. `http://prog21.dadgum.com/38.html`, 2009.

[48] Robert Harper. Dynamic Languages are Static Languages. `https://existentialtype.wordpress.com/2011/03/19/dynamic-languages-are-static-languages/`, 2011.

[49] Robert Harper. *Practical Foundations for Programming Languages.* Cambridge University Press, 2nd edition, 2016.

[50] Haskell Community. Introduction to IO. `https://wiki.haskell.org/Introduction_to_IO`.

[51] Haskell Community. IO Inside. `https://wiki.haskell.org/IO_inside`.

[52] Haskell Wiki Community. Rank-N Types. `https://wiki.haskell.org/Rank-N_types`.

[53] Geoffrey Hayes. How One Missing Var Ruined our Launch. `http://blog.safeshepherd.com/23`, 2011.

[54] Fred Hébert. *Learn You Some Erlang for Great Good!* No Starch Press, 2013.

[55] David Herman, Luke Wagner, and Alon Zakai (eds.). asm.js Working Draft. `http://asmjs.org/spec/latest/`.

[56] Rich Hickey. Clojure. `http://clojure.org/`.

[57] Rich Hickey. Clojure Core Library Source Code. `https://github.com/clojure/clojure/blob/master/src/clj/clojure/core.clj`.

[58] C. A. R. Hoare. Hints on programming language design. Technical report, Stanford University, Stanford, CA, USA, 1973.

[59] C. A. R. Hoare. Communicating sequential processes, 1985.

[60] Tony Hoare. Null References: The Billion Dollar Mistake. *InfoQ*, August 2009.

[61] Allen Holub. Why Getter and Setter Methods are Evil. *Java World*, 2003.

[62] Paul Hudak, John Hughes, Simon Peyton Jones, and Philip Wadler. A history of Haskell: Being lazy with class. In *Proceedings of the Third ACM SIGPLAN Conference on History of Programming Languages*, HOPL III, pages 12–1–12–55, New York, NY, USA, 2007. ACM.

[63] IEEE Task P754. *ANSI/IEEE 754-1985, Standard for Binary Floating-Point Arithmetic.* IEEE, New York, NY, USA, August 1985.

[64] IETF Trust. The JavaScript Object Notation (JSON) Data Interchange Format. RFC 7159, Internet Engineering Task Force, 2014.

[65] Intel. Intel®64 Architecture. http://www.intel.com/content/www/us/en/architecture-and-technology/microarchitecture/intel-64-architecture-general.html.

[66] ISO. Common Language Infrastructure (CLI). ISO/IEC 23271:2012, International Organization for Standardization, Geneva, Switzerland, 2012.

[67] ISO. Information technology – programming languages – ada. ISO/IEC 8652:2012, International Organization for Standardization, Geneva, Switzerland, 2012.

[68] JBoss. Javassist. http://jboss-javassist.github.io/javassist/.

[69] Kris Jenkins. What Is Functional Programming? http://blog.jenkster.com/2015/12/what-is-functional-programming.html, 2015.

[70] Joyent, Inc. Node.js. http://nodejs.org.

[71] Alan C. Kay. The Early History of Smalltalk. In Thomas J. Bergin, Jr. and Richard G. Gibson, Jr., editors, *History of Programming Languages—II*, pages 511–598. ACM, New York, NY, USA, 1996.

[72] Khan Academy. Intro to JS: Drawing & Animation. https://www.khanacademy.org/computing/computer-programming/programming.

[73] Kx Systems. *K Reference Manual.* Kx Systems, Inc., 1998.

[74] Angelika Langer. Java Generics FAQs. http://www.angelikalanger.com/GenericsFAQ/JavaGenericsFAQ.html.

[75] Barbara Liskov. A History of CLU. In Thomas J. Bergin, Jr. and Richard G. Gibson, Jr., editors, *History of Programming Languages—II*, pages 471–510. ACM, New York, NY, USA, 1996.

[76] Cristina Lopes. *Exercises in Programming Style.* Taylor & Francis, 2014.

[77] Lua.org, PUC-Rio. About Lua. http://www.lua.org/about.html.

[78] Lua.org, PUC-Rio. The Lua 5.3 Reference Manual. http://www.lua.org/manual/5.3/.

[79] Robert C. Martin. The Barbarians are at the Gates. https://blog.8thlight.com/uncle-bob/2011/12/11/The-Barbarians-are-at-the-Gates.html, 2011.

[80] Yukihiro Matsumoto. Ruby 1.9, Google Tech Talk. https://www.youtube.com/watch?v=oEkJvvGEtB4, 2008.

[81] Alex McCaw. *The Little Book on CoffeeScript.* O'Reilly Media, 2012.

[82] Steve McConnell. *Code Complete, Second Edition.* Microsoft Press, Redmond, WA, USA, 2004.

[83] Paul Merolla, John Arthur, Rodrigo Alvarez-Icaza, Andrew Cassidy, Jun Sawada, Filipp Akopyan, Bryan Jackson, Nabil Imam, Chen Guo, Yutaka Nakamura, Bernard Brezzo, Ivan Vo, Steven Esser, Rathinakumar Appuswamy, Brian Taba, Arnon Amir, Myron Flickner, William Risk, Rajit Manohar, and Dharmendra Modha. A million spiking-neuron integrated circuit with a scalable communication network and interface. *Science*, 345(6197):668–672, August 2014.

[84] Microsoft. .NET. http://www.microsoft.com/net.

[85] Microsoft. TypeScript—JavaScript that Scales. http://www.typescriptlang.org/.

[86] Robin Milner. A theory of type polymorphism in programming. *Journal of Computer and System Sciences*, 17(3):348–375, 12 1978.

[87] Don Monroe. Neuromorphic computing gets ready for the (really) big time. *Communications of the ACM*, 57(6):13–15, June 2014.

[88] Natasha the Robot. Swift 2.0: Why Guard is Better than If. `https://www.natashatherobot.com/swift-guard-better-than-if/`.

[89] Michael Nielsen. Lisp as the Maxwell's Equations of Software. `http://www.michaelnielsen.org/ddi/lisp-as-the-maxwells-equations-of-software/`.

[90] Chris Okasaki. *Purely Functional Data Structures*. Cambridge University Press, New York, NY, USA, 1998.

[91] Oracle. Java Tutorials: Lesson: Generics. `https://docs.oracle.com/javase/tutorial/java/generics/`.

[92] John K Ousterhout. Scripting: Higher-level programming for the 21st century. *Computer*, 31(3):23–30, March 1998.

[93] OW2 Consortium. ASM. `http://asm.ow2.org/`.

[94] Palo Alto Research Center, Inc. Milestones: Xerox PARC History. `http://www.parc.com/about/`.

[95] Tim Peters. PEP 20 – The Zen of Python. `http://python.org/dev/peps/pep-0020/`.

[96] Simon Peyton Jones et al. The Haskell 98 language and libraries: The revised report. *Journal of Functional Programming*, 13(1):0–255, Jan 2003.

[97] Benjamin C. Pierce. *Types and Programming Languages*. MIT Press, Cambridge, MA, USA, 2002.

[98] Rob Pike. Concurrency is not Parallelism. `https://talks.golang.org/2012/waza.slide`, 2012.

[99] Pivotal Software, Inc. Spring Framework. `http://projects.spring.io/spring-framework/`.

[100] Maria Popova. How Ada Lovelace, Lord Byron's daughter, became the world's first computer programmer. *Brain Pickings*, December 2014.

[101] Python Software Foundation. Python. `http://python.org`.

[102] Axel Rauschmayer. *Speaking JavaScript*. O'Reilly Media, Inc., 2014.

[103] Axel Rauschmayer. *Exploring JavaScript*. Leanpub, 2015.

[104] Red Hat, Inc. Hibernate. `http://hibernate.org/`.

[105] The Rust Community. The Rust Programming Language. `https://www.rust-lang.org/`.

[106] The Rust Project Developers. The Rust Programming Language. `https://doc.rust-lang.org/book/`.

[107] Jean Sammet. *Programming Languages: History and Fundamentals*. Prentice-Hall series in automatic computation. Prentice-Hall, 1969.

[108] Zed Shaw. *Learn Ruby the Hard Way: A Simple and Idiomatic Introduction.* Addison-Wesley Professional, 2014.

[109] Clay Shirky. Ontology is Overrated: Categories, Links and Tags. `http://www.shirky.com/writings/ontology_overrated.html`, 2005.

[110] Michele Simionato. The Python 2.3 Method Resolution Order. `https://www.python.org/download/releases/2.3/mro/`.

[111] Garrett Smith. Erlang The Movie II: The Sequel. `http://www.gar1t.com/blog/erlang-the-movie-ii-the-sequel.html`, March 2013.

[112] Simon St. Laurent. *Introducing Erlang: Getting Started in Functional Programming.* O'Reilly, 2013.

[113] Daniel Steinberg. Daddy, Are We There Yet? A Discussion with Alan Kay. *OpenP2P*, April 2003.

[114] Steve Dekorte. Io Programming Guide Introduction. `http://iolanguage.org/guide/guide.html#Introduction`.

[115] Bjarne Stroustrup. A history of C++: 1979–1991. *SIGPLAN Notices*, 28(3):271–297, March 1993.

[116] S. Tucker Taft. The ParaSail Programming Language. `http://parasail-lang.org`.

[117] The Internet Society. The Base16, Base32, and Base64 Data Encodings. RFC 4648, Internet Engineering Task Force, 2006.

[118] Dave Thomas, Andy Hunt, and Chad Fowler. *Programming Ruby 1.9 & 2.0: The Pragmatic Programmers' Guide.* Pragmatic Bookshelf, 2013.

[119] Franklyn A. Turbak and David K. Gifford. *Design Concepts in Programming Languages.* The MIT Press, 2008.

[120] Inc. Unicode. Blocks-9.0.0.txt. `http://www.unicode.org/Public/9.0.0/ucd/Blocks.txt`, 2016.

[121] The Unicode Consortium. Unicode. `http://unicode.org/`.

[122] Bill Venners. The Philosophy of Ruby. `http://www.artima.com/intv/rubyP.html`, 2003.

[123] Bret Victor. The Future of Programming. `http://worrydream.com/dbx/`, 2013.

[124] Larry Wall, Tom Christiansen, and Jon Orwant. *Programming Perl, Third Edition.* O'Reilly Media, Inc., 3 edition, 2000.

[125] Berry Warsaw, Jeremy Hylton, David Goodger, and Nick Coghlan. PEP Purpose and Guidelines. `https://www.python.org/dev/peps/pep-0001/`.

[126] Eric Wastl. PHP Sadness. `http://www.phpsadness.com/`.

[127] Web Hypertext Application Technology Working Group. Web Wokers (HTML Living Standard). `https://html.spec.whatwg.org/multipage/workers.html`.

[128] WebAssembly Community Group. WebAssembly. `https://webassembly.github.io/`.

[129] Peter Wegner. Concepts and paradigms of object-oriented programming. *SIGPLAN OOPS Messenger*, 1(1):7–87, August 1990.

[130] Ken Whister and Asmus Freytag. The Unicode Character Property Model. Technical Report 23, Unicode Consortium, 2016.

[131] Arthur Whitney. K. *Vector*, 10(1):74–79, July 1993.

[132] Wikipedia. Brainfuck.

[133] Wikipedia. C++.

[134] Wikipedia. Grace Hopper.

[135] Wikipedia. Hindley-Milner Type System.

[136] Wikipedia. Programming Paradigm.

[137] Wikipedia. Red-Black Trees.

[138] Wikipedia. Simula.

[139] Wikipedia. Standard Streams.

[140] Wikipedia. Static Program Analysis.

[141] Steve Yegge. Execution in the Kingdom of Nouns. `http://steve-yegge.blogspot.com/2006/03/execution-in-kingdom-of-nouns.html`, 2006.

Index

9781498738460